Annual International Scientific Conference on Geoinformatics (GI 2021)

Supporting Sustainable Development by GIST

E3S Web of Conferences Volume 227 (2021)

Tashkent, Uzbekistan
27 – 29 January 2021

Editors:

L. Foldvary
I. Abdurahmanov

ISBN: 978-1-7138-2441-1

Printed from e-media with permission by:

Curran Associates, Inc.
57 Morehouse Lane
Red Hook, NY 12571

Some format issues inherent in the e-media version may also appear in this print version.

For additional information, please contact EDP Sciences – Web of Conferences
at the address below.

EDP Sciences – Web of Conferences
17, Avenue du Hoggar
Parc d'Activité de Courtabœuf
BP 112
F-91944 Les Ulis Cedex A
France

Phone: +33 (0) 1 69 18 75 75
Fax: +33 (0) 1 69 28 84 91

contact-edps@webofconferences.org

Additional copies of this publication are available from:

Curran Associates, Inc.
57 Morehouse Lane
Red Hook, NY 12571 USA
Phone: 845-758-0400
Fax: 845-758-2633
Email: curran@proceedings.com
Web: www.proceedings.com

TABLE OF CONTENTS

ANNUAL INTERNATIONAL SCIENTIFIC CONFERENCE ON GEOINFORMATICS – GI 2021

Co-funded by the Erasmus+ Programme of the European Union

TIIAME

Erasmus+ DSinGIS

Óbuda University

UNIVERSITY of SALZBURG

IAMO Leibniz Institute of Agricultural Development in Transition Economies

ANNUAL INTERNATIONAL SCIENTIFIC CONFERENCE ON GEOINFORMATICS – GI 2021

"Supporting sustainable development by GIST"

Tashkent Institute of Irrigation and Agricultural Mechanization Engineers (TIIAME), Tashkent, Uzbekistan

January 27-29, 2021

ONLINE

Scientific editors: Dr. Lorant Foldvary, Obuda University (OU), Hungary

Mr. Ilhom Abdurahmanov, Tashkent Institute of Irrigation and Agricultural Mechanization Engineers (TIIAME), Uzbekistan

Statement of Peer review

In submitting conference proceedings to *Web of Conferences*, the editors of the proceedings certify to the Publisher that

1. They adhere to its **Policy on Publishing Integrity** in order to safeguard good scientific practice in publishing.
2. All articles have been subjected to peer review administered by the proceedings editors.
3. Reviews have been conducted by expert referees, who have been requested to provide unbiased and constructive comments aimed, whenever possible, at improving the work.
4. Proceedings editors have taken all reasonable steps to ensure the quality of the materials they publish and their decision to accept or reject a paper for publication has been based only on the merits of the work and the relevance to the journal.

Title, date and place of the conference

ANNUAL INTERNATIONAL SCIENTIFIC CONFERENCE ON GEOINFORMATICS – GI 2021
"Supporting sustainable development by GIST", January 27-29, 2021
TIIAME, Tashkent, Uzbekistan

Proceedings editor(s):

Dr. Lorant Foldvary, Obuda University (OU), Hungary
Mr. Ilhom Abdurahmanov, Tashkent Institute of Irrigation and Agricultural Mechanization Engineers (TIIAME), Uzbekistan

Date and editor's signature

2020.12.07.

E3S Web of Conferences **227**, 01001 (2021) https://doi.org/10.1051/e3sconf/202122701001
GI 2021

Object-based approach for urban land cover mapping using high spatial resolution data

Malgorzata Verone Wojtaszek [1,*], *Levente* Ronczyk [2], *Zokhid* Mamatkulov[3], and *Mamanbek* Reimov[3]

[1]OU Alba Regia Technical Faculty, Pirosalma str. 1-3, 8000 Székesfehérvár, Hungary
[2]UP Faculty of Sciences Institute of Geography, Ifjúság str. 6. 7624 Pécs, Hungary
[3]Tashkent Institute of Irrigation and Agricultural Mechanization Engineers (TIIAME), Koriy Niyaziy str., 39, 100000, Tashkent, Uzbekistan.

Abstract. This paper deals with object-oriented image analysis applied for an urban area. Very high-resolution images in conjunction with object-oriented image analysis have been used for land cover detection. Using the eCognition software with object-oriented methods, not only the spectral information but also the shape, compactness and other parameters can be used to extract meaningful objects. The spectral and geometric diversity of urban surfaces is a very complex research issue. It is the main reason why additional information is needed to improve the outcome of classification. The most consistent and relevant characteristic of buildings is their height. Therefore, elevation data (converted from LIDAR data) are used for building extraction, segmentation and classification. The study deals with the problem, how to determine the most appropriate parameters of segmentation, feature extraction and classification methods. The data extraction includes two phases, the first part consists the following steps: data pre-processing, rule set development, multi-scale image segmentation, the definition of features used to map land use, classification based on rule set and accuracy evaluation. The second part of the data process based on classical raster analysis GIS tools like focal and zonal function.
Keywords: OBIA, GIS, remote sensing, land cover, urban environment, zonal statistics

1 Introduction

The loss of natural and semi-natural ecosystems and hereby induced negative change in the environment are among the most actual problems nowadays. In the last decades the rapid increase of large cities and the transformation of urban landscapes are the grates challenges to human welfare and a viable global environment. Urbanization is the most visible anthropogenic force on the earth, affecting its surface, atmosphere, water, its biodiversity and the life of people. There needed reliable data to assess the ecosystem health and biodiversity of many urban areas. The town planning and city management requires wide spectrum of topographical, environmental, infrastructure and several other

* Corresponding author: wojtaszek.malgorzata@amk.uni-obuda.hu

information as well. Remote sensing provides an important data source for many fields of research. Data available from satellites presents an opportunity to collect information relevant to urban environments at various of spatial and temporal scales. Urban land cover and land use data play an important role in ecological and environmental investigations. In several studies remote sensing data are used as a source, but the results are not quite satisfying for detailed land use detection in urban areas [6, 8, 14-17,]. Many studies have been carried out to find an appropriate method to classify the high resolution remote sensing data [3, 7, 12-13]. The difficulty of mapping land cover in urban environment is not due to the classification methods (algorithms). The problem is related to the creation and selection of features. It is very difficult to find out the representative features or training samples for man-made objects. Recent sensor development including high spatial and spectral resolution like WorldView sensors have the potential for more detailed and accurate mapping of urban land cover and land use [4]. Number of study have focused on the issues of spectral properties of urban materials and their representation and mapping from remote sensing data. The results indicate highly complex and diverse spectral properties of urban land cover types. The spectral reparability of urban land cover types is strongly dependent on spectral sensor characteristics [18]. The development of classification methodologies for analysis of high resolution satellite imagery is relevant to studies related with mapping of urban environments [11-12].

High-resolution satellite images provide an important data source for many fields of research. The difficulty of mapping land cover in urban environment is not due to the classification methods (algorithms). The problem is related to the creation and selection of features. It is very difficult to find out the representative training samples or features for man-made objects. The present study aims at finding an appropriate classification method in order to map urban land cover based on high resolution remote sensing data. This paper deals with object-oriented image analysis applied for an urban area. Very high-resolution aerial photos and satellite images in conjunction with object-oriented image analysis have been used for land cover detection. Using the eCognition software with object-oriented methods, not only the spectral information but also the shape, compactness and other parameters can be used to extract meaningful objects. The spectral and geometric diversity of urban surfaces is a very complex research issue. It is the main reason why additional information is needed to improve the outcome of classification. The most consistent and relevant characteristic of buildings is their high. The elevation data are suitable for building and the different vegetation levels extraction, segmentation and classification.

2 Materials and methods

The study area (Székesfehérvár) is situated in central Hungary and one of the largest settlements in Transdanubia (Fig.1). It is the county seat, the cultural and economic centre of Fejér County. According to the historical documents the place has been inhabited since the 5th century BC, but Székesfehérvár was first mentioned in 1009, as Alba Civitas. After World War II the city was subject to industrialization. Parallel with industrialization the population was also growing. By the 1970s Székesfehérvár had swelled to more than 100,000 inhabitants (in 1945 it had only about 35,000.) The city and its surroundings were the most dynamically growing regions of Hungary in the 1990's. A number of new housing estates were built, but the city centre was able to preserve its Baroque character. It is inhabited by 101,943 people [5], the density of population is 594-person/sq km and ranks among the middle-sized cities of the country. With a total area of 170.9 sq km it is the second largest county centre in the Transdanubian region.

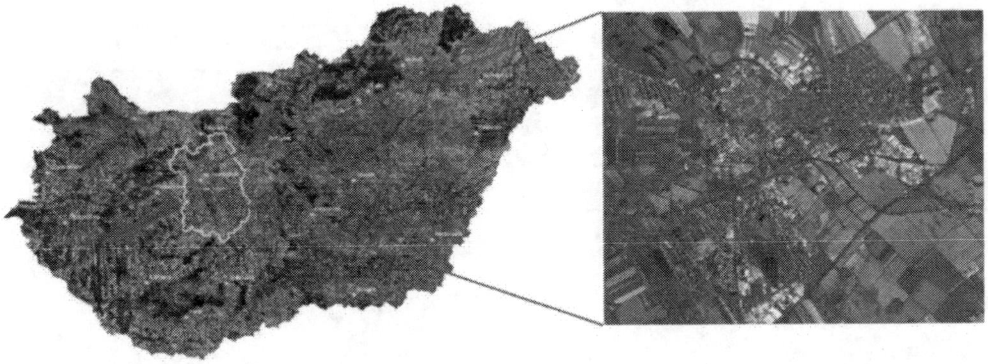

Fig. 1. Study area: Székesfehérvár (Hungary)

2.1 Data source

The study used the following remote sensing data:

- WorldView2 satellite image (2011, 2017)
- Airborne photographs (2008, 2009, 2015)
- LIDAR data (2008)

WorldView-2 developed by DigitalGlobe, USA, in 2009, is the first high resolution multispectral commercial satellite having eight spectral sensors from the visible to near-infrared range. Spatial resolution of the Pan is 0.5 m and MS sensors is 2.0 m. WorldView-2 image was acquired 2011 and 2017.

In 2008 the inner part of the city Székesfehérvár was laser scanned and in parallel with the laser scanning a digital airborne photographs were taken in visible and in NIR spectra. The images cover an area of 5 km2. The air photo made in 2009 covered all study area. Both of the airborne datasets were acquired in similar spatial resolution of approximately 0.5 m. The Lidar points measured directly from the field, their point original density is 6-8 point/m2, and the absolute horizontal accuracy is 0,5 m, the vertical accuracy is ±0,15 m (1 Table, Fig. 2).

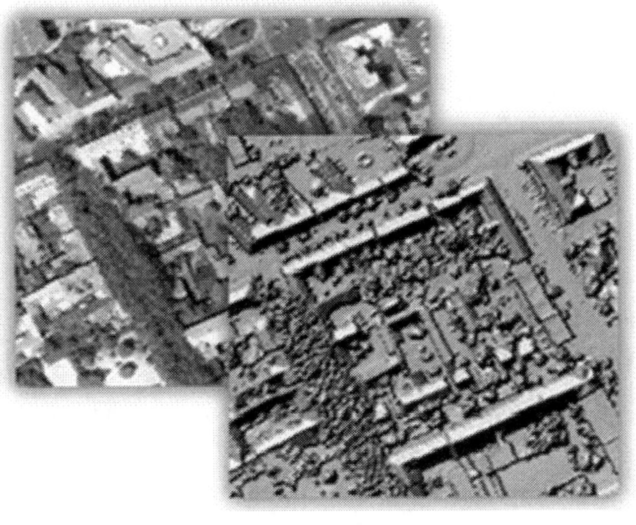

Fig. 2. Airborne photograph and Digital Surface model based on LIDAR

Table 1. The specification of remote sensing data

Datasets	Spectral resolution	Other specification
WorlView2 (2011, 2017)	Multispectral (400-1040 nm), 8 bands, Panchromatic	Spatial Resolution: PAN – 0,5 m MS – 2 m, Quantization Level: 11 Bits
Aerial photographs (2008, 2009, 2015)	visible/infrared	Spatial Resolution: 0,5 m
LIDAR (2008)		LIDAR point cloud FE/LE Absolute accuracy: horizontal < ±0.50 m; height < ±0.15 m

2.2 Methodology

One of the main objectives of this study was to develop a methodology to generate land cover classes of urban area from high spatial resolution satellite images through object-based classification. Using the eCognition software not only the spectral information but also the shape, compactness and other parameters can be employed for mapping. To extract buildings a combination of spectral (WorldView2) and LIDAR data contains elevation information were applied. The most relevant information about buildings is their different elevation compared to their surroundings. We used an object-oriented approach, which allowed context consideration during the classification process. A consideration of spatial context is very important for correct identification and classification of objects.

The aim of this papers is to describe the potential of object-based image analysis (OBIA) for urban land cover detection and mapping impervious surface within parcels using very high spatial resolution imagery.

2.3 Object-oriented image classification

The land cover classification system presented in this paper includes only the more generalized levels (first and second level). The classification system is capable of further refinement. Defining the categories, the next main view-points were considered: to give names to categories by simply using accepted terminology, to use remote sensing as the primary data source. The process considered spectral value and spatial characteristics of objects and it is included the following steps:

- Segmentation
- Feature extraction
- Object classification
- Refinement of classification based on spectral value and spatial characteristics of objects
- Final identification of categories
- Accuracy assessment

2.4 Segmentation and feature extraction

Segmentation of images is one of the main steps that have to be solved in the object-based classification [1-2]. Segmentation is a process by which pixels are grouped into segments according to their spectral similarity and other criteria (shape, area, position etc.) [9]. At the beginning of segmentation, the heterogeneity criterion is thresholded by the user. This is done by choosing a scale factor and by fixing the weights of the colour and shape criteria, and the smoothness and compactness criterions as well. The multiresolution segmentation algorithm was run with different scale factors (10, 20, 30, and 40) while all other parameters were held constant (shape: 0.1 and compactness: 0.5; shape: 0.4 and compactness: 0.5). In general, a scale factor of 10 a shape factor of 0.1 and compactness of 0.5 produced the best results. The algorithm spectral difference was used to refine existing segmentation results by merging spectrally similar image objects.

One of the critical steps in image analysis is to determine the most relevant features and algorithms to be used in classification [1]. The eCognition works with objects and information of whole objects like as the spectral signature, the shape and size or context is available. All these attributes can be applied and combined for classification. Features usually define the upper and lower limits of a range measure of characteristics of image objects. Image objects within the defined limits are assigned to a specific class, while those outside of the feature range are assigned to a different class (or left unclassified). Table 2. summarizes the features used to aid the classification process.

Table 2. The features used to aid the classification process

Class	Main features	Other features
Vegetation (trees, brushy, grass, barren)	NDVI	Vegetation (trees, brushy, grass, barren)
Building	Height information	Spectral features, NDVI, context (Rel. Border to building), size information
Other man-made (road, other)	Spectral information (NDWI, NDVI)	size: area, length to width ratio
Parking place (big area)	Spectral information (NDVI, NDWI)	Shape: rectangular fit, shape index; size: area, length to width ratio, context: rel. Border to parking place)
Water*	Spectral information (NDWI)	
Shadows*	Spectral information (NDWI)	
* At the end of classification the segments of shadow class were reclassed to the road and other man-made classes according to the shape. The water in not included in presented study area.		

3 Results and discussion

3.1 Rule set based classification

The classification of urban areas was done in two levels. In the first one the main categories of land cover were determined by taking advantage of spectral properties of WorldView2 image. In this process some normalized difference index was created according to the spectral mean data of determined classes in different bands. Normalized Difference Vegetation Index (NDVI) was used to select vegetation and no-vegetation and to investigate different kind of vegetation. Other indices also were developed and utilized during the classification (e.g. NDWI Normalized Difference Water Index). In the second level during the building extraction the elevation data was used as the most relevant characteristic of buildings. Additionally, the spectral and context features were employed to clarify the classification. Table 2 and Fig. 3 summarize the features, which were utilized to reveal the appropriate land cover classes.

https://doi.org/10.1051/e3sconf/202122701001

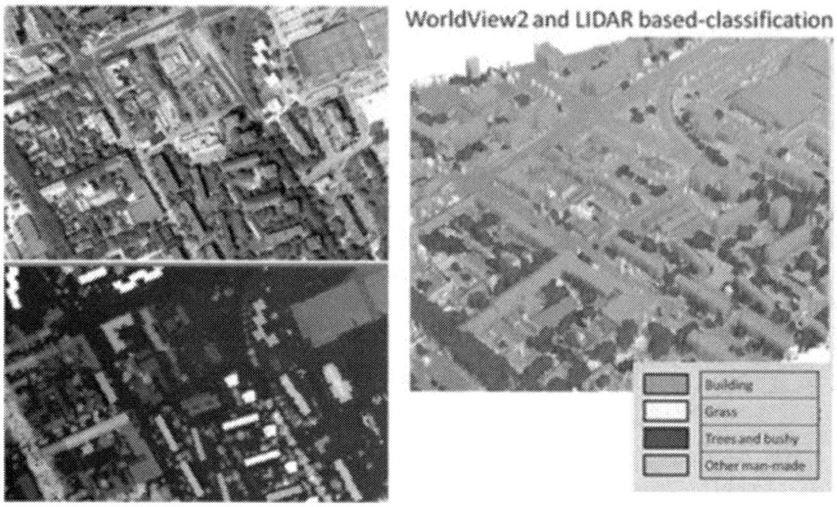

Fig. 3. Input data and the results of classification draped on DSM (Digital Surface Model based on LIDAR points)

3.2 Performing accuracy assessment

The results of the land cover classifications derived from remotely sensed data are compared by an accuracy assessment. An accuracy assessment analysis was performed by producing an error matrix that compared the interpreted land cover map and a map contains the result of ground truth investigation. The most important outputs from the accuracy assessment included an error matrix, errors of omission and commission, both for all classes and on a per category basis. The assessment indicates the producer's and user's and overall accuracy (Table 3).

Table 3. Accuracy assessment

Class	LIDAR and WorldView based classification	
	Producer's accuracy %	**User's accuracy %**
Building	88	99
Parking place	96	99
Road	67	92
Other man-made	97	26
Vegetation (trees-bush)	98	54
Grass	94	51
	Overall accuracy: 82	

3.3 Mapping impervious surface

In this process other kind of segmentation (super object) were used to determined objects representing the parcels and all types of parcels were identified at super object level. To identity impervious surface within the parcels a new segmentation level (sub object) was added and the parcel area has been split up into image objects suited for identifying impervious surface. In the classification process the objects were assigned into two classes, vegetation and impervious surface. The final step of this work focus on categorizing parcels based on the degree of impervious surface. Four classes have been distinguished using the "classification" algorithm. The class distinction was based on imperviousness, in steps of 25% (Fig.4). The process is included the following steps:

1. Identify parcels (super object level)
    - Segmentation representing the parcels (cheesboard segmentation)
    - Classify the parcels (thematic layer attributes)
2. Identify impervious surface within the parcels (subobject level)
    - A segmentation representing impervious surface (multiresolution segmentation)
    - Classify impervious surface
3. Categorize parcels (classification, class description)

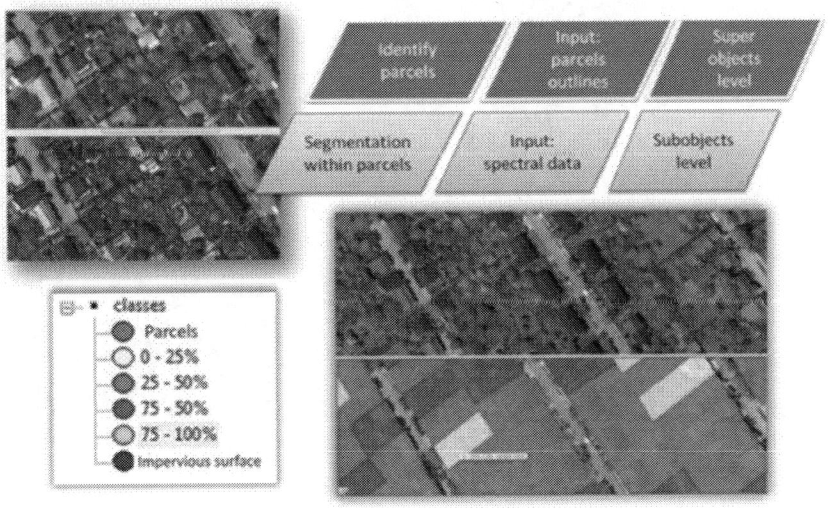

Fig. 4. Mapping impervious surface within the parcels

3.4 Urban landscape extraction using high spatial resolution imagery and GIS techniques

In this process a raster analysis GIS tools like focal and zonal function we used for urban landscape extraction. It included the following steps:

1. Segmentation
    - Cheesbroad segmentation using thematic layer (Level 1)
    - Cheesbroad segmentation 1 hectare units (Main Level)
    - Multi resolution segmentation using mainly (0.9) the spectral information (Working Level)
2. Object classification
    - Spectral values of 8 bands, indexes
    - Position
    - Relation to super object

In the classification process the objects were assigned into two classes, vegetation and water. Water bodies and vegetated area composed the class of Blue & Green Area (BGA). The rest of the object was denoted as non-vegetated surface. To reach this classification firstly the vegetated area and lakes were separated and merged into one common class. The portion of BGA in one hectare was the final result of the classification and it was export out from the eCognition software as raster file. Two basic raster analysis tools were run on the exported data. Focal functions process data using different neighbourhoods cells. During the application of the zonal statistics tool of ArcGIS 9.3 software the polygons of district dataset calculated the statistics of raster layers were produced by neighbourhoods tool. The summary statistics of the zonal function is suitable to reveal the following information:

- Majority – the BGA value that appears most often in a given district,
- Range – range of the BGA values; homo- or heterogeneity of the BGA values in a given district
- Std – Standard deviation of BGA values in a given district

The outcome of our research confirms the original assumption that from the city center BGA value started to increase. The city center has the lowest BGA value, and there was not any cell which was covered more than 90% by BGA. The majority of the pixels BGA value is between 50-20 % (Fig. 5).

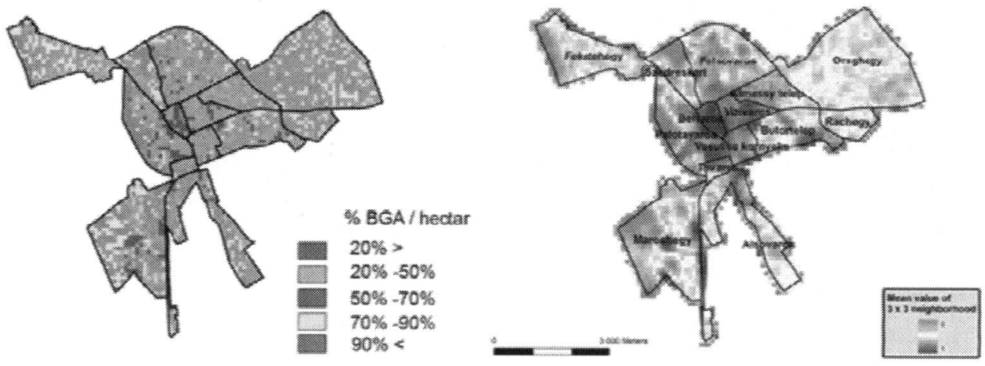

Fig. 5. BGA % in one hectare

Employing zonal statistic functions other aspects of urban landscape were analyzed. According to urban morphological zone the statistical parameters of the district also were dissolved (Table 4) Five categories were used according to the Urban Atlas of European Union (URBAN ATLAS 2010). Value one demonstrate more than 90% of VBA value, and category 5 shows more than 80% of sealed or anthropogenic surfaces.

Table 4. Different statistical parameters of the urban morphological zones

Morphological zone	Maximum value	Range value	Mean value	STD value	Major value	Minority value
Downtown	4	2	3,5	0,5	4,0	2
Village like	5	5	1,8	1	1,6	5
Traditional	4,8	4,8	2,4	0,7	2,8	3
House with garden	5	5	1,8	0,75	2,0	5
Building estate	4,75	4,5	2,3	0,9	2,2	4

An additional result was calculated the neighborhood of 1 hectare BGA area around a given cell, to highlight the accessibility of natural environment of examined cell. Our results are able to explain the reason behind recent population trends of Székesfehérvár. Using focal function and calculating the maximum BGA value in 5 cells distance highlights, that the most of the historical center of the city does not have any area, which can provide 1 ha green area for the young inhabitants. This is one of the main motivation, why the young families leave the downtown and separating their living and working environment.

4 Conclusion

An urban ecosystem presents a wide structural diversity and consequently spectral variability therefore the process of classification needs not only spectral information but other information like context or geometry as well. The fusion of remote sensing techniques (imagery and aerial LIDAR) offers an efficient way to map and update GIS ready land cover information. Using enhanced data source and object based image analysis accurate results could be obtained.

Accuracy measures have shown that the use of elevation parameters leads to significant improvements in comparison to only spectral bands classifications. However, the best classification results were reached with the application of combined dataset of spectral and elevation features, the further studies are needed in this field. The accuracy of this method can be increased by integrated using high resolution aerial photos or hyperspectral data. Our study presented a multifaceted method to evaluate an index, which is able to represent one of the most important elements of the urban landscape. Object based image analysis and further GIS functions supplied useful information of inner pattern of Székesfehérvár and clarify some reason of present population trend.

References

1. T. Blaschke, ISPRS J. Photogramm. Remote Sens. **65**, 2 (2010)
2. C. Burnett, T. Blaschke, Ecol. Modell. **168**, 233 (2003)
3. I. Fekete, D. Dezső, I. László, K. Ócsai. *Information Technology in higher education is an electronic publication* (Debrecen, 2008)
4. S. Hartling and others, doi: 10.3390/s19061284 (Published online, 2019)
5. KSH 2014, *Gazetteer of the Republic of Hungary* (Online: http://portal.ksh.hu/docs/hun/hnk/hnk_2011.pdf, 2015)
6. D. Lu, S. Hetrick, E. Moran, Photogramm. Eng. Rem. S, **76**, No 10, 1159 (2010)
7. S. W. Myint, P.Gober, A. Brazel, S. Grossman-Clarke, Q. Weng, Remote Sens. Environ. **115 (5)**, 1145 (2011)
8. R. Pu, S. Landry. Yu, Qian, Int. J. Remote Sens. **32 (12)**, 3285 (2011)
9. E. Schöpfer, S. Lang, J. Strobl*: In: Rashed, T. – Jürgens, C. (Eds): Remote Sensing of Urban and Suburban Areas,* 181 (Springer, Dordrecht, 2010)
10. R. Valbuena, T. F. de Sevilla, F.Mauro, C. Pascual, Abril, Martín, A. G. S., J. A. Manzanera, *Lidar and true-orthorectification of infrared aerial imagery of high Pinus sylvestris forest in mountainous relief* (Edinburgh, UK http://oa.upm.es/3104/2/INVE_MEM_2008_53075.pdf, *2008*)
11. M. Verőné Wojtaszek, L. Ronczyk, *Object based classification of urban land cover extraction using high spatial resolution imagery* (2012)
12. Q. Weng, J. Remote Sens. Environ. **117**, 34 (2012)
13. Y. Zhang, J. Photogramm. Remote Sens., **54 (1)**, 50 (1999)

14. W. Zhou, G. Huang, A. Troy, M. L. Cadenasso, J. Remote Sens. Environ, **113**, 1769 (2009)
15. S.Abdurakhmanov, Diss. Autoref. (TIIAME, Geodesy and Geoinformatics, 2020)
16. L. Gulyamova, J. Geograph. Societ. Uz. Spacial Vol., 28 (2018)
17. M. Sultanov, Research. Diss. (NUUz, 2018)
18. M. Herold, S. van der Linden, P. Hostert, D. Roberts, J. Urban Remote Sens. 137 (2006)

Land Use and Land Cover Change in Khorezm, Uzbekistan

Sherzod Rakhmonov[1,*], *Uktam* Umurzakov[1] , *Kosimdjon* Rakhmonov[1], *Iqbol* Bozarov[1] and *Ozodbek* Karamatov[2]

[1]Tashkent Institute of Irrigation and Agricultural Mechanization Engineers (TIIAME), Department of Geodesy and Geoinformatics, 100000 Tashkent, Uzbekistan
[2]Kastamonu University, Department of Economics, 37150 Kastamonu, Turkey

Abstract. This article depicts on discussions about land use and land cover change distribution in Khorezm province, Uzbekistan between 1987 and 2019. For the study Landsat 5 TM and Landsat 8 OLI respectively used to detect land use changes in the study area. Khorezm region affected by Aral Sea shrinkage having received salt wind from northeast of the region. Moreover, population increased within study period, making population density intense. Research is carried out to detect reflection of ecology and density in land use. RS techniques - maximum likelihood employed to classify land use to generate land cover distribution map. In total seven class selected such as agricultural land, built up, bare land, lowland, saline land, sand and waterbody. The research of Khorezm region for 32 years has been thoroughly studied and found out that agricultural land, built up and saline land increased tremendously while lowland and bare soil are decreased accordingly. The result map can be used for decision makers and government bodies for future long term urban and regional planning.

1 Introduction

Land Use and Land Cover (LULC) terms are two separate terms which corresponds to physical properties of land surface electromagnetic elements and human eye perception of land cover [1,2]. Therefore, it is not separated in use and the term broadly used among remote sensing scientists. LULC changes represents result of human activity influence on surrounding area [3-7]. Multi-spectral data is widely used among RS researchers to detect and learn timely temporal changes of human-environment interaction as well as natural disaster on land use [8-12].

US Geological Survey has launched satellites with remote sensors to orbit. Geological Survey provides remotely sensed data publicly available to anyone via earthexplorer.usgs.gov. RS researchers use freely available geodata with good resolution and timely updates. Researcher all over the world use the geoportal to carry out research to detect land cover changes on a timely base [13,14]. RS offers different approaches and

* Corresponding author: sherzod.rakhmonov@tiiame.uz

methodologies such as maximum likelyhood, principal component analysis and interactive supervised classification to work on geodata.

Various articles related to remote sensing research results on Central Asian countries are studied in this research. Yin et al [14] delivered results of a forest cover map for Central Asian countries. Interactive supervised classification applied to MODIS and Landsat geodata to generate a forest map. Kraemer et al. [15] analyzed Kostanay region in Kazakhstan to detect LULC changes. Multi-spatio-temporal Landsat geodata is analyzed using Support Vector Machine method to develop agricultural land using geoinformation between 1953-2010. Edlinger et al. [16] applied RS technique with Landsat MSS and TM geodata for 1972-2000. Research result compared with local geodata delivered quite good result which detected irrigated land in Kashkadarya region, Uzbekistan.

Aral Sea has been decreased starting from 1970s. Almost vanished Aral Sea resulted in water scarcity and strong northeast wind delivering salt and dust to surrounding area, including Khorezm province. Khorezm province is located 400 km away from Aral Sea and hence chosen for study for this research to detect how ecology resulted to the site.

Khorezm is located in the northwestern part of Uzbekistan and bordered with Turkmenistan on southwest. It was established on January 15, 1938 with 6,0 thousand km^2 area. The population is about 1.8 mln. people, consisting of mostly Uzbeks, as well as Turkmen, Russian, Kazakh, Tatar, Korean, Karakalpak and other nationalities. The average population density is 300 people per 1 km^2 (2017). Khorezm has 10 districts, 3 cities, 7 towns, 100 village - communities. Its capital is Urgench city.

2 Materials and methods

2.1 Study Area

Khorezm is located between 41°01' and 42°05' north and 60°11' and 62°15' east. The territory of the region is placed on the left bank of the Amudarya downstream, on the northern part of the Khorezm has lowland oasis at an average height of 100 m. Khorezm's west and southwest sides borders with the sandy Karakum Desert. The fossil minerals include limestone, sand, clay and others. The climate is sharp continental. The winter is moderately cold and light snow. The average temperature in January is -5°C, the lowest temperature is -32°C. The summer is hot and dry, the average temperature July is 30°C and the highest temperature is 45°C. Vegetation period is 200-210 days a year. The rainfall is 78-79 mm per year, it mostly rains in March-April. North and northeast winds blow in the area. The Amudarya River is the only river that runs through the valley. Water is used to irrigate crops through large canals. In the south of the region there are small salt lakes, marshes and salts. Meadow soils is spread over river valleys and sands to the west. The irrigated land is mainly high salinized. Poplar, willow, oleander, owl, sandy saxaul and other species grow on the shores of the Amu Darya river.

Regions main agriculture consists of cotton and grain. Horticulture, vegetable growing, gardening, cattle breeding and silkworm breeding also developed. Cotton, grain, potatoes, vegetables are planted on the fields. Khorezm region ranks number one in the country for rice cultivation. Cattle, sheep and goats, poultry are raised in commune and private farms. There are plants and factories in Urgench and Khiva cities.

The Tashkent-Moscow, Dushanbe-Moscow-Tashkent railway runs through the region and as well as Urgench-Tashkent. The length of the railway is 128 km long. The Khazarasp railway bridge over the Amudarya River was put into operation recently. The length of roads in the region is 2750 km. The trolleybus route from Khiva to Urgench was launched in 1997. Buses run from Urgench, Khiva to Tashkent, Samarkand, Navoi, Bukhara, Nukus and other cities. Urgench international and domestic airports have regular air communication with foreign countries. Central Asia - Center, Bukhara - Uralsk, Turkmenistan - Russia international gas main pipelines run through the region.

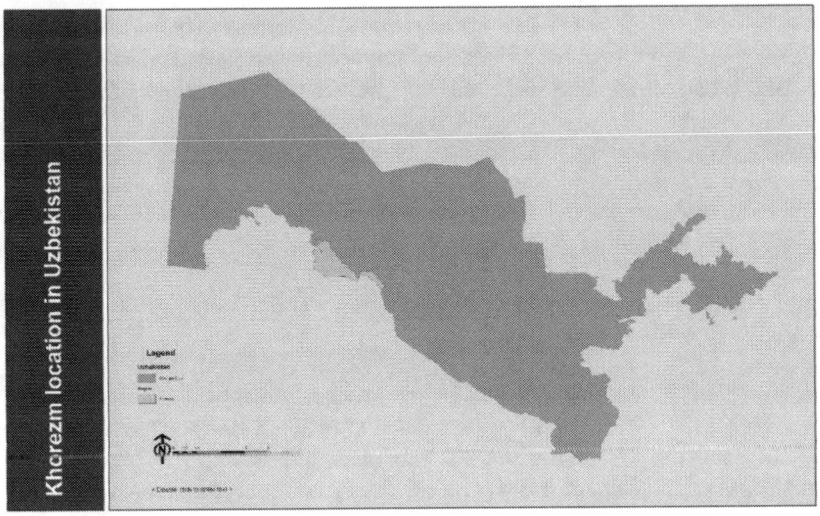

Fig. 1. Location of the study area.

There are 538 secondary schools in the region (including 28 lyceums and gymnasiums), 57 professional colleges and secondary specialized educational institutions, Urgench University, Urgench branch of Tashkent Medical Academy. 36 hospitals and 211 medical institutions, clinics, outpatient clinics, rural hospitals. Urgench branch of the State Scientific Center for Emergency Medical Care and its subdivisions are established at all district central hospitals.

2.2 Data Preparation and Processing

Landsat 5 TM data of 20 July 1987 and Landsat 8 data of 12 July 2019 derived from Geological website (earthexplorer.usgs.gov). The spatial resolution of Landsat 5 TM and 8 OLI is 30 meter. The study area is mostly plane and lowland oasis. Ground reference interpretation was done via visual gaining information from Google maps, Google Earth and as well as from local cadastral data. ArcGIS 10.3 software was used to process the Landsat geodata. The preprocessing included assigning coordinate system to data, radiometric correction, layer stacking of separate bands of datasets, mosaic the stack of bands and generating polygons after process.

Radiometric correction is a technique to correct or reduce errors in digital numbers of remote sensed data. This technique is, due to inconsistencies during different times of the geodata collection, important when comparing two datasets of different time of period. Radiometric errors must be corrected to obtain a true ground radiance and reflectance values. Radiometric correction is applied using values taking into account effects of satellite sensors, angle of sun, topography and atmosphere [19].

Maximum likelihood algorithm is popular and was used effectively in assessing satellite imageries [17]. At the end, seven types of LULC classes were derived from study area: agricultural land, bare soil, waterbodies, built-up areas, lowland, sandy area, saline land. In order to get accurate results, 18 training samples for each land use were collected and generated sample signature. Finally, to find out change detection results a post-classification method was applied, and the LULC change direction and change detection matrix between initial and final states was evaluated.

3 Results and Discussions

All classes were named generally and they may further be divided into subclasses, such as Agricultural Land, which also included crop fields and fallow fields (cotton field, grain, garden and rice field); Bare Soil, which included rock, stony and unused land plots; Waterbodies, which included both rivers, lakes and reservoir; Built-Up areas, which included residential, commercial and industrial subclasses; Saline Land which includes salinized land, land where water left saline land behind; Lowland which includes hilly areas that does not have vegetation and finally the Sand that are covered with sand. Selected signature fits nomenclature system adopted in Uzbekistan.

In Fig. 2 and 3 the spatial distribution of LULC for 1987 and 2019 years is provided. The classified image for 1987 is divided into seven classes, with 51.5% (3098,54 km^2) of the area covered with agricultural land, 2.1% (123,86 km^2) with built up, 12.4% (746,17 km^2) with bare land, 7.5% (453,79 km^2) with lowland areas, 20.2% (1216,21 km^2) with sandy land, 5.5% (332,47 km^2) with water and 0.8% (50,78 km^2) with Saline Land. About 60.8% (3661,08 km^2) of the area of the 2017 image is classified as agricultural land, 10.1% (610.8 km^2) as built-up areas, 1.8% (109,23 km^2) as bare land, 0.4% (23,47 km^2) as lowland areas, 0.4% (730,06 km^2) as sand, 4.8% (288.01 km^2) as water and 9.9% (6021,82 km^2) as built-up areas. We can see that agricultural land gained 562.54 km^2 (+18.2%), built up area increased by 486.94 km^2 (+393.1%), and salinized land by 548.39 km^2 (+1079.9%). Bare land decreased by 636.9 km^2 (-85.4%), lowland areas by 430.32 km^2 (-94.8%), sandy land by 486.15 km^2 (-40%), and finally waterbody by 44.46 km^2 (-13.4%).

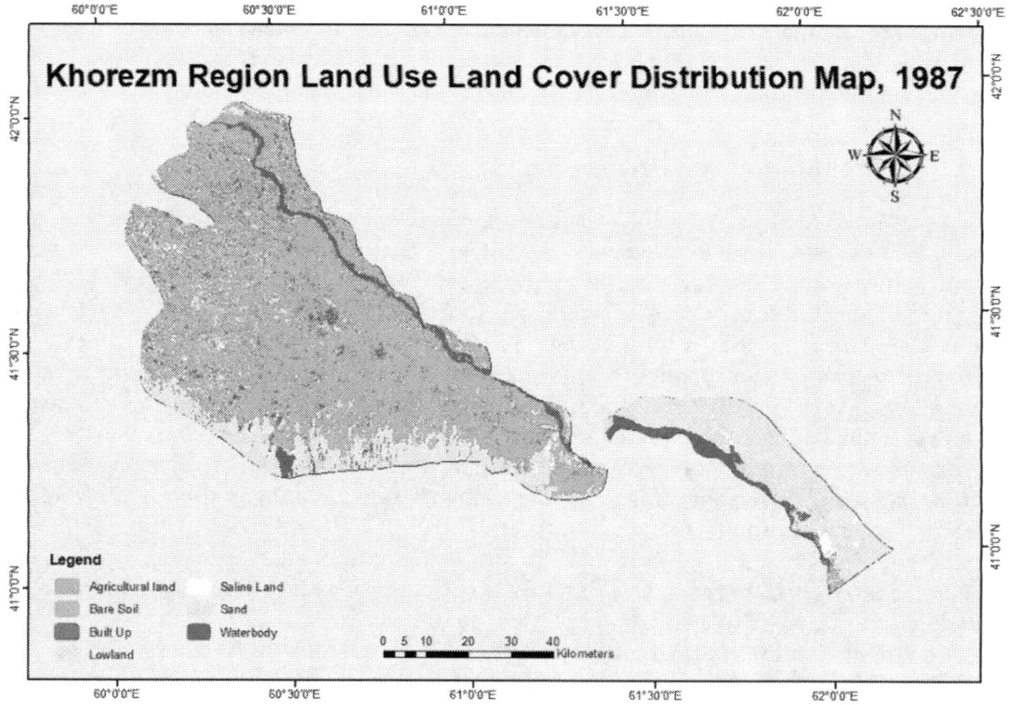

Fig. 2. LULC spatial distribution of Khorezm region, 1987.

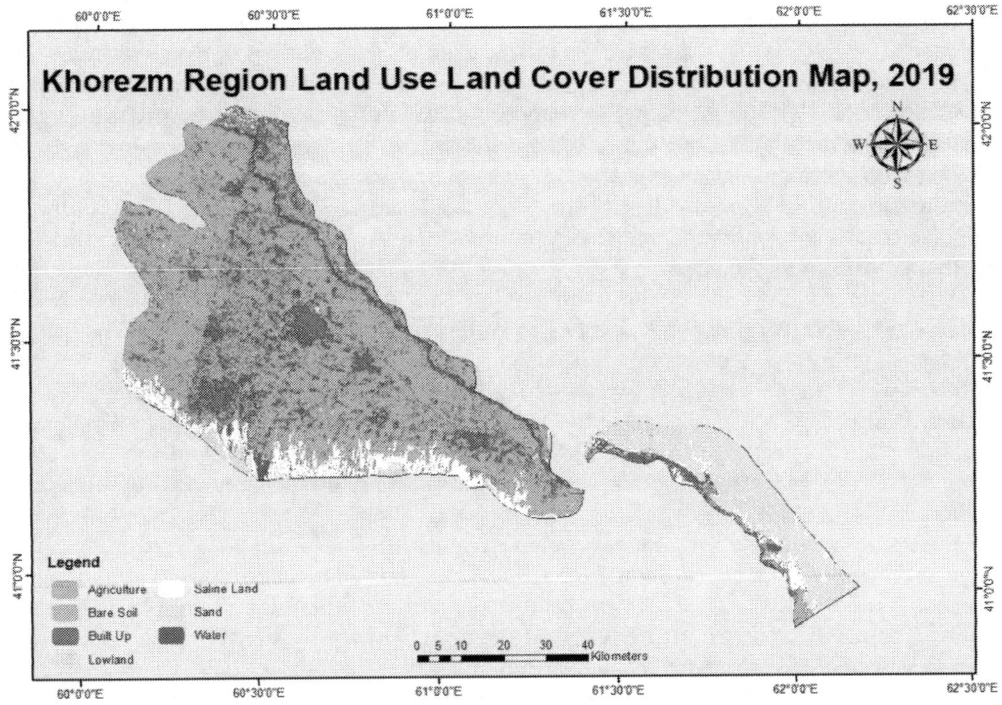

Fig. 3. LULC spatial distribution of Khorezm region, 2019

Above numbers can be discussed as population increased by 60% (1991-2017) therefore built up area increased accordingly, people gained irrigated land at the expense of hilly areas and bare land to increase agricultural land. Area of agricultural land was increased from 3098,54 km^2 to 3661,08 km^2, built up area from 123,86 km^2 to 610,8. But Salinized land increased by 11 fold from 50.78 km^2 to 599.17 km^2. This can be explained as Aral Sea's natural disaster resulted in Khorezm region. Since Uzbekistan, including Khorezm has been facing water scarcity therefore water surface area was decreased by 44.46 km^2 (-13.4%) in 2019. See table 1.

Table 1. Land use and land cover change (1987-2019).

#	Land use type	1987		2019	
1	Agricultural Land	3098,54	51,5	3661,08	60,8
2	Built Up	123,86	2,1	610,8	10,1
3	Bare Land	746,17	12,4	109,23	1,8
4	Lowland	453,79	7,5	23,47	0,4
5	Sand	1216,21	20,2	730,06	12,1
6	Water	332,47	5,5	288,01	4,8
7	Saline Land	50,78	0,8	599,17	9,9
		6021,82	**100,0**	**6021,82**	**100,0**

We can see on Fig.3 land in south and southwest of western Khorezm was salinized as well as some areas in east of eastern Khorezm. Agricultural class with increased dynamically. The class rose by 18.2% (562.54 km^2). Increase can discussed as increase in population reflected in the agricultural areas increase, mainly because of subsidence farming: About 75% of the Khorezm region's population lives in houses with small irrigated land attached to it. People settled within 32 years in place where it was bare land. It is also needed to mention that vast areas of rice plots disappeared in 2019.

Built-up area was increased in mostly where bare land is. The Built up area, with an increase of 393.1% (486.94 km^2) for the total area of the region. This is obvious since built-up areas expansion is related to the number of population and the population increased from 1 066 000 to 1 776 700 in the period between 1991 and 2017. Population density in the Khorezm province is 30.0 per km^2, and from 1991 to 2017 the population increased by 66.7%, with an annual average rate of 2.0 %. Main income of The Khorezm province's economy is tourism. Because Khiva and Urgench touristic sites are located in Khorezm, and also resorts and children's camps are built in last 30 years within the region, which is another reason for this increase.

The water class on the image mainly represents Amudarya River and lakes, muddy water, suspended water and salty water. In order to get accurate image analysis, we got separate samples for muddy water, suspended water and salty water. Because they scatter differently for each one. Later we united all of them into one as waterbody. By the classified images, a slight change in the waterbody class is detected with a 13.4% decline (44.46 km^2). This result is mainly related with less water coming from Amudary River and water scarcity in general. As we can see that some lakes are disappeared leaving saline land behind.

Bare land area is occurred highest decline, minus 85.4% (636.94 km^2) from the initial area. From Table 2 we can see that the bare soil class has changed to forest and agriculture classes. As the study area befalls the protected zone of Uzbekistan from 1992, the main effort is being paid for the afforestation of the area. Most probably this is the one of the main reasons behind the decline of the bare soil class.

4 Conclusions

Remote Sensing tool – ArcMap was used together with Landsat imagery to find out land use and land cover distribution change in Khorezm region. Processed geodata with maximum likelihood classification brought good results for how the land use distribution changed over 32 years. Research find out the significant increase of agricultural land and built up area. Due to rise of population reflected in increasing built up as well as agricultural land. People pushed pressure on natural bare land to gain more space. Also saline land is increased tremendously (11 fold) because of natural disaster (Aral Sea shrink). Strong wind moved salt from former bottom of the Sea to neighboring area, including Khorezm. The most gains at the expense of bare land and lowland areas.

The land use and land cover maps generated from the research will be further used for decision maker from government bodies to think and take action towards softening natural hazards. It might be utilized by local khakimiyat and cadastral bodies to enhance urban and region planning. The LULC distribution maps for 1987 and 2019 alerts ecology specialists that Aral Sea has affecting not only Karakalpakistan but also Khorezm region.

References

1. E. López, G. Bocco, M. Mendoza, E. Duhau. Predicting land-cover and land-use change in the urban fringe. Landscape and Urban Planning, **55 (4)**, 271 (2001)
2. P.H. Verburg, K. Neumann, L. Nol. Challenges in using land use and land cover data for global change studies: land use and land cover data for global change studies. Global Change Biology, **17 (2)**, 974 (2011)
3. Yang, G. Sun. Landsat-Based Land Cover Change in the Beijing-Tianjin-Tangshan Urban Agglomeration in 1990, 2000 and 2010. ISPRS International Journal of Geo-Information, **6 (3)**, 59 (2017)

4. P. Gong, J. Wang, L. Yu, Y. Zhao, Y. Zhao, L. Liang, J. Chen. Finer resolution observation and monitoring of global land cover: first mapping results with Landsat TM and ETM+ data. International Journal of Remote Sensing, **34 (7)**, 2607 (2013)
5. S. Fuchs, V. Röthlisberger, T. Thaler, A. Zischg, M. Keiler. Natural Hazard Management from a Coevolutionary Perspective: Exposure and Policy Response in the European Alps. Annals of the American Association of Geographers, **107 (2)**, 382 (2017)
6. J.S. Rawat, M. Kumar. Monitoring land use/cover change using remote sensing and GIS techniques: A case study of Hawalbagh block, district Almora, Uttarakhand, India. The Egyptian Journal of Remote Sensing and Space Science, **18 (1)**, 77 (2015)
7. Ruiz-Luna, C.A. Berlanga-Robles. Land use, land cover changes and coastal lagoon surface reduction associated with urban growth in northwest Mexico. Landscape Ecology, **18 (2)**, 159 (2003)
8. M.G. Turner, C.L. Ruscher. Changes in landscape patterns in Georgia, USA. Landscape Ecology, **1 (4)**, 241 (1988)
9. Y. Chen, X. Li, X. Liu, B. Ai. Analyzing land-cover change and corresponding impacts on carbon budget in a fast developing sub-tropical region by integrating MODIS and Landsat TM/ETM+ images. Applied Geography, **45**, 10 (2013)
10. F. Yuan, K.E. Sawaya, B.C.Loeffelholz, M.E. Bauer. Land cover classification and change analysis of the Twin Cities (Minnesota) Metropolitan Area by multitemporal Landsat remote sensing. Remote Sensing of Environment, **98 (2-3)**, 317 (2005)
11. X. Wang, D. Zheng, Y. Shen. Land use change and its driving forces on the Tibetan Plateau during 1990-2000. CATENA, **72 (1)**, 56 (2008)
12. X. Li, S. Zhao, H. Yang, D. Cong, Z. Zhang. A Bi-Band Binary Mask Based Land-Use Change Detection Using Landsat 8 OLI Imagery. Sustainability, **9 (3)**, 479 (2017)
13. O.R. Abd El-Kawy, J.K. Rød, H.A. Ismail, A.S. Suliman. Land use and land cover change detection in the western Nile delta of Egypt using remote sensing data. Applied Geography, **31 (2)**, 483 (2011)
14. H. Yin, A. Khamzina, D. Pflugmacher, C. Martius. Forest cover mapping in post-Soviet Central Asia using multi-resolution remote sensing imagery. Scientific Reports, **7 (1)** (2017)
15. R. Kraemer, A.V. Prishchepov, D. Müller, T. Kuemmerle, V.C. Radeloff, A. Dara, A. Terekhov, M. Frühauf. Long-term agricultural land-cover change and potential for cropland expansion in the former Virgin Lands area of Kazakhstan. Environmental Research Letters, **10 (5)**, 054012 (2015)
16. J. Edlinger, C. Conrad, J. Lamers, G. Khasankhanova, T. Koellner. Reconstructing the Spatio-Temporal Development of Irrigation Systems in Uzbekistan Using Landsat Time Series. Remote Sensing, **4 (12)**, 3972 (2012)
17. D. Jovanovic, M. Govedarica, F. Sabo, Z. Bugarinovic, O. Novovic, T. Beker, M. Lauter. Land cover change detection by using remote sensing: A case study of Zlatibor (Serbia). Geographica Pannonica, **19 (4)**, 162 (2015)
18. Statistics of Uzbekistan by region. Derived from stat.uz on February 4, 2020.
19. USGS, 2020. Landsat Surface Reflectance. Accessed in April, 2020 (https://www.usgs.gov/core-science-systems/nli/landsat/landsat-collection-1-surface-reflectance)

Features of the organization of agroclusters in the structure of land management projects

Abdugani Mukumov[1,*], *Komil* Xujakeldiev[2], *Fayzullo* Xamidov[3], *Sharafatdin* Narbaev[4] and *Khudoyberdi* Abdivaitov[1]

[1]Tashkent Institute of Irrigation and Agricultural Mechanization Engineers (TIIAME), Land Use Department, 39 Kari Niyazov Street, Uzbekistan
[2]Karshi Engineering Economic Institute (KMI), Geodesy and Cadastre Department, 225 Mustakillik Avenue, Uzbekistan
[3]Bukhara Branch of the Tashkent Institute of Irrigation and Agricultural Mechanization Engineers (BBTIIAME), 200105, Bukhara city, Gazli shokh street, 32, Uzbekistan
[4]Tashkent Institute of Irrigation and Agricultural Mechanization Engineers (TIIAME), State Cadasters Department, 39 Kari Niyazov Street, Uzbekistan

Abstract. The features of the organization of production units - agro-clusters, which are one of the components of land management projects which aimed at organizing the rational and efficient use of available land resources are described in this article. The term *"cluster"* is a French word that means *"bundle", "collection"* in Uzbek. It can be taken as "the geographical proximity of enterprises and institutions cooperating with each other in a particular field". The development of value-added production in the agricultural sector and the development of this market requires ensuring product quality standards, full use of scientific and scientific achievements in the processing process, development of existing research institutes and using the potential of geographical location. In our view, the solution of this problem can be found in the agricultural production system through the organization of cluster production, which is used by developed countries in America, Europe and Asia.
Keywords: agriculture, land management, project, agro-cluster, labor resources, land fund, land plot, productivity.

1 Introduction

In the conditions of formation of a competitive innovative economy in the Republic of Uzbekistan, including in the agricultural sector, ensuring the country's independence and food security, improving the living standards and quality of the rural population, poverty eradication, rational and efficient use of natural resources and environmental protection for future generations are the top priorities of the state.

It is known that the agricultural sector of the country has a great potential for the production and export of fruit and vegetables. The analysis shows that in the years when fruit and vegetables grown in the country exceed domestic demand, the average price of the

* Corresponding author: abduganimuratov59@gmail.com

product in the domestic market decreases, and farmers who grow crops suffer. In addition, a low market prices hinder the release of products into the market, resulting in the loss of most products before they reach the consumer. Among many other factors, a poor sales system is a major obstacle to falling average market prices or increasing losses. This, in turn, has worsened the economic situation of agricultural enterprises, leading to problems in providing the country's population with quality food products. It is well known that ensuring sustainable agricultural production requires not only a short-term approach to increasing production, but also a long-term approach to processing and balanced distribution of products throughout the supply chain.

2 Materials and methods

According to the sources [1], the basis of the "cluster" theory is Alfed Marshall's "Principles of Economics", written in the late nineteenth century. According to its scientific findings, the territorial harmonization of specialized entities is based on the following:

- availability of qualified labour resources;
- growth of supplier and ancillary industries;
- existence of specialization in different links of production processes of different enterprises.

In recent years, the cluster theory has become popular by setting different goals and applying them in practice. Some views prioritize clustering quality over geographic location, others prioritize networking, and still others prioritize innovation.

According to British theorists (J. Dunning, K. Briman, J. Humphrey), a cluster is an institutional theory that defines the basis of the economy as a system of interacting institutions. In this case, the cluster itself is considered a modern institution. Scandinavian scientists (B. Lundwal, B. Johnson, B. Asheim, A. Isaacson) say that the evolution of the cluster has gone through a number of stages.

The theory of clusters has been studied by Russian scientists [2], and projects have been developed for its implementation [3, 4].

In our view, the development of cluster theory can be described in two ways (Figure 1).

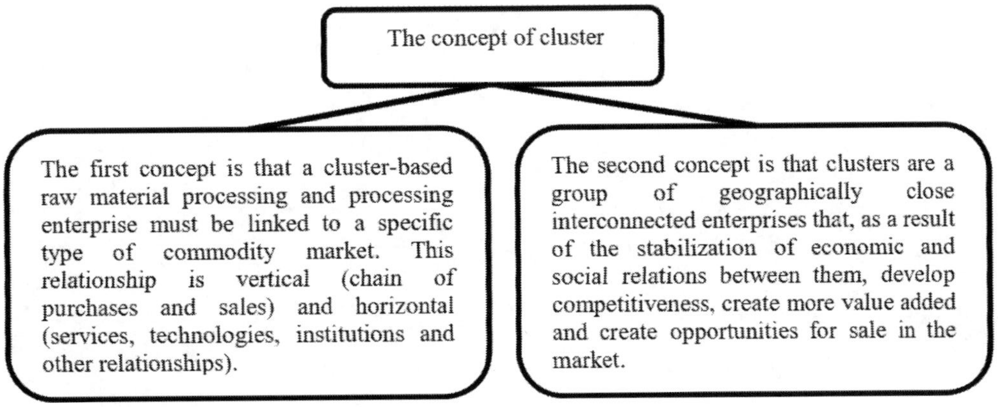

Fig. 1. Description of the cluster concept

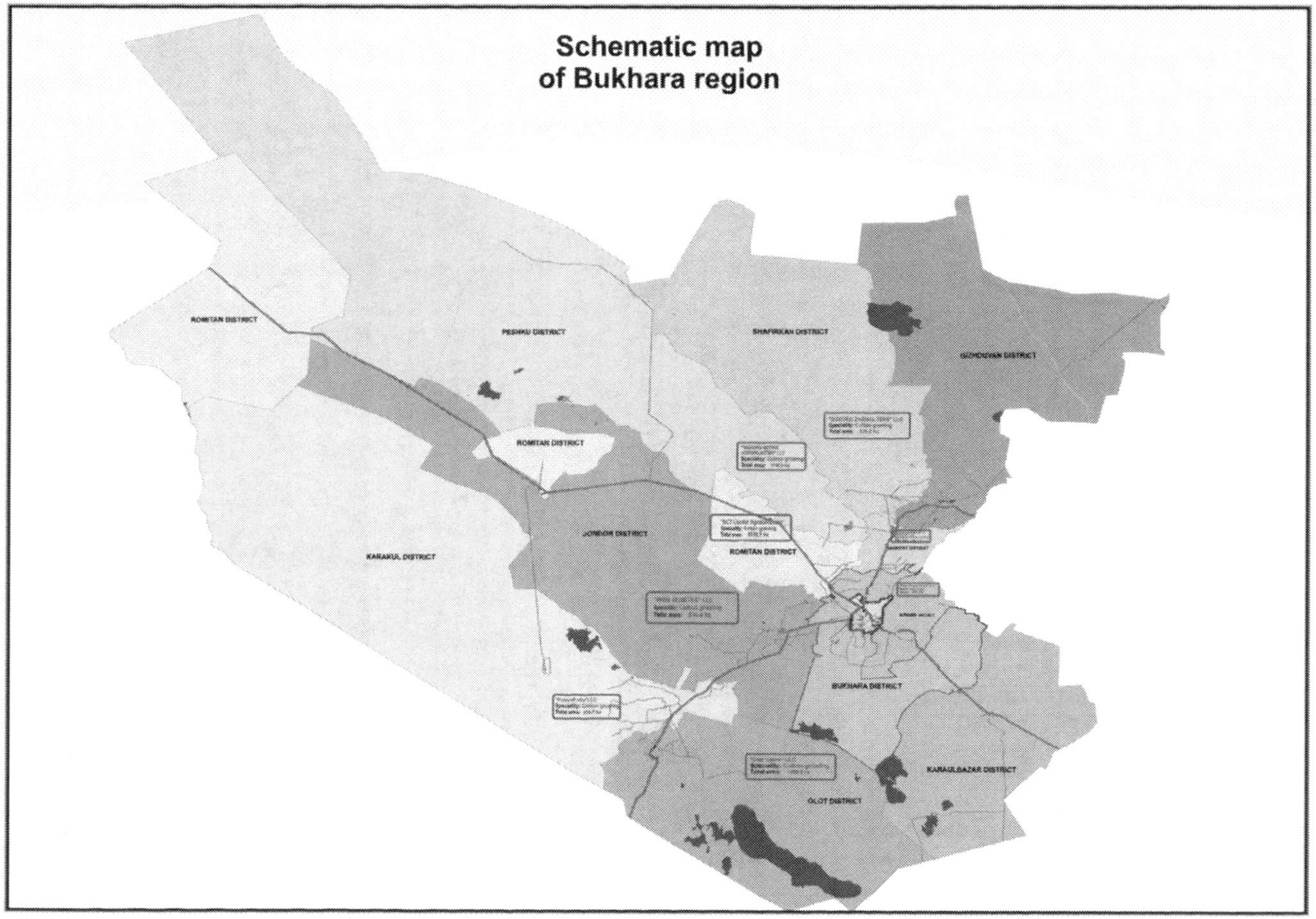

Fig. 2. Scheme of location of cotton agro-clusters in the region

The descriptions of the concept of clusters emphasize that the interaction between enterprises and organizations ensures the economic efficiency of clusters. By further refining and expanding these theoretical definitions, the *"agrocluster-village improving the competitiveness of agricultural products in the domestic and foreign markets, the formation and development of infrastructure in rural areas, increasing employment and incomes of the rural population and improving the environment through the integration of production, processing and sales of agricultural products into a single chain and the use of high technology innovations is a business entity"* engaged in focused activities.

In the Bukhara region, agro-clusters are mainly organized in the following areas: cotton, livestock, horticulture, vegetable growing.

It has the highest number of cotton clusters and is organized regionally in Jondor, Romiton, Vobkent, Alat, Shafirkan, Kagan, Karakul districts of the region (Figure 2).

It is known that the cotton complex is an important element of the socio-economic sphere of the region. At the same time, the potential of this industry is not fully used. One of the problems hindering the development of this sector of the agro-industrial complex is the lack of storage and processing facilities, the lack of modern technologies and the availability of agricultural machinery, and others.

Fruit and vegetable agro-clusters are the second largest agro-clusters in the region. Territorially, they are located in Bukhara, Karavulbozor and Gijduvan districts of the region.

The advantage of agro-clusters in this direction is that they are located on the basis of existing orchards and vineyards and close to large settlements. It is known that the area of agro-clusters in this area is not very large. The main reasons for this are that the processes of planting, processing and harvesting are still largely manual, the capacity of storage and processing enterprises is low, the crop is not guaranteed by the state, and so on [5].

The agro-cluster is the only one in the region located in Karakul district of the region.

As part of the regional land fund (Table 1), the agro-cluster is organized on an area of 361.2 hectares, with hayfields and pastures covering 2541.3 thousand hectares. Factors contributing to this situation include a low yields of forage crops in natural pastures (1.5-2.0 s / ha) for year-round grazing of livestock, and degradation of pastures.

Table. 1. Regional land fund

№	Land type name	Land area		Cultivated lands		Lands occupied by trees		Unused lands		Pastures	
		Total	Including irrigated	Total	Including irrigated	Total	Including irrigated	Total	Including irrigated	Total	Including irrigated
1	Types of agricultural land	2769.0	227.6	201.2	201.2	19.5	19.5	6.9	6.9	2541.3	-

Due to the lack of water resources in the 676,106,000 hectares of land in Alat, Gijduvan, Peshku and Jondor districts of the region, it is not possible to establish year-round cultural pastures that require large capital expenditures. However, there is an opportunity to organize agro-clusters in 4 livestock areas, specializing in karakul due to the organization of rotational grazing.

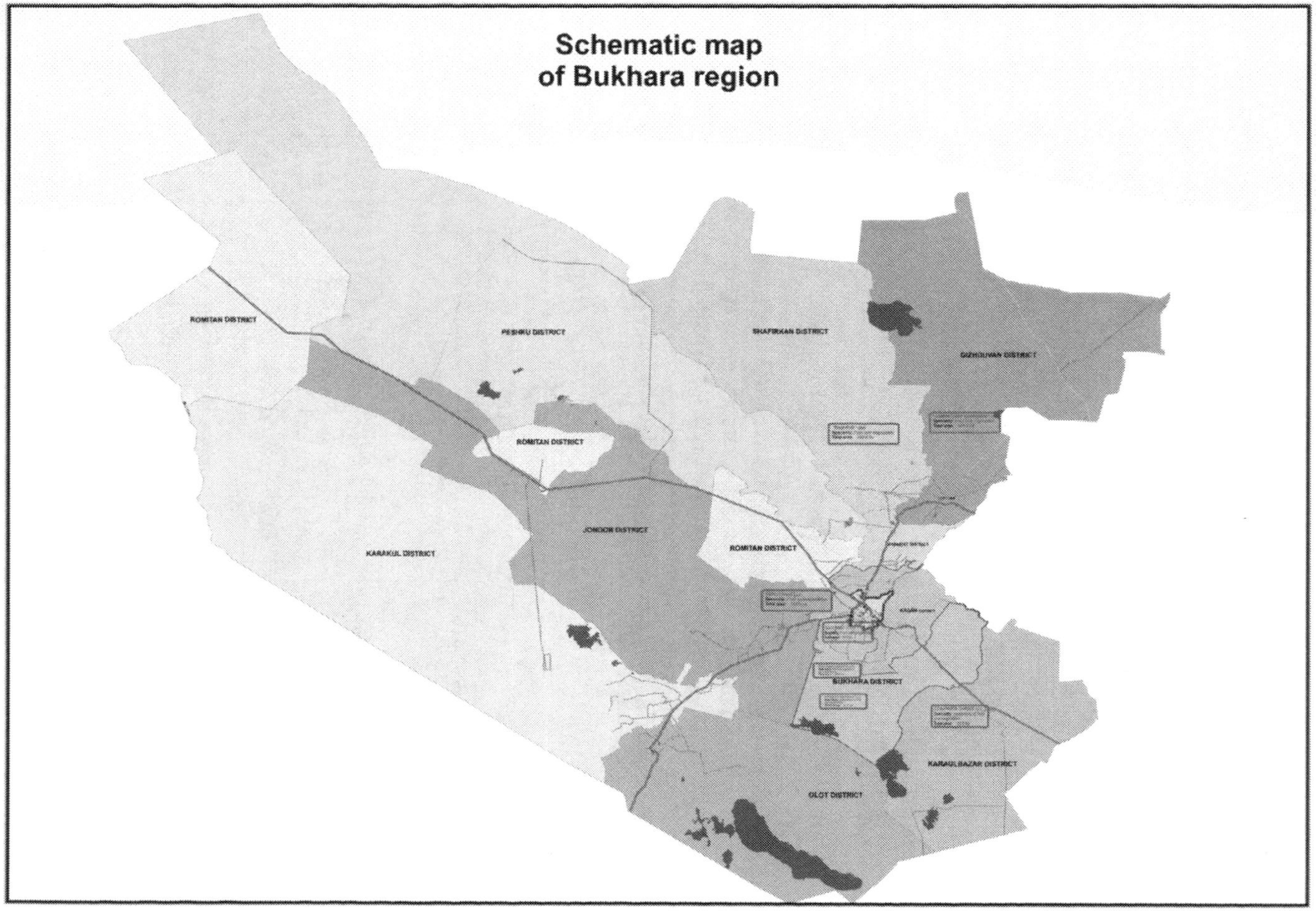

Fig. 3. Scheme of location of agro-clusters in the field of fruit and vegetable growing in the region

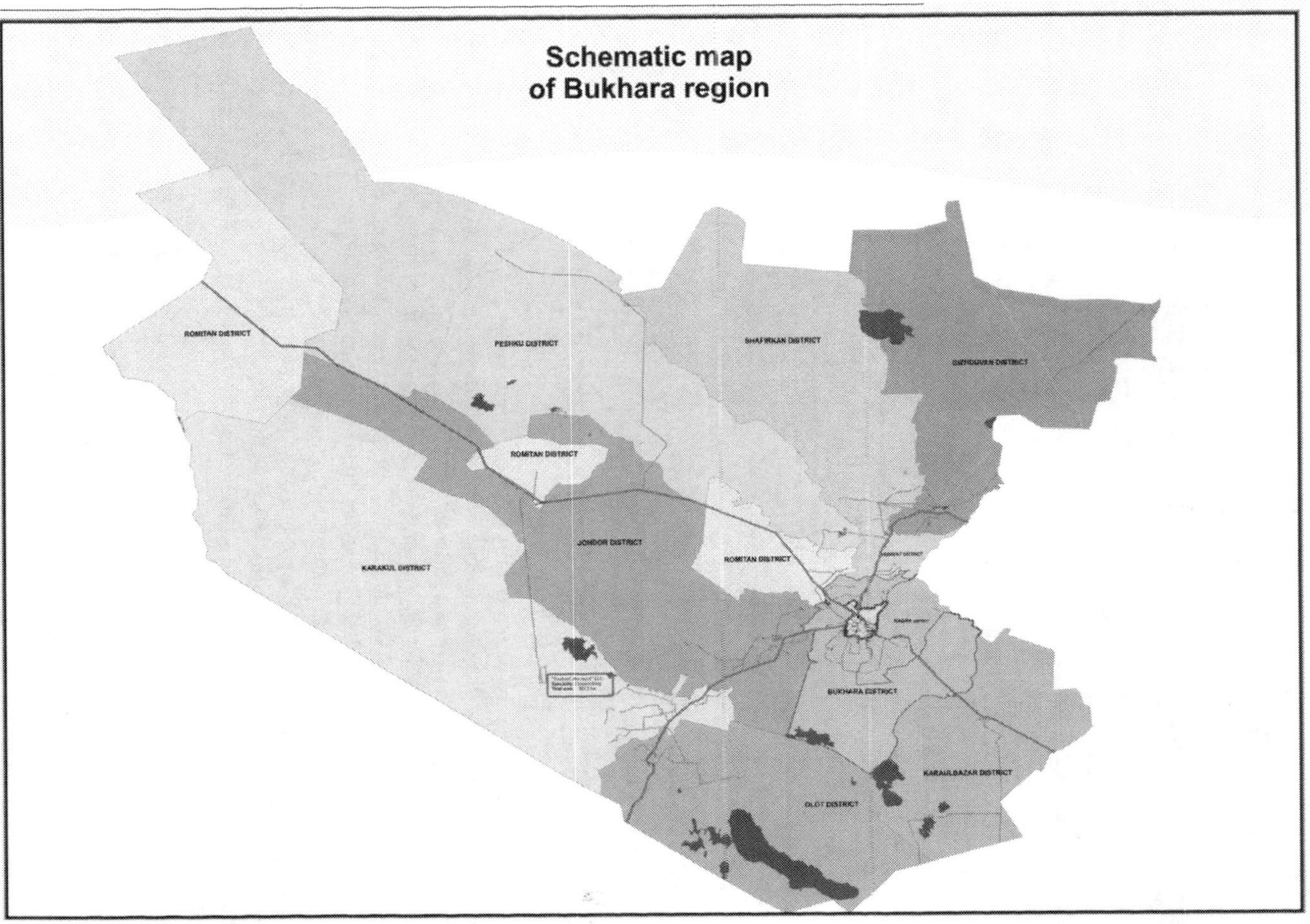

Fig. 4. Scheme of the location of agro-clusters in the field of livestock in the region

https://doi.org/10.1051/e3sconf/202122701003

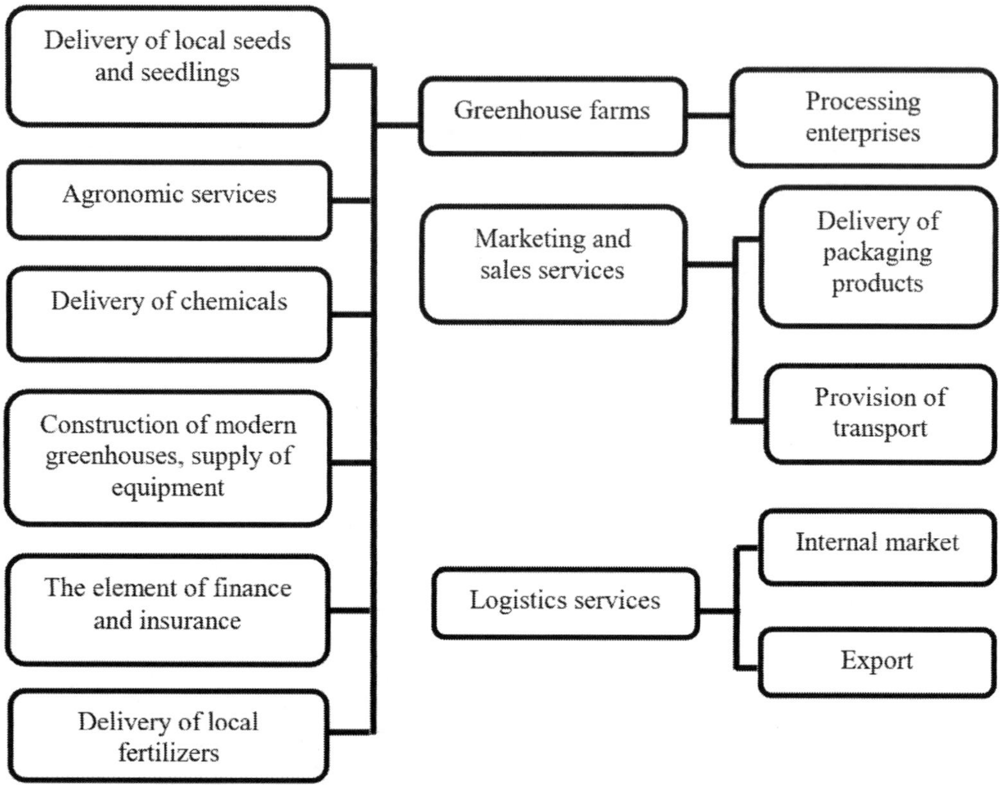

Fig.5. Developed scheme of the agro-clusters in Bukhara region

3 Results

It is known that the application of cluster theory has its own characteristics in different countries and industries, but it also has a number of features that are common to all countries. In our opinion, they are as follows:

- enterprises (firms) have the opportunity to increase labour productivity and production efficiency due to their direct connection to suppliers, qualified specialists, information supply, service and training centres. In enterprises located in clustered areas, the increase in labour productivity by 1.5 times and salary by up to 30% has been confirmed in practice.
- financial opportunities will be created to encourage scientists and specialists who have achieved high results by conducting effective research activities.

One of the important features of the formation of clusters in our country, in contrast to other countries, is the role of the state.

The development of the agricultural sector on a cluster basis requires the organization of ancillary suppliers, service providers and processors [6]. Therefore, the following scheme has been developed in Bukhara region to create an agro-cluster to increase agricultural production, improve its composition, introduce innovative and resource-saving technologies in the industry (Figure 5).

An important aspect of the organization of the agro-cluster is to strengthen the trust of its members through the implementation of joint projects that combine the processes of joint cultivation, processing - sale – research, [7, 8, 9]. With this in mind, the following scheme is proposed for the creation of a chain of value added of food products in agro-clusters organized for the cultivation of livestock and agricultural products (Figure 6).

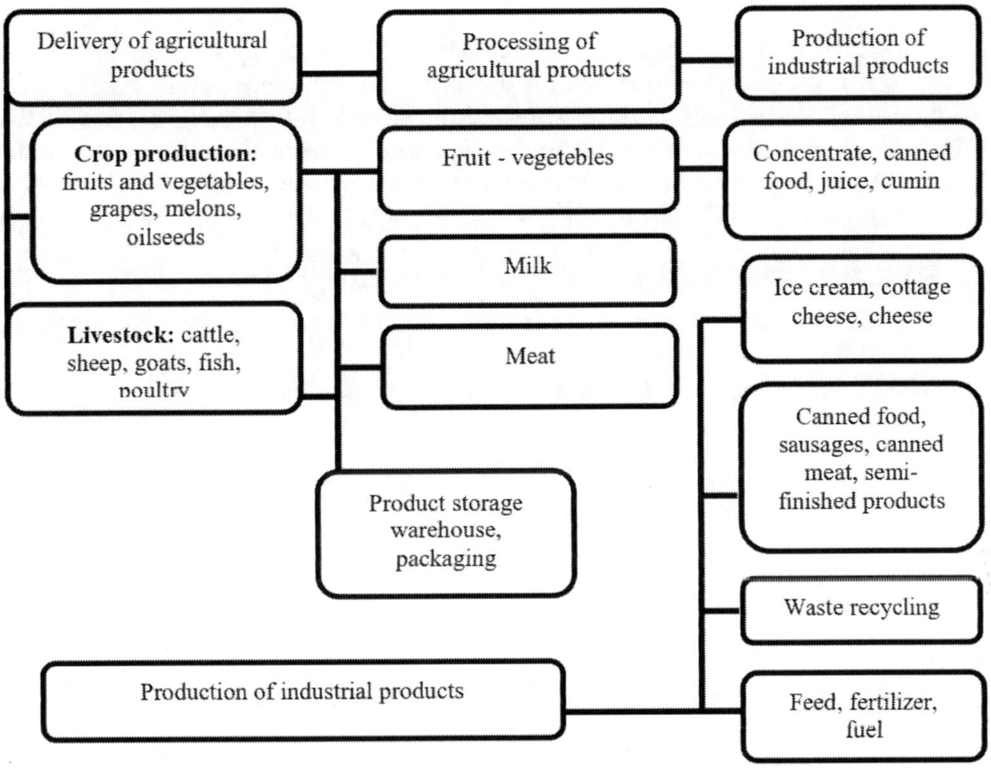

Fig. 6. Proposed scheme for the creation of a chain of value added of food products in agro-clusters

In our opinion, the implementation of the recommended scheme for creating a value chain of food products in agricultural clusters (Fig. 6) will serve as a mechanism for strengthening the confidence of its participants through the implementation of joint projects that combine the processes of joint cultivation of agricultural and other products, their processing-sale-research.

The authors would like to thank Sadulla Avezbaev for his suggestions in this paper.

References

1. M. A. Rakhmatov, V. Z. Zaripov. Cluster integration, innovation and economic growth. Tashkent, Zamin Nashr, 165 p (2018)
2. I. V. Andronova, Yu. P. Bachinina, *Cluster approach in the ensuring of the competitiveness of the region.* Monograph – Tyumen: Tyumen state oil and gas University, 120 p (2010)
3. S. Avezbaev, S.N.Volkov, *Automated design systems in land management.* Textbook. Toshkent, 168 b (2010)
4. S. N. Volkov, *Automated design systems in land management.* Textbook. Moscow, 667 p (2018)
5. A. S. Chertovitsky, A. K. Bazarov, *Land use Economics.* Textbook. Tashkent, TIIM, 220 p (2009)
6. U. P. Umurzakov, *Improving the efficiency of using the resource potential of the agricultural sector of the economy of Uzbekistan.* Monograph. Tashkent, "Fan", 211 p (2005)

7. T. V. Papaskiri, *Geoinformation systems and computer-aided design technologies in land management.* Educational and methodological guide (4th edition, revised and supplemented) - Moscow: publishing house "New printing technologies", 249 p (2013)
8. T. V. Papaskiri, *Automation of land management design and land management (efficiency and organization).* Land management, cadastre and land monitoring, Moscow: Panorama publishing house, Athena publishing House, **5**, Pp. 12-22 (2014)
9. F. R. Hamidov, The role of land management in organization of rational use and protection. European Sciences review Scientific journal № 11–12 (November–December), (2017)
10. Land Records and the Role of Government. — London: Land and Property Economics, p.10 (2001)
11. Google Maps. - Access mode: https://www.google.com/maps

An overview of land fund categories distribution in Surkhandarya, Uzbekistan

Khudoyberdi Abdivaitov[1,*], *Sanjarbek* Muratov[2], *Abdurashid* Altiyev[1], *Sevarakhon* Abdivaitova[3], *Musurmon* Norkulov[4] and *Barnokhon* Khalilova[1]

[1]Tashkent Institute of Irrigation and Agricultural Mechanization Engineers (TIIAME), Land Use Department, 39 Kari Niyazov Street, Uzbekistan
[2]State Inspection of Geodesy Supervision under the State Committee on Land Resources, Geodesy, Cartography and State Cadastre of the Republic of Uzbekistan, Osiyo Street
[3]Tashkent Institute of Irrigation and Agricultural Mechanization Engineers (TIIAME), English language Department, 39 Kari Niyazov Street, Uzbekistan
[4]Uzbek State Scientific Research Institute of Land Management "Uzdavyerloyiha"

Abstract. Uzbekistan has a wide range of land use categories, because of its best geographical location. The distribution of the land fund according to its specific purpose is always one of the most important issues in the country. This article provides an analysis of the use of the Republican land fund for the last decade. Explanations of the concept of the land fund have been studied and detailed clarifications have been given for each of its categories. Surkhandarya region was selected as the main study area and the land fund of the last two years was analysed. During the work, statistical land-use data which provided from the state committee on land resources, geodesy, cartography and state cadastre of the Republic of Uzbekistan were used. Besides that, using the importance of geographic information system (GIS) and remote sensing techniques was reviewed for rational land use.

1 Introduction

The term land fund is used in all land legislation acts of Uzbekistan. An analysis of the literature reviewed during the study shows that the term has been explained by a number of scholars, including professor S.Avezbayev, who defined it as follows: The land fund is the total of all lands in a given area, within its boundaries (country, province, district, etc.), which for management, ownership, estate, use, and objects of rent [1]. Another scientist, K.Rakhmonov, gave a slightly different definition: Land fund - the part of the Earth's surface, in particular the bottom of water bodies, located within the state borders of the Republic of Uzbekistan, the administrative borders of the autonomous Republic of Karakalpakstan, regions, districts, cities [2].

The term of land fund is basically used by Commonwealth of Independent States and it is defined as total land area within the boundaries of individual land users or administrative divisions [3].

* Corresponding author: abdivaitov90@list.ru

The issue of land distribution is a key part of economic development analysis. According to Hossein Azadi and others, the following objectives of land distribution are mentioned:

- to improve the living standards in rural areas
- to redistribute of land by improving new land registration
- to redistribute income through organising a new job
- to increase the productivity of agricultural products [4].

However, we think that a number of other points can be added, such as, through the proper distribution of land, it is possible to increase the efficiency of its use, to reduce the illegal occupation of land and etc.

The land fund of the Republic is regularly distributed and redistributed among all sectors of the economy for use in clearly defined purposes by categories. The distribution of the land fund should take into account the zoning of the land in terms of the natural adaptation of agriculture, the natural conditions of the area and a number of other important features. Therefore, the organization of rational and efficient use of land resources is always important and at the same time complex issue. The country's economy can be improved by addressing this issue as accurately as possible. At the same time, the process of rational distribution and redistribution of land resources of the Republic is not always carried out in accordance with the purpose, which shows that there are gaps in this regard. In addition, the lack of work done using Geographic information system (GIS) and remote sensing data requires to do research on this topic.

Land fund categories are a constant process of continually changing the size of the area. One of the reasons for this is that Uzbekistan has a high population growth rate compared to many other countries in Asia and the world. When land is allocated to the population for different purposes, it transfers from one category to another. Moreover, the population of this country has been significantly increased during the last decade, and the primary source of income of most of the people is directly connected to agriculture activities. Various factors such as geographical location, different climate conditions, soil fertility, timely exchange of seasons, proper road and power line networks, and water resources, provide this country to a potential of land use in a wide range of agriculture and industrial purposes. Today the importance of agriculture still plays one of the main role in the government's economy in Uzbekistan [5]. In comparison with the other four Central Asian countries, (i.e. Kazakhstan, Kyrgyzstan, Tajikistan, and Turkmenistan) agriculture has significantly developed in Uzbekistan at the beginning years of independence [6]. But the previous land use policy has caused a huge number of negative effects on the fertility of the soil in agricultural areas of Uzbekistan. In spite of a few land reforms that have been implemented until now, it can be easily defined the numerous issues in this field. Creating such a national data set and land categorisation make it possible to present a better land market system for the government and local people. Therefore, it is possible to increase the efficiency of land use only through the most accurate distribution and redistribution of the land fund of the republic among all sectors of the economy.

Observing changes in land categories is a body part of learning for the global environmental sciences research. The literature review provides the importance of remote sensing and GIS for numerous domains such as natural and social science [7]. Using remote sensing and GIS software is the best methods in order to know the spatial distribution and evolution of geographical phenomena related to land-use and land-cover changes [8]. Today, using various types of remote sensing data is the best option to evaluate, monitor and analyse other characteristics of land categories [9]. In addition, using the remote sensing data and analysis techniques can give a huge number of advantages such as for widely monitoring, detecting changes in time in land cover and land use [10]. Thus the significance of this study consists of a detailed analysis of the land fund and its categories at the country level.

1.1 Overview of republican land fund categories distribution

As of January 1 of each year, each department of land resources and state cadastre prepares a "Report on existing landowners, land users of irrigated agricultural land types and their distribution" and writes a comment letter for this report. The data in the report are obtained on the actual, territorial boundaries of arrays, subsidiary farms of organizations and institutions, other ministries, agricultural enterprises and other landowners, as well as land users, land categories and types. A report made a year in advance is used as a basis for a report for the current year. The analysis of data from the district and regional land resources and state cadastre organizations showed that data on 8 categories of land resources are not prepared at the district's level. Based on the information provided by all districts and regions, it is summarized by the state committee for geodesy and cadastre, and as a result, the Republican land fund is formed (Figure 1).

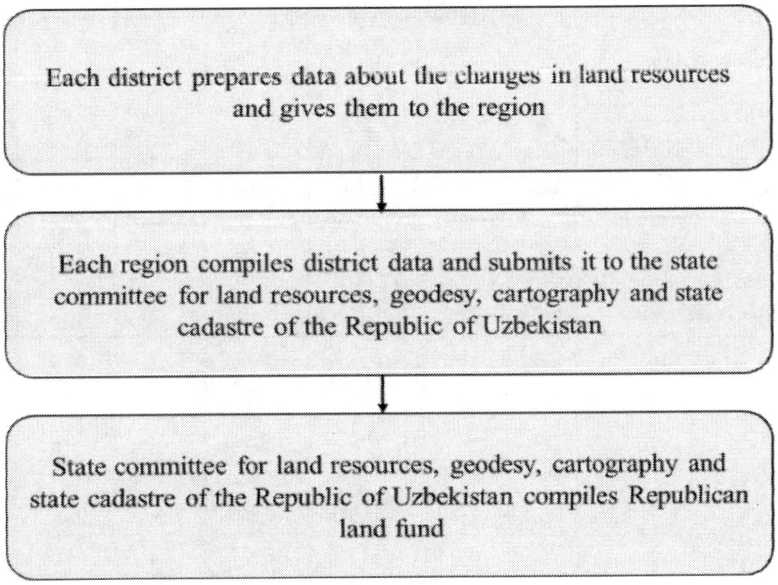

Fig. 1. Formation of Republican land fund (Note: There are 170 districts and 12 regions, 1 autonomous republic, capital city in Uzbekistan).

As of January 1 of each year, the state committee for land resources, geodesy, cartography and state cadastre of the Republic of Uzbekistan prepares a National Report on the state of land resources of the Republic. Because this Committee is a specially authorized state organization that implements the unified state policy in the field of land resources, geodesy, cartography and state cadastre. The report is compiled on the basis of information on all changes in the lands of enterprises, institutions, farms and organizations in the country and as a result of summarizing the annual land reports for the country, approved by the decisions of the district, city and regional governors. According to the National Report, the total land fund area is 448924 km^2, in 2019. 45.08 % of the total area i.e., 202633 km^2 of land is used for agricultural purposes in the country (Table 1) [11]. Irrigated land is 43066 km^2 or 9.6% of the total land area.

Table 1. Distribution of the land fund of the Republic of Uzbekistan by categories, km^2 (Analysis of changes for the period 2010-2019).

№	Categories of land fund	Unit of measurement	2010 year		2019 year		Difference 2010-2019 years, (+/-)	
			Total land area	Including, irrigated lands	Total land area	Including, irrigated lands	Total land area	Including, irrigated lands
1	Agricultural lands	km^2	214532	42177	202363	41912	-12169	-265
		%	48.31	9.5	45.08	9.34	-3.23	-0.16
2	Lands of settlements	km^2	2204	483	2214	510	10	27
		%	0.50	0.1	0.49	0.11	-0.01	0.01
3	Lands for industry, transportation, communications defense and lands for other purposes	km^2	19958	120	8663	123	-11295	3
		%	4.49	0.03	1.93	0.03	-2.56	0
4	Lands for nature protection, healthcare and recreational purposes	km^2	759	6	7104	6	6345	0
		%	0.17	0.001	1.58	0.001	1.41	0
5	Lands for historical and cultural heritage	km^2	11	0	145	0	134	0
		%	0.002	0	0.03	0	0.028	0
6	Lands of the forest fund	km^2	94623	307	111995	446	17372	139
		%	21.31	0.07	24.95	0.09	3.64	0.02
7	Lands of water fund	km^2	8211	46	8369	47	158	1
		%	1.85	0.01	1.87	0.01	0.02	0
8	Lands of reserve	km^2	103805	18	108071	22	4266	4
		%	23.37	0.004	24.07	0.005	0.7	0.001
	Total lands:	km^2	**444103**	**43157**	**448924**	**43066**	**4821**	**-91**
		%	**100**	**9.72**	**100.0**	**9.59**	**0**	**-0.13**

The land fund of the Republic of Uzbekistan has its own characteristics in accordance with the purpose and procedure of land use, which are divided into 8 categories in accordance with article 8 of the Land Code of the Republic of Uzbekistan. The distribution of the land fund by categories as of January 1, 2019 is shown in table 1 above [12]. The table shows that from 2019 to 2010, changes can be seen in all categories of the country's land fund. The following is a description of the republican land fund by categories and changes in their areas (Table 1).

Changes in the Republican land fund for 2010-2019 are discussed for each category. These data show that the distribution of land resources and their categories varies from region to region, with the share of agriculture, forestry and water resources, as well as reserve land being high.

The first category of land fund includes lands allocated to agricultural enterprises and organizations, institutions and citizens engaged in the cultivation of agricultural products or for this purpose [13]. There is a significant decrease in the amount of this category of land, i.e. by 12169 km^2 compared to the total area of the land fund of the republic, and irrigated land by 265 km^2.

The second category of land fund includes rural settlements located within the administrative boundaries of cities and towns and not included in the lands used by agricultural, forestry enterprises, institutions and organizations. Relative growth can be observed in the land area of settlements.

The third category of lands of the land fund includes railway, inland water transport, automobile, air and pipeline transport, which are given to industrial enterprises, including mining, energy enterprises for permanent use for production and construction of ancillary buildings and structures; communications, radio, television and information, military units of the armed forces, border, Internal Affairs and railways, military educational institutions, enterprises, institutions and organizations, land for defence purposes and other purposes for the location of communication lines and related facilities lands given to legal entities for use. The area of land used for industry, transport, communications, defence and other purposes decreased by 11295 km^2 compared to the total area of the land fund of the republic.

The fourth category of land resources includes lands occupied by specially protected natural areas, which have natural healing properties, as well as lands used for public recreation and tourism. The main part of this category of land is occupied by nature reserves and national parks. State nature reserves are nature protection, research and environmental education institutions whose purpose is to preserve and study natural processes and events, the genetic fund of flora and fauna, certain groups of plants and animals, and typical and unique ecosystems. Between 2019 and 2010, the area of the land fund increased by 6345 km^2.

The fifth category of lands of the land fund includes historical and cultural reserves, memorial gardens, cemeteries, archaeological, historical and cultural monuments, lands given to the relevant institutions and organizations for permanent use and are specially protected. This category of land increased by 134 km^2.

The sixth category of land resources includes lands covered with forests, as well as lands allocated for forestry needs, even if they are not covered by forests. Considering the nature protection features of forest areas and the long process of reforestation, it is necessary to pay more attention to the problem of their protection and reproduction. For this reason, since 1991, forested pastures and hayfields used by agricultural enterprises and organizations have been transferred to the category of forest lands in accordance with the established procedure. This creates some favourable conditions for expanding the area of forest lands and, on this basis, for preventing soil degradation and deforestation for economic purposes. Among the categories of land, the largest increase in land area is for these purposes, an increase of 17372 km^2.

Lands of the seventh category of the land fund are allocated to enterprises, institutions and organizations for water management needs in the region occupied by water basins, rivers, lakes, reservoirs, hydraulic and other water management structures, as well as along the banks of water basins and other water bodies lands. Lands of the water fund include lands occupied by water bodies, structures, and lands of protection zones adjacent to water bodies allocated for water management needs. Lands of this category increased by 158 km^2.

The last eighth category of the land fund includes all (including underwater) reserve lands that have not been transferred to legal entities and individuals for the purpose of possession, use and rent of lands and are not included in other land categories. The difference is 4266 km^2.

2 Materials and methods

2.1 Study area

Surkhandarya is located in the extreme south-east part of Uzbekistan, between 37°00' and 39°20' north and 66°30' and 68°20' east, and is bounded on the north by the Hissor ridge, on the east by the Babatag ridge, on the northwest by the Boysun ridge, and on the west by the Kuhitang ridge. The region is divided into 14 districts. As of January 1, 2020, according data to the regional department of land resources and state cadastre, the land area of the districts is distributed as follows (Fig. 2). The Surkhandarya region is the hottest and longest growing region in the country due to the fact that it is surrounded by high mountains from the north, where cold air masses come from. In the relief of Surkhandarya, along with the mountains, there are also plains. The southern slopes of the Hissor ridge, which surround the northern part of the region, face the Surkhandarya. Its highest peak is 4643 m above sea level. The surface of the region gradually decreases from north to south. There are lowlands in the area, between which there are narrow and deep valleys. As Surkhandarya differs from other regions by its relief, climate, natural conditions, soils and many other features, it is very important to properly organize the land use system.

2.2 Tests

During this part, we used different programs such as ArcGIS, ENVI image analysis software were used for visualising the topographic map of the study area (Fig. 3) and doing Land use and Land Cover (LULC) map (Fig. 5). Another software Global Mapper is also widely used in the worldwide geospatial community by a huge number of researchers for the last three decades [14]. Because today remote sensing techniques are the most famous methods for LULC change detection. For different aspects of land cover change, multi-temporal remote sensing (RS) based on change detection analysis is used [15-16].

For a better representation of the topographic map of Surkhandarya region, we used TIN (triangulated irregular network) tool. The tin file was created and visualised in figure 3. Data were downloaded on the website of USGS (United States Geological Survey) Earth Explorer. Then mosaic tool was used in ArcGIS and as a result made the topographic map of the study area.

For LULC map, Landsat 8 OLI (operational land imager) data of July 2018 were also downloaded on the website of USGS. The Optical Landsat 8 OLI used in this part is Level 1 C ortho-rectified imagery data. Besides that, radiometric calibration was performed to accurately represent and convert the satellite digital number data (DN) values to radiance with help ENVI 5.0 software. The atmospheric correction was applied in order to remove or reduce the influence of the atmosphere. For all the processing and post-classification steps were done using the software packages ENVI 5.0 and ArcGIS 10.6.1. Supervised classification methods and maximum likelihood algorithm were used for making LULC maps. Because at the present time maximum likelihood algorithms are prominent and were repeatedly used effectively in assessing satellite imageries [17].

In addition, ArcGIS was used for doing the map of the territorial administrative division of Surkhandarya region and the diagram of distribution of land areas in the districts was done using Microsoft Excel 2019.

https://doi.org/10.1051/e3sconf/202122701004

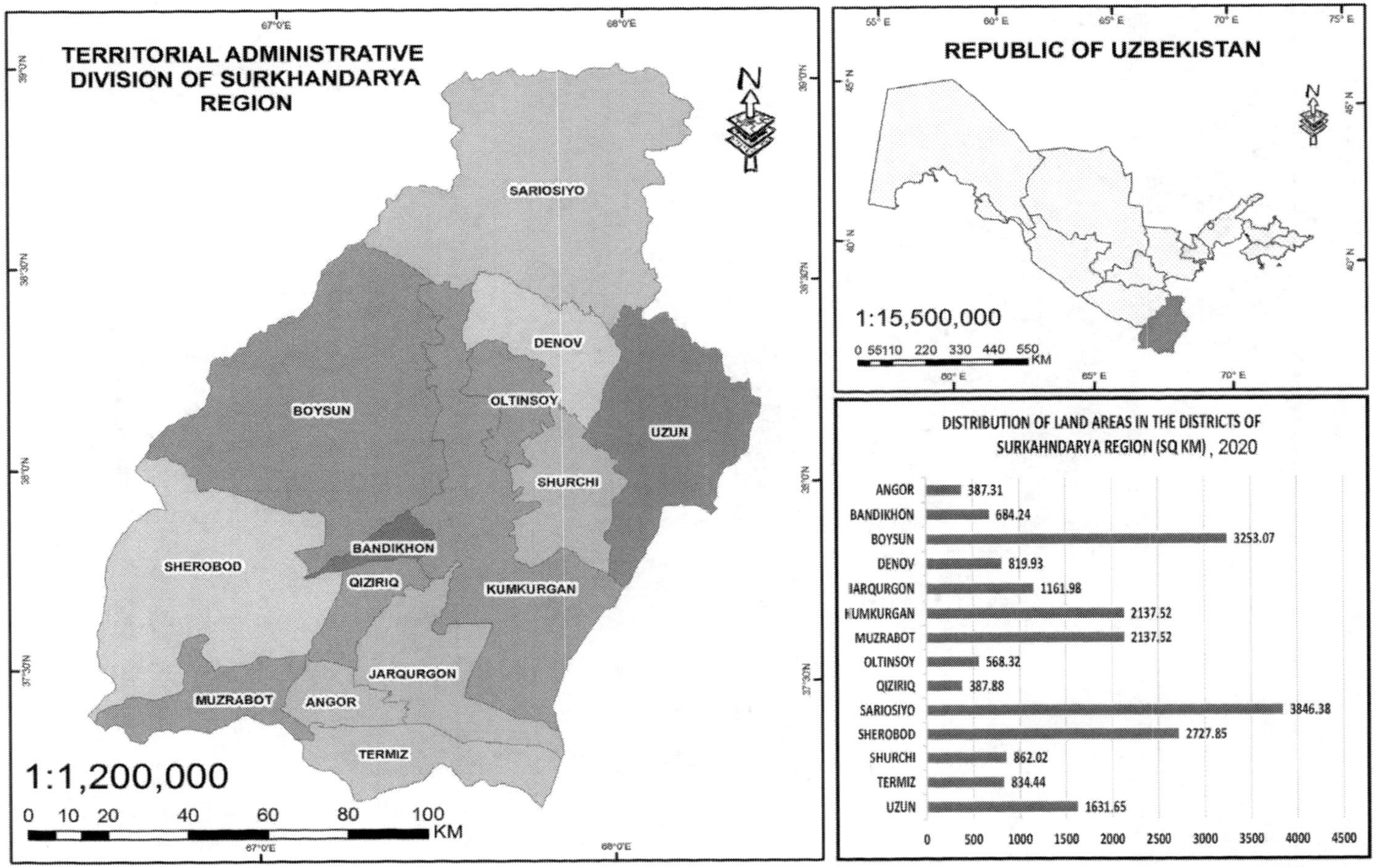

Fig. 2. Location of the study area.

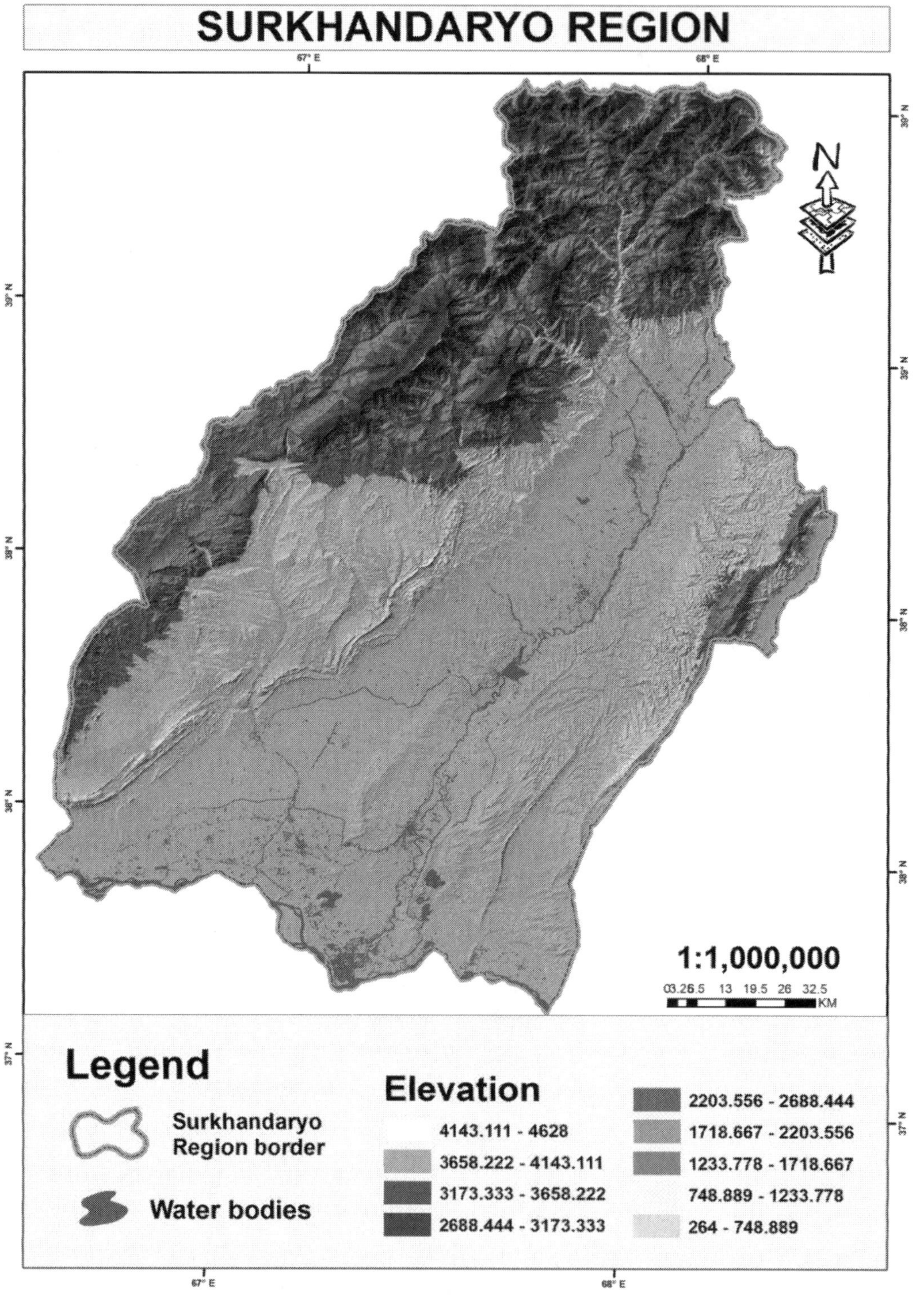

Fig. 3. The topographic map of Surkhandarya Region.

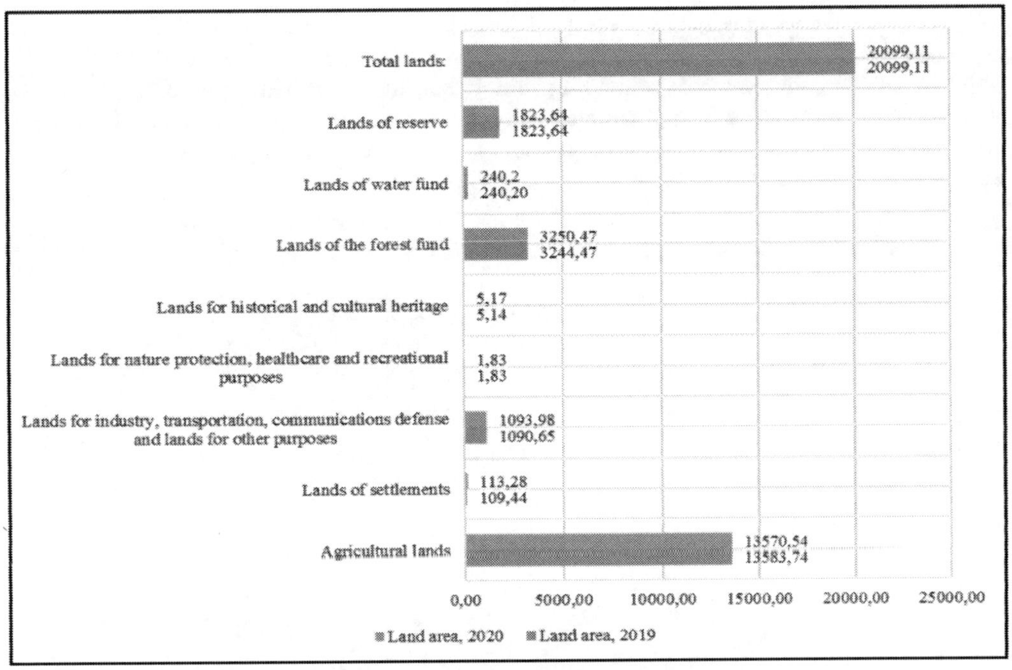

Fig. 4. Distribution of land fund of Surkhandarya region by categories (km^2).

If we analyze the changes of the land fund of the region in the last two years (2019-2020), we can observe a slight change in each category, except lands for nature protection, healthcare and recreational purposes. The main part of the land fund of Surkhandarya region is agricultural land (67.6% of the total land area of the region). Lands for industry, transportation, communications defence and lands for other purposes has slightly increased, namely 3.33 km^2 (Fig. 4).

3 Results and discussions

The classification of the land fund in the Republic of Uzbekistan does not differ much from the categories listed in Article 7 of the Land Code of the Russian Federation. That is, the fourth and fifth categories of land fund are recognized as one category in the Land Code of the Russian Federation [18]. It is called as lands of specially protected territories and objects. Other land fund categories called the same names. Therefore, the land fund is divided into eight categories in Uzbekistan, it is seven categories in the Russian Federation. It is possible to see how much the land fund needs to be properly distributed and redistributed in all sectors of the economy. Therefore, the main problem of the world land fund is to regulate the use of agricultural land. In many countries, measures are being taken to preserve the land fund and regulate its composition [19].

Correctly land distribution among in every single land users and carrying out constantly land monitoring require the best solution. In generally land resources are considered to be the important resource of the economy. That's why organizing rational and effective land use is one of the top problems in Uzbekistan now. One of the important measures of organizing rational land use is land management. The conception of land management is understood about land policy, land laws, land monitoring, state regulation, and economic development [20]. Using modern technics and technologies in doing land management work is today's demand. Soil salinity could be clearly explained in the maps by using modern GIS technologies [21]. However, it is difficult to find GIS and remote sensing works when categorizing lands for the study area. One of the main reasons for this is that the speed and quality of internet connection is not good [22]. Internet connectivity is often a problem in this country for downloading remote sensing data. This should be taken into account when working GIS applications and materials (e.g. ArcGIS Pro, ArcGIS Online, downloading free satellite imagery data sources such as USGS (United States Geological Survey) Earth Explorer, Landsat and etc.).

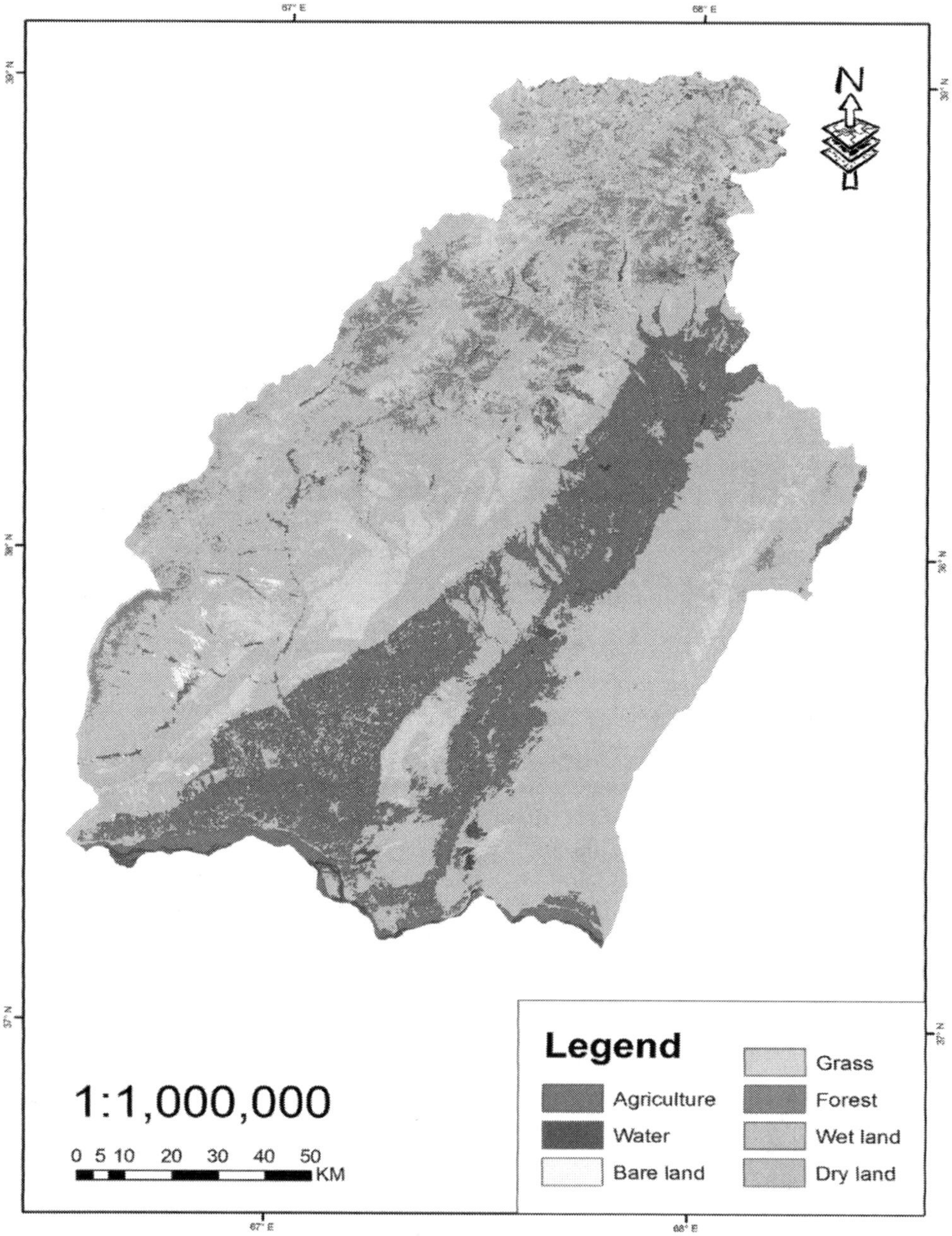

Fig. 5. Land use and land cover (LULC) map of Surkhandarya region derived from Landsat 8 OLI (operational land imager).

LULC map shows that this region has different types of LULC classes, such as agricultural lands, forest, water, grass, wetland and dry land. In general, Surkhandarya region is surrounded by mountains on three sides. It mains that this region of Uzbekistan has a huge potential on using these types of land categories, especially for the cultivation of agricultural products, expansion of forest areas, further development of cattle - raising in pasture areas and etc. Equations should be centred and should be numbered with the number on the right-hand side.

4 Conclusion

In this paper, the distribution of land fund categories in Uzbekistan have been reviewed. The use of land resources was analyzed on the example of Surkhandarya region and at the national level. Some papers were reviewed on how the importance of remote sensing and GIS for land use. The government lands of Uzbekistan still have not been evaluated by the earth observation data of Sentinel-1 & 2 for land use suitability purpose. There is a need to analyse the usefulness of these lands using the current applications of geographic information system and earth observation data. More work is required to organize of extensive use of modern GIS and remote sensing data in the assessment, observation and analysis of a number of land features in Uzbekistan. Any sustainable development in all fields should balance preparing high qualified experts, the most modern information supply, and good policy in the country.

We would like to thank Josef Strobl and Sabine Hennig from Department of Geoinformatics - Z_GIS, University of Salzburg, Austria for their suggestions during this paper.

References

1. S. Avezbayev, S.Voklov, *Land management projecting*. Textbook, Tashkent: Yangi asr avlodi, p. 784, (2002)
2. Q. Rakhmonov, Sh. Narbayev, Z. Muqumov, *Land resource management*. Textbook, TIIAME, p. 157, (2018)
3. S.Voklov. *Basics of land management*. Textbooks and manuals for students of higher educational institutions. Moscow : P. 53-59. (2015)
4. H. Azadi, E. Vanhaute, K. Janečková, P. Sklenička, D. Teklemariam, L. Feng, F. Witlox, Evolution of land distribution in the context of development theories, Land Use Policy, P. 1-9, **97** (2020). https://doi.org/10.1016/j.landusepol.2020.104730
5. I. Abdullaev, M. Giordano, A. Rasulov, "Cotton in Uzbekistan: water and welfare", *The cotton sector in Central Asia economic policy and development challenges*. The School of Oriental and African Studies. University of London, London: P. 112-128, (2007)
6. P. Bloch, "Agrarian reform in Uzbekistan and other Central Asian countries", Land Tenure Center University of Wisconsin–Madison, WORKING PAPER, NO. **49**, May (2002)
7. S. Hennig, A. Nazarkulova, Benefits and Challenges of eLearning in Central Asia. International Journal of Geoinformatics, P. 43-51, **15**, (2019)
8. E. Popovici, K. Gheorghe, Assessment of changes in land-use and land-cover pattern in Romania using CORINE Land Cover database. Carpathian Journal of Earth and Environmental Sciences, **8**, P. 195-208, (2013)

9. S. Fonji, and G. Taff, "Using satellite data to monitor land-use land-cover change in North-eastern Latvia" Springerplus **3**(1): 61. (2014)
10. J. Rogan, and D. Chen, "Remote sensing technology for mapping and monitoring land-cover and land-use change" Progress in planning **61**(4): P.301-325. (2004).
11. National Report on the State of Land Resources of the Republic of Uzbekistan, (2010-2019). State committee for land resources, geodesy, cartography and state cadastre of the Republic of Uzbekistan. https://ygk.uz/
12. Land Code of the Republic of Uzbekistan. Tashkent, Adolat, (2018). https://ygk.uz/uz/node/705
13. Land fund of the Republic of Uzbekistan, (2010-2019). State committee for land resources, geodesy, cartography and state cadastre of the Republic of Uzbekistan. https://ygk.uz/
14. H. Mohd, and Ch. Ruzinoor, Online 3D terrain visualisation using Unity 3D game engine: A comparison of different contour intervals terrain data draped with UAV images, IOP Conference Series: Earth and Environmental Science, **37,** (2016)
15. X. Li, S. Zhao, H. Yang, D. Cong, Z. Zhang. A Bi-Band Binary Mask Based Land-Use Change Detection Using Landsat 8 OLI Imagery. Sustainability, **9** (3), p.479, (2017).
16. O. R. Abd El-Kawy, J. K. Rød, H. A. Ismail, A. S. Suliman. Land use and land cover change detection in the western Nile delta of Egypt using remote sensing data. Applied Geography, **31** (2), p.483, (2011).
17. A. C. Abino, S.Y. Kim, M. N. Jang, Y. J. Lee, J. S. Chung. Assessing land use and land cover of the Marikina sub-watershed, Philippines. Forest Science and Technology, **11** (2), p.65, (2015).
18. Land Code of the Russian Federation of October 25, (2001). https://cis-legislation.com/document.fwx?rgn=3104
19. A. Altiyev. Development of mechanisms of regulating the system of using land resources, Doctoral dissertation, (2019)
20. S. Enemark, Building land information policies. Proceedings of Special Forum on Building Land Information Policies in the Americas. Aguascalientes, Mexico, Citeseer, (2004)
21. T. Gorji, A. Tanik, E. Sertel "Soil salinity prediction, monitoring and mapping using modern technologies" Procedia Earth and Planetary Science **15**: P. 507-512, (2015)
22. O. Akbarov, B. Markus, and A.Podor, Development of Advanced Education in Geoinformatics for Enabling Sustainable Development in Uzbekistan. Proceedings FIG Congress, P. 1-15, (2014)

Influence of Adverse Ecological Factors on the Incidence of Malignant Neoplasms

Izimbet Turdimambetov[1], *Amet* Madreymov[2], *Lorant* Foldvary[3], *Medetbay* Oteuliev[1,*], *Murat* Kurbanov[2], *Kamila* Utarbaeva[1], *Kuatbay* Bekanov[4],

[1]Karakalpak State University, Geography and Hydrometeorology Department, 230 112, Nukus, Uzbekistan
[2]Karakalpakstan Medical Institute, 230100, Nukus, Uzbekistan
[3]Obuda University, Alba Regia Technical Faculty, Institute of Geoinformatics, 8000, Szekesfehervar, Pirosalma str.1-3, Hungary
[4]National University of Uzbekistan, Cartography Department, 100174, Tashkent, Uzbekistan

Abstract. An analysis of the leading sanitary-ecological factors of the environment affecting the incidence rates among the population of the Republic of Karakalpakstan in 2009-2018 found that between the indicators of the primary incidence of malignant neoplasms and water samples of open reservoirs that are not standard in chemical composition, strong direct correlations in the Northern (rxy = 0.89) zone, average direct correlations in the Western (rxy = 0.67) and Central (rxy = 0.57) zones were established. Similarly, strong connection was found in the Chimbay (rxy = 0.73) district, moderate correlation in Nukus (rxy = 0.44) and Ellikkala (rxy = 0.66) districts. The moderate correlation of the dynamics of malignant neoplasms with indicators of chemical pollution of tap water in the Chimbay (rxy = 0.33) district, well water in Kanlykul (rxy = 0.32) and atmospheric air in the city of Nukus (rxy = 0.41) was revealed and in the Republic of Karakalpakstan (rxy = 0.39). The solution to the problem lies in providing for the population of the selected districts with centralized drinking water that meets the state standard.

Keywords: malignant neoplasms, open reservoir, primary oncological incidence, primary incidence rate, South Aral Sea region.

1 Introduction

At Karakalpak State University, together with the Medical Institute of Karakalpakstan and the Republican Center for Sanitary and Epidemiological Welfare (abbreviated RC SEW; former abbreviation RC SSES) a research is being carried out to study and assess the most important environmental pollution in the context of cities, regions and conditionally allocated zones, to determine their impact on human health, as well as identify the most contaminated areas, and to identify risks which may lead to diseases.

At the end of the 20th century and at the beginning of the 21st century, as a result of a sharp reduction in the water flow from the Amudarya and Syrdarya rivers into the Aral Sea,

* Corresponding author: medetbay.oteuliev@gmail.com

the actual disappearance of the Aral Sea, the ecological balance was disturbed in the Aral Sea region, the external environment was degraded, an extreme condition for people's habitation arose [1, 2, 3, 4].

Polluted atmospheric air, soil and water serve as a source of pollution of plant and livestock products, later used by the population as products of food [2, 3, 4,5].

One of the diseases that needs to be studied in connection with harmful ecological factors in the Republic of Karakalpakstan (hereinafter: RK) is malignant neoplasms (hereinafter: MN).

The purpose of this study is to identify and evaluate the leading sanitary-ecological factors of the environment that affects human health and appears on incidence level of malignant neoplasms among the population of cities, districts and conditionally allocated zones of the RK.

2 Data

The data for the study was f-12-health reports from the Health Institute of the Ministry of Health (hereinafter: MH) of the Republic of Uzbekistan (hereinafter: RUz) and the Ministry of Health of the Republic of Karakalpakstan, the results of studies of the laboratory complex of the Republican Center of the State Sanitary and Epidemiological Surveillance (hereinafter: RC SSES) of the MH of the RK for 2009-2018.

The dynamics of the incidence of MN with a diagnosis for the first time (incidence per 100 000) in the RUz and the RK, tends to decrease slowly (Table 1).

Table 1. Incidence rates (intensive incidence rates) per 100 000 population

Years	Uzbekistan	Karakalpakstan
1991	76.2	79.0
1995	68.5	63.1
2000	71.7	65.2
2005	63.7	66.1
2010	67.7	67.3
2015	67.7	69.6
2018	71.0	67.2

According to the statistical reports of oncological dispensaries, in 2010 in the RUz, intensive incidence rates for malignant neoplasms amounted to 67.7 (per 100 000 population; from now on the unit of incidence is not indicated but considered to be "per 100 000 population"), among rural residents 79.0 cases, among women 73.4 cases. In 2018, these values were 71.0, 82.6 and 82.9, respectively.

Breast cancer was the No. 1 form of MN in RUz in 2010 (11.9 cases), followed by stomach cancer (9.5) and circulatory and lymphatic tissue (8.6). In the RK, forms of MN (in descending order) are esophageal cancer (17.6), stomach cancer (10.4) and cervical cancer (10.0) [6].

In 2018, the form of oncological incidence in the whole RUz most often was breast cancer (10.9), followed by stomach cancer (5.7), and cervical cancer (5.0). In the RK in 2018, (in decreasing order) esophageal cancer has accounted for 12.0, stomach cancer 10.4, breast cancer 9.8 and cervical cancer 8.7 cases.

Compared with 2009, in 2018 in the RUz there was an increase in the incidence of breast cancer by 45.3%, in the RK by 22,5%.

According to the MH of the RUz [9], the incidence rates of MN in 2012 amounted to 156.0 in Kazakhstan, 114.0 in Kyrgyzstan, 72.8 in Tajikistan, and 64.5 in Uzbekistan.

In 2010, cancer mortality in Uzbekistan amounted to 34.5, while in Karakalpakstan 43.0, therefore the number of cases in the RK 24.6% above than the RUz average. And in 2018, cancer mortality in Uzbekistan amounted to 41.1, in Karakalpakstan 51.4, indicating again relevantly (25%) more cases in the RK.

The above regional differences in the primary oncological incidence and the excess of the incidence rate of Karakalpakstan over Uzbekistan prompted us to disclose their causes.

Given the uneven distribution of incidence by territory and by time, the territory of the RK is conditionally divided into 4 zones: the Western zone (Muynak, Kungrad, Kanlykul and Shumanay districts), the Northern zone (Takhtakupyr, Karauzyak, Chimbay, Kegeyli districts), the Central zone (Nukus city, Khodjeyli, Takhiatash and Nukus districts), and also the Southern zone (Amudarya, Beruni, Ellikkala and Turtkul districts).

In order to increase the visibility of the tendencies, the studied 10-year period is divided into 2 five-year periods: 2009-2013 and 2014-2018.

When studying the dynamics of the incidence of MN with a first established diagnosis, in the second five-year period compared with the first, an increase in the incidence rate was found in the Western (by 1,5%), Northern (by 1,8%), Central (by 7,6%) and Southern (5,1%) zone (Table 2).

Table 2. Primary incidence rates of MN in the RK per 100 000 population [7]

City, districts	2009	2010	2011	2012	2013	2014	2015	2016	2017	2018
Muynak	**84.2**	77.5	**96.9**	79.0	**92.1**	**87.8**	**83.3**	66.2	78.2	74.0
Kungrad	67.7	67.2	48.5	64.4	63.7	59.6	66.0	65.5	66.7	55.6
Kanlykul	**86.0**	43.1	57.1	**106.5**	64.4	57.2	50.1	**87.0**	**97.2**	74.0
Shumanay	32.0	46.0	59.5	49.6	41.6	41.2	72.4	57.2	52.8	56.0
Western zone	67.5	58.5	65.5	74.9	65.5	61.5	68.0	69.0	73.7	64.9
Takhtakupyr	**90.0**	50.8	**81.4**	59.4	64.6	67.0	66.7	71.6	60.6	**125.6**
Karauzyak	48.8	57.3	**80.4**	65.4	56.7	54.0	75.1	59.1	79.3	72.8
Chimbay	69.1	**82.8**	74.9	70.3	65.9	67.0	**82.5**	**93.5**	**92.7**	74.2
Kegeyli	63.1	71.3	74.9	70.7	78.2	61.0	62.5	71.3	70.1	78.5
Northern zone	67.8	65.6	77.9	66.5	66.4	62.3	71.7	73.9	75.7	87.8
Nukus city	73.3	79.7	75.7	59.7	**80.4**	74.9	**81.7**	67.2	**80.9**	70.6
Nukus district	69.6	67.1	79.2	**83.7**	74.3	**82.8**	70.3	**91.1**	**98.5**	63.8
Khodjeyli	50.1	65.5	**84.6**	69.9	68.4	75.9	**84.3**	58.7	**100.0**	63.1
Takhiatash	64.7	60.9	65.7	78.2	x	x	x	x	79.7	62.2
Central zone	64.4	68.3	76.3	72.9	74.4	77.9	78.8	72.3	**89.8**	64.9
Amudarya	58.5	71.7	60.6	55.3	62.9	41.0	46.7	53.9	56.3	67.7
Beruni	68.1	74.2	67.9	62.6	62.8	58.1	63.8	63.8	60.9	67.8
Ellikkala	41.5	59.5	43.2	21.2	61.8	45.8	66.9	52.5	74.0	57.0
Turtkul	65.0	52.6	53.1	45.9	**85.4**	76.1	64.7	**85.6**	**84.5**	59.5
Southern zone	59.4	65.0	56.0	47.0	69.1	55.3	60.5	64.0	68.9	63.0
RK	63.7	67.3	68.5	59.8	69.9	63.6	69.6	67.8	76.1	67.2
RUz	68.4	71.0	65.9	64.5	66.2	65.7	67.7	66.7	70.2	71.0

Note: x-Takhiatash was included in the Khodjeyli district

The average the primary incidence rate of MN for RK over the first five-year period (2009-2013) is 65.8. High rates were observed in Muynak (85.9), in Nukus (74.8), in Chimbay (72.6), in Kegeyli (71.6), in Kanlykul (71.4) districts and the city of Nukus (73.8). In the next five-year period, the regional (RK) average was 68.9; relatively high primary incidence rates of MN were recorded in Chimbay (82.0), Nukus (81.3), Takhtakupyr (78.3),

Muynak (77.9), Khodjeyli (76.4), Kanlykul (73.1) districts and in the city of Nukus (75.1) (Table 2). Relatively low primary incidence rates of MN were detected in the first five-year period in Shumanay (45.7) and Ellikkala (45.4) districts.

The highest incidence rates of MN were recorded in Muynak district (84.2, 96.9, and 92.1 in 2009, 2011 and 2013, respectively), in Kanlykul (86.0 and 106.5 in 2009 and 2012). The lowest rates were observed in Ellikkala district (21.2 in 2012) and in Shumanay (32.0 in 2009).

3 Tests and results

We presume that the dynamics of the level of oncological incidence for regions under similar living conditions, lifestyle and nutritional features should be similar; significant influences are exerted by harmful external environmental factors resulting from the Aral ecological disaster. Therefore, we attempt to connect the dynamics of the primary incidence of MN with the main harmful sanitary-ecological factors that formed in the South Aral Sea region with variations of chemical pollution of drinking water and atmospheric air (Fig. 2).

At the end of 2018, 62% of the population of the Republic was provided with tap water, the rest uses well water (mainly water from tubular wells) and water from open reservoirs.

First of all, we compared the levels of primary oncological incidence (Table 2) with indicators of non-normative water samples of open reservoirs by chemical composition (by hardness and mineralization) (Table 3).

The strength of the relationship between the primary incidence rates of oncological diseases (Table 2) and the indicators of water samples from open reservoirs, exceeding the maximum permissible concentration (MPC) of chemicals and indicators (Table 3) was analized for each zones and districts. As an example, the time series are plotted on Fig 1. for the Western zone of the RK.

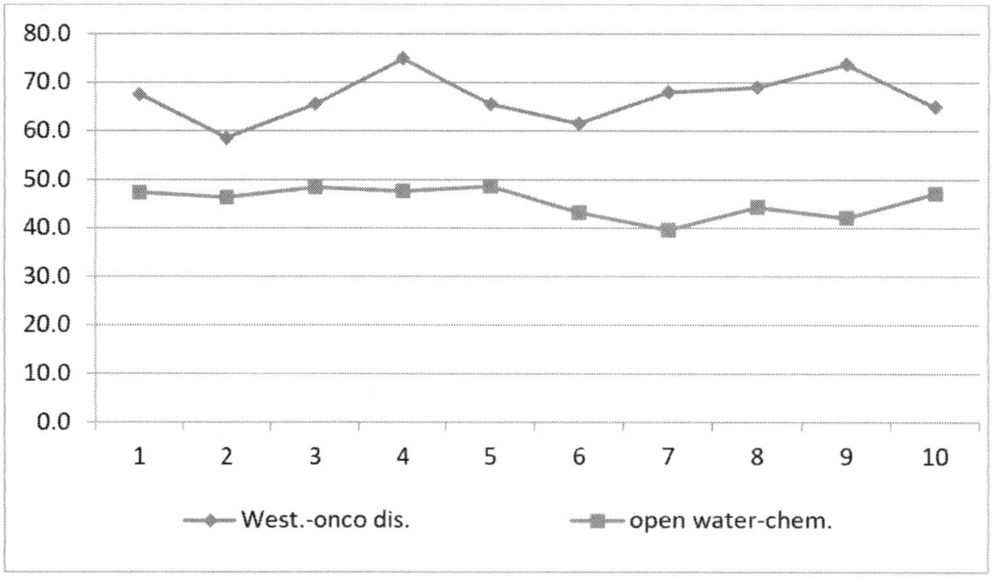

Fig. 1. The MN incidence indicators in the western zone of the RK per 100 000 population and % of water samples of open reservoirs with excess of the MPC of chemicals substances.

MAP OF THE PRIMARY INCIDENCE OF MALIGNANT NEOPLASMS IN THE REPUBLIC OF KARAKALPAKSTAN

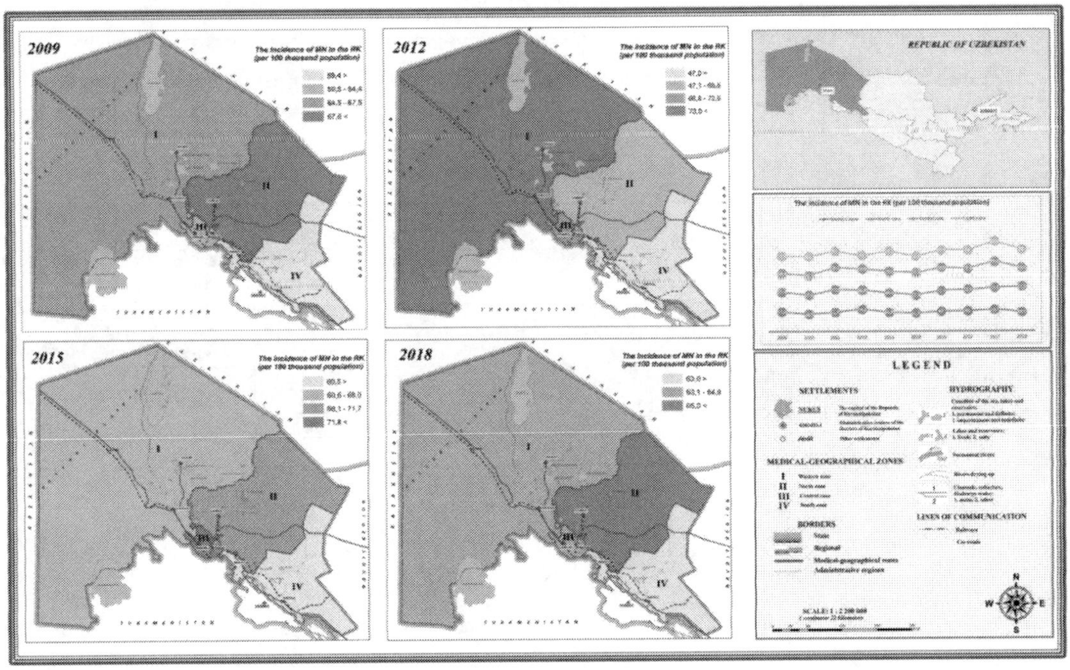

Fig. 2. Map of the primary incidence of malignant neoplasms

Table 3. The percentage of open water reservoirs samples, which found the maximum permissible concentration (MPC) of chemical substances and chemical indicators for the 2009-2018 [8]

№	Districts	2009	2010	2011	2012	2013	2014	2015	2016	2017	2018
1	Muynak	60.8	24.3	19.4	8.3	16.7	36.9	29.9	17.0	18.2	10.6
2	Kungrad	25.4	30.1	37.3	48.8	41.3	7.6	0.9	28.3	20.2	32.0
3	Kanlykul	73.5	100.0	100.0	100.0	100.0	100.0	96.4	100.0	100.0	100.0
4	Shumanay	29.4	30.8	36.9	33.3	36.2	28.2	31.3	31.7	30.1	45.8
	Western	**47.3**	**46.3**	**48.4**	**47.6**	**48.6**	**43.2**	**39.6**	**44.3**	**42.1**	**47.1**
5	Takhtakupyr	100.0	72.5	100.0	100.0	100.0	100.0	100.0	100.0	100.0	100.0
6	Karauzyak	46.0	42.9	31.0	35.8	31.2	23.3	36.8	22.5	26.0	27.3
7	**Chimbay**	**4.8**	**15.3**	**13.6**	**6.3**	**1.5**	**9.6**	**19.8**	**18.8**	**12.6**	**15.5**
8	Kegeyli	11.1	20.0	0.0	0.0	26.3	36.4	71.4	75.0	94.0	76.6
	Northern	**40.5**	**37.7**	**36.2**	**35.5**	**39.8**	**42.3**	**57.0**	**54.1**	**58.2**	**54.9**
9	Nukus city	23.1	12.2	37.5	27.3	31.4	24.7	41.9	40.8	45.1	41.4
10	**Nukus district**	**75.0**	**50.0**	**77.8**	**52.2**	**100.0**	**77.8**	**13.0**	**95.0**	**100.0**	**84.3**
11	Khodjeyli	20.7	28.9	12.9	10.3	14.3	21.9	30.2	31.5	49.5	51.4
12	Takhiatash	16.3	0.0	43.3	36.4	x	x	x	x	75.0	21.0
	Central	**33.8**	**22.8**	**42.9**	**31.6**	**48.6**	**41.5**	**28.4**	**55.8**	**67.4**	**49.5**
13	Amudarya	72.2	89.9	97.3	5.8	30.6	78.0	81.1	77.5	73.6	63.5
14	Beruni	69.0	51.0	82.0	83.6	90.3	59.1	63.3	78.0	26.7	69.0
15	**Ellikkala**	**20.0**	**83.3**	**95.8**	**0.0**	**85.7**	**25.7**	**40.9**	**37.8**	**100.0**	**100.0**
16	Turtkul	100.0	82.9	100.0	100.0	100.0	91.7	70.0	87.0	100.0	100.0
	Southern	65.3	76.8	93.8	47.4	76.7	63.6	63.8	70.1	75.1	83.1
	RK	41.9	39.4	43.5	31.3	38.3	48.7	45.4	55.8	47.9	49.9

The strength of the relationship between the curves on Fig 1. (and in other cases), correlation coefficient according to Pearson [10] has been determined. The calculation was carried out on a computer program Formulas. Though the time series is "short", i.e. including 10 samples only (k=10), to decide on the dependence of oncological cases and water quality measures, a hypothesis test should be conducted. The hypothesis-test for correlation coefficient can be performed by a Student's t-test with degrees of freedom $k-2$ [11], where the null-hypothesis is that the two time series are uncorrelated, i.e.

$$H_0: r_{xy} = 0 \tag{1}$$

For 'large' number of samples, it is known that

$$\bar{t} = \frac{r_{xy}\sqrt{k-2}}{\sqrt{1-r_{xy}^2}} \tag{2}$$

follows a Student's t-distribution with degrees of freedom $k-2$. If

$$|\bar{t}| \leq t(p, k-2) \tag{3}$$

then the null-hypothesis should be accepted, thus the two time series are independent at a confidence level p. The values of t-distribution at k-2=8 freedom for probability of 0.8, 0.9 and 0.95 are 1.397, 1.806 and 2.306, respectively. Accordingly, the dependence can reliably by assumed, when the null-hypothesis is rejected, i.e.

$$|\bar{t}| > t(p, k-2) \tag{4}$$

Note, however, that the results may be distorted due to the small number of samples. According to [12], in case of small samples, the distribution may differ.

The correlation coefficient for the Western zone for 2009-2018 was -0.1488, for 2009-2013 it is 0.3843, and for 2014-2018 it became -0.3378. The indicator $\bar{t}$ = -0.426, 1.1774 and -1.0151, respectively for the 2009-2018, 2009.2013 and 2014-2018 periods have not met the criteria t, so for this case the hypothesis should be rejected. All the resulted correlations and values are listed in Table 4. In the Table those cells has been emphasized by green color, where correlation has been accepted: light green is used for cases accepted on 0.8 probability level (the value of t exceeds 1.397), darker green shows the cases, which are accepted on 0.9 probability level (the value of t exceeds 1.806), while the darkest green shows those cases, which could have been accepted on 0.95 probability level (the value of t exceeds 2.306).

The results in Table 4 suggest no systematic correlation by time or by location. The strong correlation in Chimbay is mainly a result of the strong correlation during the 2009-2014 period, which cannot be observed anymore. A notable (a quite unfavorable) feature is that in Muinak, Kalykul and in Amudarya there was no correlation observed in the 2009-2013 period, but it became relevant in the 2014.2018 period.

Then we compared the levels (incidence) of primary oncological diseases (Table 2) with the indicators of tap water samples, the percentage of which exceeded the MPC of chemical substances and indicators (Table 5).

Rationing of drinking tap water is based on the application of the State Standard of Uzbekistan (UzSS) 950-2011 "Drinking water. Hygienic Requirements and Quality Control" (SS-state standard) updated every 10 years. The district laboratories of the sanitary-epidemiological service measured the hardness, mineralization (dry residue) in the water sample. This dry residue in our regions mainly consisted of nitrates, sulfates and chlorides. No pesticide contamination of water and food has been detected over the past 10 years. They have ceased to be used in agriculture for 25 years.

The results of the comparison are shown on Table 6. In the Table light green is used for cases accepted on 0.8 probability level (the value of t exceeds 1.397), darker green shows the cases, which are accepted on 0.9 probability level (the value of t exceeds 1.806), while the darkest green shows those cases, which could have been accepted on 0.95 probability level (the value of t exceeds 2.306).

The levels of primary oncological diseases depended on chemical contamination (Table 2 and Table 5) at several locations in the Western, Northern and Southern regions during the 2014-2018 period. It is quite unfavourable situation as at most locations no strong correlation was observed before. As for the Central region, no relevant correlation was observed for the 5 year periods (apart from Khojeli during the 2009-2013 period), but the correlation seems to be convincing for the 10 year period (Fig.3).

Table 4. Correlation between the primary incidence rates of MN in the RK per 100 000 population (Table 2) and the percentage of open water reservoirs samples, which found the maximum permissible concentration (MPC) of chemical substances and chemical indicators (Table 3)

Zones, district	rxy (2009-2018)	$\bar{t}$ (2009-2018)	rxy (2009-2013)	$\bar{t}$ (2009-2013)	rxy (2014-2018)	$\bar{t}$ (2014-2018)
Muinak	0.2471	0.7213	-0.0777	-0.1350	0.8213	2.4935
Kungrad	-0.2128	-0.6161	-0.2397	-0.4276	-0.3321	-0.6098
Kanlykul	-0.1785	-0.5131	-0.3261	-0.5975	0.6518	1.4886
Shumanai	0.2709	0.7961	0.6825	1.6174	0.1619	0.2842
Western	-0.1488	-0.4255	0.3843	0.7210	-0.3378	-0.6216
Takhtakupyr	0.3781	1.1553	0.6401	1.4431	NaN	NaN
Karauzyak	-0.2943	-0.8711	-0.7027	-1.7107	0.3244	0.5940
Chimbay	0.7331	3.0486	0.9429	4.9032	0.4969	0.9918
Kegeyli	-0.0029	-0.0082	0.2663	0.4785	0.6031	1.3096
Northern	0.5447	1.8373	0.3683	0.6861	0.6462	1.4666
Nukus city	0.0479	0.1357	-0.1278	-0.2232	0.1588	0.2786
Nukus district	0.4356	1.3689	-0.0196	-0.0340	0.5696	1.2003
Khojeli	0.2143	0.6205	-0.4818	-0.9523	0.1796	0.3162
Takhiatash	NaN	NaN	NaN	NaN	NaN	NaN
Central	0.4872	1.5779	0.6225	1.3777	0.2840	0.5130
Amudarya	-0.0230	-0.0650	0.5423	1.1180	-0.8874	-3.3341
Beruni	0.0364	0.1032	-0.9282	-4.3208	0.4736	0.9314
Ellikkala	0.6560	2.4582	0.7962	2.2793	0.5923	1.2733
Turtkul	0.0104	0.0293	0.2799	0.5050	0.1988	0.3513
Southern	0.5125	1.6880	0.5201	1.0547	0.6028	1.3085
RK	0.4109	1.2749	0.6506	1.4839	-0.2496	-0.4465

https://doi.org/10.1051/e3sconf/202122702001

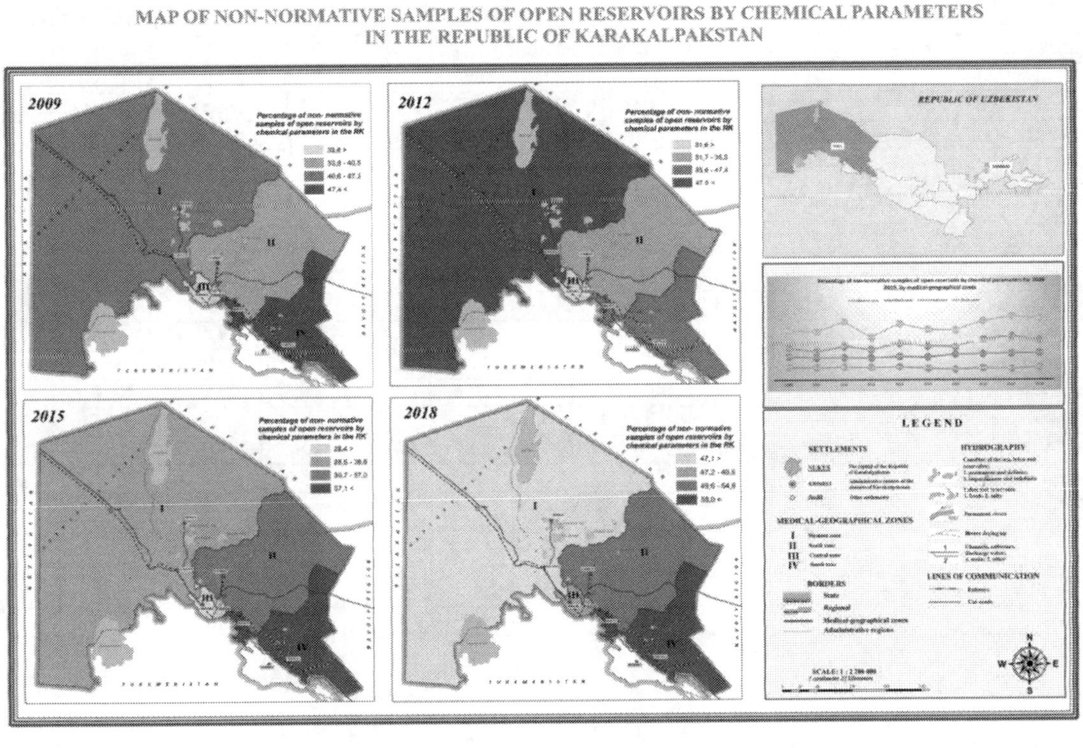

Fig. 3. Map of non-normative samples of open reservoirs by chemical parameters

Table 5. Percentage of tap water samples exceeding the MPC of chemical substances and indicators for the RK

Zones, districts	2009	2010	2011	2012	2013	2014	2015	2016	2017	2018
Muynak	43.8	16.0	0.0	16.9	20.9	26.2	36.4	7.3	15.8	11.6
Kungrad	39.2	43.6	36.7	42.9	26.8	18.4	8.9	14.1	13.1	14.6
Kanlykul	53.8	33.3	50.0	28.2	63.2	60.9	54.8	51.5	47.0	51.5
Shumanay	29.3	21.5	25.1	29.3	35.4	34.4	37.3	27.2	21.2	29.0
Western	41.5	28.6	37.3	29.3	36.6	35.0	34.3	25.0	24.3	26.7
Takhtakupyr	13.4	7.1	11.2	23.5	30.7	29.9	13.6	18.2	9.9	16.7
Karauzyak	29.0	25.4	21.5	16.6	21.2	9.6	14.3	8.6	22.2	14.7
Chimbay	**37.4**	**57.2**	**21.0**	**32.3**	**13.3**	**23.0**	**35.0**	**42.8**	**16.2**	**12.6**
Kegeyli	16.1	11.4	3.8	0.7	12.1	12.7	20.5	6.5	7.2	20.4
Northern	24.0	25.3	14.4	18.3	19.3	18.8	20.8	19.0	13.9	16.1
Nukus city	37.8	1.1	20.8	6.1	30.5	29.8	20.2	14.4	24.1	34.5
Nukus district	31.3	8.3	32.8	11.0	12.3	12.9	10.2	32.4	51.5	33.4
Khodjeyli	75.9	51.9	40.4	19.5	29.6	29.3	16.6	21.8	26.8	52.7
Takhiatash	25.3	7.6	38.5	20.3	x	x	x	x	2.2	14.8
Central	42.6	17.2	33.1	14.2	24.1	24.0	15.7	22.9	26.2	33.9
Amudarya	18.6	36.1	24.0	0.0	26.6	33.9	31.6	33.3	37.8	43.1
Beruni	40.0	30.7	35.6	40.7	37.7	58.5	54.7	47.3	34.0	70.9
Ellikkala	34.9	25.9	33.3	5.2	15.1	12.3	19.6	12.2	10.4	21.4
Turtkul	31.3	15.9	39.2	22.6	25.9	37.3	29.7	18.5	14.5	19.6
Southern	31.2	27.2	33.0	22.8	26.3	35.5	33.9	27.8	24.2	38.8
RK	34.5	24.7	28.0	18.5	28.6	30.6	26.6	23.1	23.2	31.6

Table 6. Correlation between the primary incidence rates of MN in the RK per 100 000 population (Table 2) and the percentage of tap water samples exceeding the MPC of chemical substances and indicators (Table 5)

Zones, district	rxy (2009-2018)	$\bar{t}$ (2009-2018)	rxy (2009-2013)	$\bar{t}$ (2009-2013)	rxy (2014-2018)	$\bar{t}$ (2014-2018)
Muinak	0.1244	0.3546	-0.3788	-0.7090	0.8378	2.6581
Kungrad	-0.0190	-0.0537	0.2168	0.3846	-0.5742	-1.2148
Kanlykul	-0.2995	-0.8878	-0.2257	-0.4013	-0.8216	-2.4960
Shumanai	0.0662	0.1877	-0.4028	-0.7623	0.2817	0.5084
Western	-0.2713	-0.7972	0.0440	0.0762	-0.6590	-1.5176
Takhtakupyr	-0.0523	-0.1482	-0.1448	-0.2535	-0.0036	-0.0062
Karauzyak	-0.0932	-0.2647	-0.5817	-1.2387	0.8832	3.2617
Chimbay	0.3272	0.9792	0.7644	2.0532	0.4125	0.7843
Kegeyli	-0.2162	-0.6262	-0.3501	-0.6474	0.0173	0.0300
Northern	-0.5922	-2.0789	-0.7309	-1.8549	-0.4897	-0.9728
Nukus city	0.2195	0.6363	0.2793	0.5039	0.0153	0.0265
Nukus district	0.4684	1.4997	0.0350	0.0606	0.5422	1.1177
Khojeli	-0.5487	-1.8562	-0.6561	-1.5060	-0.3651	-0.6793
Takhiatash	NaN	NaN	NaN	NaN	NaN	NaN
Central	-0.3329	-0.9984	-0.3419	-0.6302	-0.4195	-0.8003
Amudarya	0.1331	0.3798	0.8940	3.4550	0.8616	2.9406
Beruni	-0.2626	-0.7698	-0.8148	-2.4346	0.4826	0.9543
Ellikkala	0.0116	0.0327	0.3306	0.6068	0.0366	0.0634
Turtkul	-0.1510	-0.4320	0.0975	0.1698	-0.2858	-0.5166
Southern	-0.0795	-0.2255	0.1921	0.3390	-0.6927	-1.6635
RK	-0.0425	-0.1203	0.4293	0.8232	-0.6836	-1.6222

Comparison of the levels (incidence) of primary oncological diseases (Table 2) with the indicators of well water samples exceeding the MPC in terms of chemical indicators (Table 7) is shown on Table 8. In the Table light green is used for cases accepted on 0.8 probability level (the value of t exceeds 1.397), darker green shows the cases, which are accepted on 0.9 probability level (the value of t exceeds 1.806), while the darkest green shows those cases, which could have been accepted on 0.95 probability level (the value of t exceeds 2.306).

According to the results, the correlation seems to be stronger by time in every district of the Western region, in contrary, when the average values for the Western region are analysed, the tendency is opposite: the correlation is stronger in the 2009-2013 period.

In the Northern region correlation at 0.8 probability level can be observed at several districts during the 2009-2013 period, which has been absent for the 2014-2018 period, apart from Karauzyak, where the correlation became notably stronger.

In the Central region there is an obvious correlation (at least at 0.8 probability level) for both 5 year periods, but it cannot be detected for the whole 10 year long time series.

Not any dependence in the Southern region of the RK has been observed.

Table 7. Specific gravity of non-normative well water samples by chemical indicators (in%)

№	Zones, districts	2009	2010	2011	2012	2013	2014	2015	2016	2017	2018
1	**Muynak**	100.0	100.0	98.8	97.8	97.7	80.7	94.2	99.0	100.0	100.0
2	**Kungrad**	31.0	42.2	40.0	33.8	21.8	40.0	23.2	38.0	21.6	47.0
3	**Kanlykul**	62.6	50.0	55.0	52.2	64.0	53.6	70.9	65.4	73.2	94.3
4	**Shumanay**	39.6	40.8	40.6	41.7	50.8	51.7	43.7	40.7	33.5	40.4
	Western Zone	58.3	58.3	58.6	56.4	58.6	56.5	58.0	60.8	57.1	70.4
5	**Takhtakupyr**	100.0	80.0	100.0	100.0	100.0	99.2	99.2	100.0	100.0	99.4
6	**Karauzyak**	54.1	33.8	33.6	34.4	50.3	53.8	93.2	87.8	93.1	92.8
7	**Chimbay**	34.4	34.9	25.9	25.5	16.5	16.2	25.5	22.2	11.2	19.5
8	**Kegeyli**	40.5	42.0	13.2	8.9	16.7	38.8	69.0	35.9	46.9	55.6
	Northern Zone	57.3	47.7	43.2	42.2	45.9	52	71.7	61.5	62.8	66.8
10	**Nukus city**	89.9	82.6	95.0	100.0	98.9	99.7	95.7	100.0	99.7	99.4
11	**Khodjeyli**	72.6	86.5	92.4	92.1	69.4	84.8	92.9	80.6	70.3	85.2
	Central Zone	81.3	84.6	93.7	96.1	84.2	92.3	94.3	90.3	85	92.3
13	**Amudarya**	**88.7**	**76.2**	**55.9**	**44.8**	**57.7**	**54.0**	**68.6**	**65.1**	**67.2**	**56.3**
14	**Beruni**	58.6	57.1	80.5	69.0	54.2	72.5	65.1	58.0	51.6	92.4
15	**Ellikkala**	88.2	76.6	82.8	26.2	34.3	27.5	33.8	45.0	37.2	72.6
16	**Turtkul**	100.0	100.0	100.0	90.9	100.0	78.6	63.7	80.0	100.0	100.0
	Southern Zone	83.9	77.5	79.8	57.7	61.6	58.2	57.8	62.0	64.0	80.3
	RK	63.1	60.6	60.5	49.5	54.4	46.5	60.5	59.9	55.0	67.2

Note: there are no wells in Nukus district and Takhiatash

4 Conclusions

The present investigation has made use of spatial and statistical analysis tools, so establishment of an analytical relation between the causes of high mortality from MN and the mentioned environmental factors would require further studies.

According to Tables 4, the strong relation between the incidence rates of MN in the RK per 100 000 population (Table 2) and the percentage of open water reservoirs samples, which found the maximum permissible concentration (MPC) of chemical substances and chemical indicators could have been established for the 10 year (2009-2018) period in Ellikkala and Chimbay. In both cases the correlation is stronger in the first half of the period (2009-2013). For the second period (2014-2018), strong correlation was observed in Muinak and Amudarya.

The correlation between the primary incidence rates of MN and the percentage of tap water samples exceeding the MPC of chemical substances and indicators (see in Table 6) was not convincing on 0.95 confidence level for the 10 year long period (2009-2013), but for the half periods, there have been relevant correlations in Amudarya and Beruni (2009-2013) and Muinak, Kanlykul, Karauzyak and Amudarya (2014-2018).

Table 8. Correlation between the primary incidence rates of MN in the RK per 100 000 population (Table 2) and the percentage of tap water samples exceeding the MPC of chemical substances and indicators (Table 5)

Zones, district	rxy (2009-2018)	$\bar{t}$ (2009-2018)	rxy (2009-2013)	$\bar{t}$ (2009-2013)	rxy (2014-2018)	$\bar{t}$ (2014-2018)
Muinak	-0.2940	-0.8701	-0.3195	-0.5841	-0.7659	-2.0632
Kungrad	-0.5124	-1.6876	-0.2989	-0.5425	-0.8571	-2.8820
Kanlykul	0.1690	0.4849	0.1457	0.2551	0.2350	0.4187
Shumanai	-0.2400	-0.6993	-0.1301	-0.2273	-0.3058	-0.5564
Western	-0.1835	-0.5279	-0.7673	-2.0724	-0.2246	-0.3993
Takhtakupyr	0.3655	1.1108	0.6401	1.4432	-0.2296	-0.4085
Karauzyak	0.3497	1.0557	-0.7162	-1.7776	0.8024	2.3286
Chimbay	-0.1128	-0.3211	0.6338	1.4193	-0.0209	-0.0362
Kegeyli	-0.3897	-1.1968	-0.6064	-1.3208	-0.0464	-0.0804
Northern	0.4817	1.5549	-0.2714	-0.4884	0.6020	1.3058
Nukus city	-0.3438	-1.0355	-0.4819	-0.9525	-0.6054	-1.3175
Khojeli	0.1164	0.3315	0.6422	1.4512	-0.3801	-0.7119
Central	0.1234	0.3516	0.6840	1.6242	-0.6312	-1.4096
Amudarya	0.2027	0.5855	0.3835	0.7193	-0.0355	-0.0615
Beruni	0.1029	0.2927	-0.1043	-0.1816	0.5210	1.0572
Ellikkala	-0.0573	-0.1625	0.2468	0.4411	-0.0347	-0.0602
Turtkul	-0.1565	-0.4481	0.5204	1.0555	0.0920	0.1600
Southern	0.1262	0.3598	0.2029	0.3588	0.3247	0.5945
RK	0.2444	0.7127	0.3993	0.7544	0.1654	0.2904

As it can be observed in Table 8, the correlation between the primary incidence rates of MN and the percentage of tap water samples exceeding the MPC of chemical substances and indicators shows strong correlation (at 0.95 confidence level) only occasionally, in Kungrad and Karauzyak. A very favorable feature is that there is no correlation at all in the Sourthern region of RK.

Since the small number of samples distorts the results, also lower confidence level should be considered. At 0.8 confidence level, the correlation between primary incidence rates of MN and non-normative water samples of open reservoirs by chemical parameters (Table 4) and also for the tap water samples (Table 6) was established at several districts and regions of the RK. Therefore, it is reasonable to search for a dependence analytically as well. The results for the well water samples (Table 8) are not that convincing, but as open reservoirs may more properly reflect the actual water quality (as in open reservoirs the water is more homogeneous due to the diffusion of the chemical particles than in the water supply system), the source of the dependence is worth for an analytical investigation.

All in all, the results obtained allow us to conclude that the higher the proportion of non-normative water samples, i.e. chemical contamination of water in open reservoirs and non-normative samples of tap water and atmospheric air in specified districts, the higher primary malignant oncological incidence rates. The solution to the problem of the consumption of contaminated water by the population for drinking purposes is to maximize the coverage of the population of the allocated districts with centralized water supply and bring the quality indicators of the tap water supplied to the population in accordance with the state standard. Our results can contribute to planning the reduction of the pollution of the environment, providing a measure for a differential approach for specific territories of the RK.

We would like to thanks to the RC SEW who provided us their statistic data and Erasmus+ DSinGIS: Doctoral Studies in GeoInformation Sciences project. They helped us to arrange all our research activities needed for our study.

References

1. Ch.A bdirov, *Public health and priority issues of medical-ecological research in the conditions of the South Aral Sea region*, Medical-ecological problems of the Aral Sea region and public health, Nukus (1991)
2. Ch. A. Abdirov, N. A. Agadjanyan, A. E. Severin, *Ecology and human health*, Nukus, Karakalpakstan, 103, (1993)
3. O. A. Ataniyazova, *Aral crisis and medical-social problems of Karakalpakstan*, Nukus, Bilim, 116, (2002)
4. R. R. Reimov, L. G. Konstantinova, *Extremality of the environment and human ecology in the South Aral Sea region*, Abstracts of rep. scientific Pract. Conf. "Human ecology and regional pathology of the Aral Sea region", Nukus, (1993)
5. L. Yu. Shardakova, L. V. Usmanova, *Statistical analysis of dust storms in the Aral Sea region*, Sat. thesis. Int. scientific practical conf. "Problems of rational use and protection of biological resources of the South Aral Sea region", Nukus, (2006)
6. I. R. Turdimambetov, A. Madreimov, *Territorial differentiation of natural conditions and nosogeographic situation in the Republic of Karakalpakstan*. J. Bul. Geo. Soc. Uzb. **41**, issue 62-68.
7. Statistical collections of the Institute of Health and Medical Statistics of the Ministry of Health of the Republic of Karakalpakstan, Nukus, (2009-2018)
8. Materials of laboratory control of environmental objects of the Republican Center of SSES of the Ministry of Health of the RK.
9. World Health Organization material
10. K. Pearson, *Notes on regression and inheritance in the case of two parents*, Proceedings of the Royal Society of London. 58: 240–242 (1895)
11. W. S. G. Student, *Probable error of a correlation coefficient*, Biometrika. 6 (2-3): 302–310. https://doi.org/10.1093/biomet/6.2-3.302 (1908)
12. H. E. Soper, A. W. Young, B. M. Cave, A. Lee, K. Pearson, *On the distribution of the correlation coefficient in small samples. Appendix II to the papers of "Student" and R.A. Fisher. A co-operative study*, Biometrika. 11 (4): 328–413. doi:10.1093/biomet/11.4.328 (1917)

Investigation of the population area of snow leopard in the Central Tian-Shan Mountains

Mirlan Dyldaev[1], *Akylbek Chymyrov*[2,*], *Askat Mukabaev*[3] and *Orozbek Omurzak uulu*[1]

[1]Bishkek State University named after K. Karasaev, 27, Aitmatov street, Bishkek, 720044, Kyrgyzstan
[2]Kyrgyz State University of Construction, Transport and Architecture, 34 "b", Maldybaev street, Bishkek, 720020, Kyrgyzstan
[3]Institute of Geology, National Academy of Sciences, 30, Erkindik Ave., Bishkek, 720461, Kyrghyzstan

Abstract. The spatial analysis of population areal of snow leopards in the Central Tian Shan Mountains and monitoring of their population dynamics and reproduction is needed by nature conservation activities. The study area is located in the unique and remote transboundary region between the Kyrgyz Republic, Republic of Kazakhstan and People's Republic of China. The population areas, feeding base and movement of snow leopards within the Khan Tengri State Nature Park were realized by using camera traps and Geoinformation Systems (GIS). Camera traps have been used for recording the existence and activity of snow leopards in the high mountainous and uninhabited area. 14 adult and 5 snow leopard cubs are detected and recorded within the survey period in 2018-2019 with various individual characteristics of animals. The spatial distribution of snow leopard's population was mapped and analyzed by applying open source GIS software (Quantum GIS) and satellite positioning.
Keywords: snow leopard, GIS, GPS, camera trap, monitoring

1 Introduction

Investigation and analysis of the distribution range (areal) of the snow leopard (*Uncia uncia*) and other endangered species of large mammals are currently of great scientific interest, which is growing because of the climate change issues. The number of snow leopard declined from an estimated 600–700 individuals in the late 1980s [1] to 150–200 individuals by 2000 [2] has putting the species at high risk of extinction in the Kyrgyz Republic. Its main prey species – mountain goats *(Capra sibirica)* and argali *(Ovis ammon)* are also threatened by illegal or unsustainable hunting in many parts of the snow leopard range. Climate change impact is increasingly affecting the snow leopard population because of the glacier loss resulting in the reduced feeding area of its main prey and disturbance caused by increasing altitude of highland pasture grazing [3,4].

The National Academy of Sciences of the Kyrgyz Republic has declared that number of snow leopards has been slowly improving because of the environmental protection

* Corresponding author: chymyrov@gmail.com

activities and national legislation has been strengthened (250-300 in 2010 and 330 in 2019) [5]. But these numbers are resulted from the traditional research methodology based mainly on interview with huntsman, herders and other people in the snow leopard population area.

Geoinformation technology and digital identification research methods have been used as powerful tools of sophisticated research on analyzing the population area of these animals, their prey-resources, migration mobility and other biological features. Geoinformation systems, remote sensing and geospatial data processing software applications allow to solve a wide range of research tasks that allow developing digital maps with areal of the snow leopard [6,7].

It should be noted that the degree of knowledge of any type of animal is an objective indicator of the difficulty of working with it in a natural environment. The problematic link in the study of large mammals is predators, which include the snow leopard itself. Their secretive lifestyle, the general sparseness of the population, and the inaccessibility of habitats minimize the possibility of their direct observations [8,9]. Well-adapted to the cold, dry conditions, snow leopards roam the high mountains up to altitude of 5 000 m or more above sea level (a.s.l.), following the ibex and other wild ungulates from one river valley to another. Steep mountain slopes, rocky and rugged terrain provide ideal hunting grounds for these stealthy cats. The lifestyle of the Central Tien Shan snow leopard remains poorly understood due to the complexity of its geographical area, although according to some researchers it lives within its well defined territory [10,11,12].

Snow leopards usually live alone, except for mothers and their cubs. Mating of snow leopards occur in February-March and the pregnancy of females lasts 98-103 days, she gives birth every 2 years to a litter of three or four cubs (very rarely up to 5), puberty occurs at the age of two. A female keeps cubs hidden for the first eight weeks and the cubs continue to live with their mothers for about two years [1,13].

Hunting the snow leopard in Kyrgyzstan has been prohibited since 1948, claims are established for hunting and trapping without a license since 1959. It is included in the International Red Book and attached to the Convention on International Trade in Endangered Species of WWF. Since 1975, it is included in the List of Specially Protected Species of Kyrgyzstan and is protected throughout the country [11].

Currently, in the framework of the Bishkek Declaration on the Conservation of the Snow Leopard adopted by the 12 snow leopard range countries in October 23, 2013, Bishkek, Kyrgyz Republic, a number of strategic international and national documents are developed. The current system of national legislation has been strengthened including increased penalties for illegal shooting of snow leopards to 1.5 million Kyrgyz Soms (more than 20 thousand USD) to protect this unique wild species.

Ongoing monitoring study has made it possible to assess the current state of the habitat of the snow leopard in the territory of the Khan Tengri State Nature Park (SNP) in the Central Tien Shan mountain region. The territory of the future Natural Park is one of the major habitats of the snow leopard in Kyrgyzstan. Camera traps have been used for detecting and identification of animals and collected photographs and geospatial datasets are processed and analyzed by applying open source Geoinformation software to contribute into conservation of this unique wild species.

2 Materials and methods

2.1 Study area

The Khan Tengri State Natural Park is located in the eastern part of the Issyk-Kul region within the Sary-Zhaz river basin at junction of the Teskey Ala-Too and Kakshal-Too ridges of the Central Tien Shan (Fig. 1). Khan Tengri SNP is the largest nature park in Kyrgyzstan and the second largest in Central Asia. This region is characterized by the alternation of high mountain ranges and narrow valleys, the upper part of the valleys are moraine fields and glaciers that sources rivers. Its territory of 275 800 hectares bordered by China and Kazakhstan [9,14].

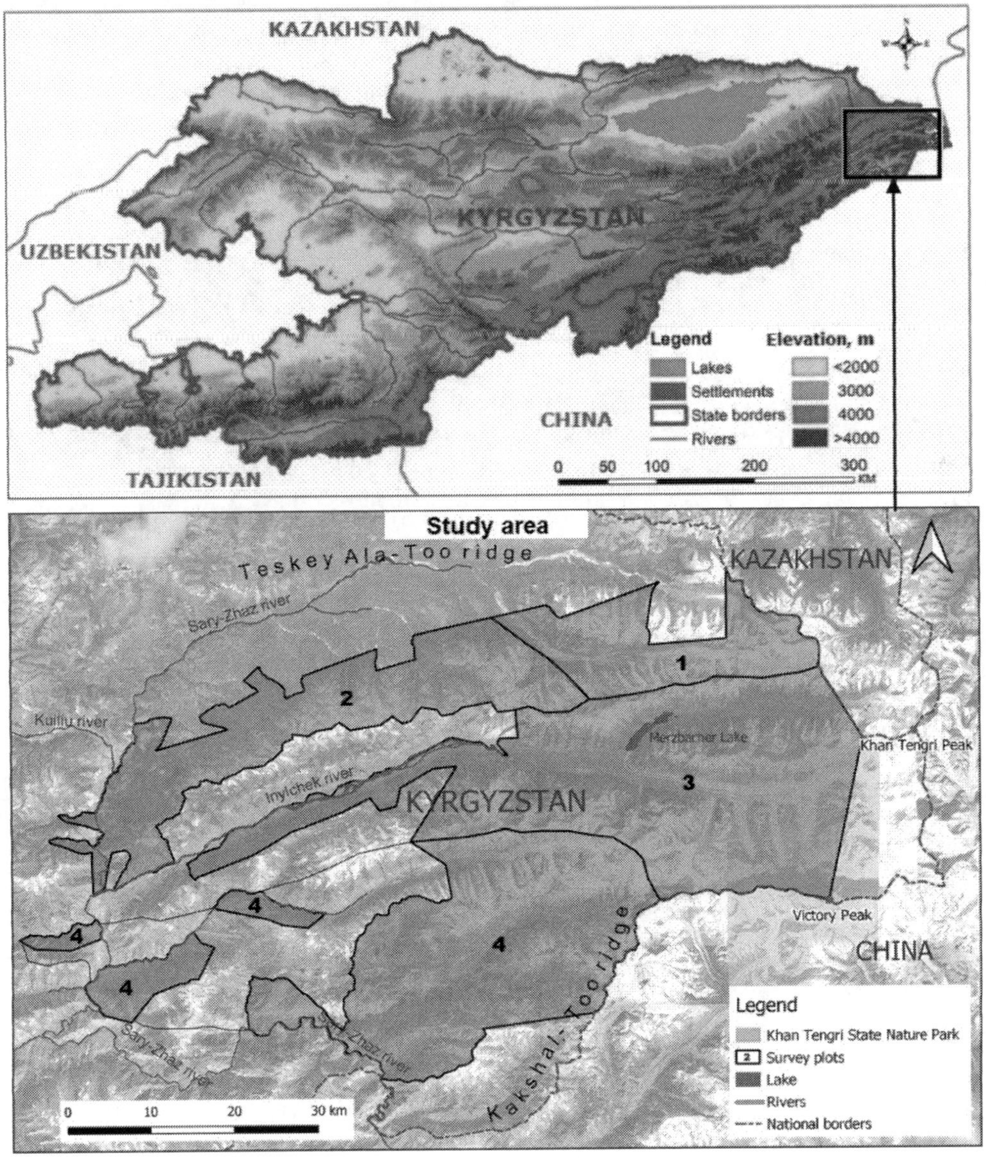

Fig. 1. Location of study area and survey plots: 1-Adyr-Tor; 2-Sary-Zhaz; 3-Inylchek; 4-Kaiyndy

The Central Tien Shan region in geomorphological terms is characterized by great contrast, since it is represented by a massive uplift of this section by mountain ranges, rock outcrops and high peaks on the one hand, on the other hand it is heavily cut and dissected by river valleys, glacial troughs and cirques. Many of these glacial valleys with a huge amount of moraine deposits form unique mountain lakes of their kind, the most mysterious of which is Lake Merzbacher in the upper reaches of the Inylchek River Valley. There are the two highest points of the Tien Shan: Victory Peak 7439 m and Khan Tengri Peak 7010 m. This mountain junction connects the largest mountain ridges of Kakshal-Too, Teskey Ala-Too and others [15,16]. Remoteness from the oceans, significant uplift and complex strongly rugged relief determined the continental climate of the region. Significant temperature fluctuations are characteristic both in the seasons of the year and during the day, low or moderate rainfall and relative dry air. Humidity increases, the duration of the frost-free period decreases. At altitudes of 3400–4000 m, frosts can be repeated throughout the entire warm season, and precipitation falls, as a rule, only in the form of snow. In the midlands, the air temperature in July is 10-15° [17]. Winter in the mountains is longer and colder with air temperature in -15-20°C in January reaching the absolute minimum -45°C.

2.2 Materials

To study the behavior, migrations and breeding of the snow leopard and other animals, camera traps were used (Fig. 2). 36 cameras have different operational modes that allow the user to customize its settings and tailor it to their specific needs (Table 1). They are used to detect and record the presence of rare species of wild animals in a particular area, to monitor their behavior, movement and abundance. The trapping-identification method using camera traps is one of the main tools at the moment for monitoring, determining the number of groups, and the migration routes of snow leopards [18,19,20,21].

Fig. 2. Reconyx Hyperfire HC500 camera trap

Table 1. Camera trap specifications

No.	Camera trap model	Camera resolution, trigger speed	Quantity, pieces
1	Reconyx Hyperfire HC500	3.1MP, 0.29 sec	16
2	Trophy Cam HD Aggressor No-Glow Trail	24MP, 0.2 sec	10
3	Cuddeback E3 Black Flash	20MP, 0.25 sec	10

Considering that the biological behavior of the snow leopard, which leads a very secretive and cautious way of life, and also that the territory of the study area is represented by high mountains, this method was chosen as the most suitable and efficient.

The grid cell method allows the effort to be spread out equally in the entire landscape and not get limited to known or easily accessible area used to cover in the study. A shape file with 5x5 km grid cells has been created using QGIS to model transects for detecting paw prints, feces, wool, skeletons of dead animals by snow leopards and were superimposed on the study area. Such gridding enables to select particular sites that snow leopards were most likely to use in order to maximize detection rates of animals [6,22].

After preliminary planning of field work using detailed topographic and satellite maps of the area, three clusters were identified (Kaiyndy, Inylchek, Adyr-Tor) for installing camera traps. When choosing a camera trap installation, preference was given to areas where either traces of the leopard's vital activity (footprints, scraping, urine marks, excrement) were found, or at the paths intersections, next to a large stone, at the base of a cliff or at the narrowing of the path along the ridge (see Fig. 3). Camera traps should be mounted to trees or rocks with sensors located 50-90 cm above the path, while accounting for possible snow accumulations, and at a distance of 3.5-4 m from the animal's presumed trajectory, preferably perpendicular [20].

Fig. 3. Paw prints and excrements of a snow leopard

When installing camera traps, the following data was recorded in the field diary: installation date and time, GPS coordinates, installation height and camera number and memory card number. In the framework of this study, traps were installed between April 2018 and May 2019, approximately each camera trap at the installation point stood on average for five months, but the duration of the operation of individual traps may depend on the river flooding and other natural hazards.

After conducting field studies, the data from the camera traps is downloaded, processed and all indicators and locations of the camera locations are entered into GIS. Data is stored, processed and analyzed in GIS program, maps are created for each species of wild animals, and transects are loaded onto each individual map cells from the GPS navigator.

In total, from April 2018 to May 2019, camera traps were installed at 52 points in the Khan Tengri State Natural Park. On average, the period of activity of each camera trap in this area was 120 (max. 162 and min. 95) days. The average installation altitude of the camera traps was 3407 (max. 3915 and min. 2384) meters a.s.l.

In total, 15 different wildlife species (snow leopard, mountain goat, argali, bear *(Ursus arctos)*, manul *(Otocolobus manul)*, gray marmot *(Marmota baibacina)*, fox *(Vulpes vulpes)*, rabbit *(Oryctolagus cuniculus)*, Himalayan snowcock *(Tetraogallus himalayensis)*, ermine *(Mustela erminea)*, weasel *(Mustela nivalis)*, wolf *(Canis lupus)*, badger *(Meles meles)*, golden eagle *(Aquila chrysaetos)* and pika *(Ochotona)*) were recorded in the research area with camera traps. Some of these animal species are listed in the Red Book of Kyrgyz Republic.

2.3 Methods

The next workflow is applied in the snow leopard monitoring investigations and mapping its distribution range, its food supply change and other unique species of animals in the territory of the Khan Tengri SNP (Fig. 4). The interviews with gamekeepers, hunters, herders and local people and field work were organized to identify the snow leopard living areal and their prey resources. The preliminary analysis of the snow leopard population area done and the monitoring sites with expected transects of animals for installation of the camera traps are identified.

Geoinformation technology is applied on collection, storage, processing, visualization and analysis of the research datasets on snow leopard monitoring. The open source Quantum GIS (QGIS) software is applied for geospatial data collection, processing and analysis in the research work.

To create and compose the mapping base of the research area using the QGIS program, cartographic shape file layers (Khan Tengri State Natural Park area, camera installation sites, hydrography, transportation, administrative borders etc.) were developed for spatial analysis and geodatabase development. Data collection, processing, visualization and analysis within the framework of QGIS allow designing and distribution of the digital and paper maps to use.

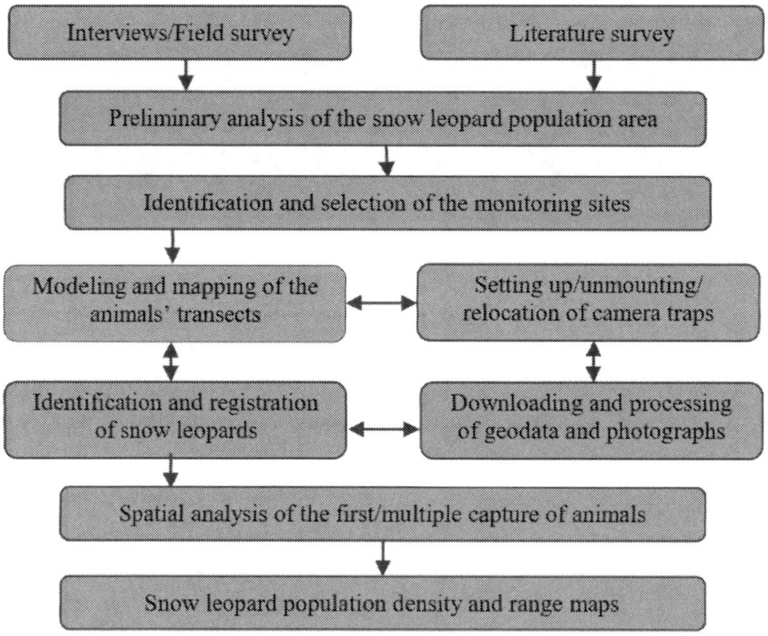

Fig. 4. Workflow of the snow leopard monitoring study

3 Results

The indicators from the traps with their coordinates are converted into a separate shape file layer with the 5x5 km cells for further research with silhouette icons of the snow leopard superimposed. These silhouettes are created using a graphical editor using QGIS to compile a thematic map with points of fixation and presence in a given area (Fig. 5).

On the basis of the obtained images in the Khan Tengri State Natural Park, snow leopard individuals were identified using coat patterns of the animal's, which are unique for each snow leopard, as well as external distinguishing features of its body shape [23]. Most

 https://doi.org/10.1051/e3sconf/202122702002

suitable for animal identification are body parts such as hind and forepaws, dorsal part of the tail, and also a pattern of dark spots throughout the body of the snow leopard (Fig. 6).

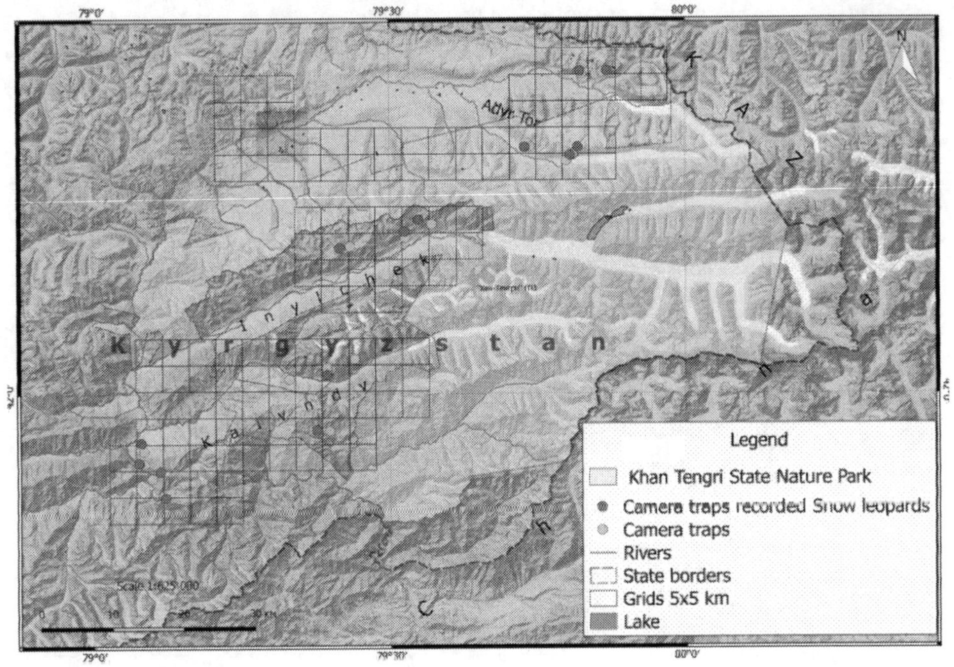

Fig. 5. Living areal of snow leopard fixed by camera traps

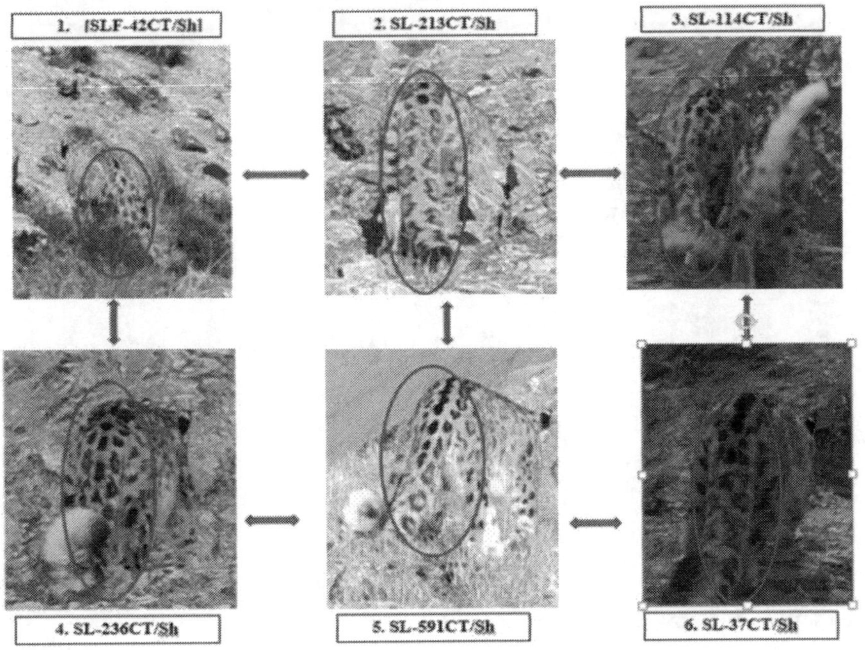

Fig. 6. Identification of six different snow leopards based on their tail patterns

As a result of photo-identification, a special passport is created for a certain type of animals, in which data on a specific individual snow leopard are recorded during the following camera traps. Snow leopard passport is compiled on the basis of images suitable for identification by their body patterns. After this procedure the individual snow leopard has been identified and an individual identification code (IIC) is assigned to it (Table 2).

In most images, it is not immediately possible to identify gender, since sexual dimorphism in snow leopards is not clearly expressed, the exception is family groups consisting of one female with kittens.

Thus, there are 16 adults and 9 snow leopard cubs identified in the Kaiyndy, Inylchek and Adyr-Tor sites in the territory of the Khan Tengri State Natural Park in the Central Tien Shan according to the results of the study conducted in 2018-2019 (Fig. 7).

Table 2. Special passport of the snow leopard

ID code	Date/ Time	Location	GPS/WGS-84 Coordinates	Altitude, m a.s.l.	Photography of the snow leopard
[SL-416CT/Sh]	03.10. 2018, 10:04	Central Tian Shan, Inylchek valley	42.18467N 79.41640E	3780	

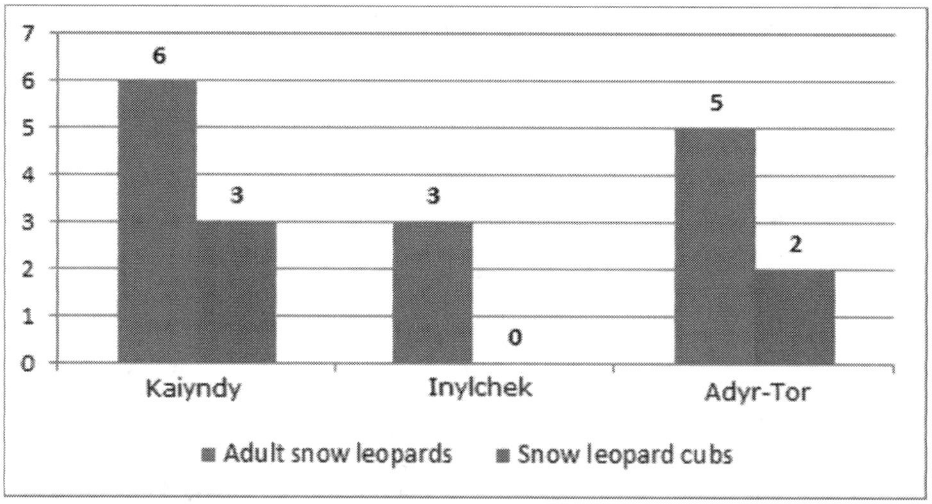

Fig. 7. Number of registered snow leopards in the survey plots

4 Conclusion

Based on the first study using new monitoring methods in the territory of the Khan Tengri State Natural Park, the approximate abundance of the snow leopard and its distribution area have been preliminarily analyzed. It should be noted that this region is unique for its species diversity, natural landscapes in many cases are not subject to active anthropogenic impact. The creation of the State Natural Park in 2016 is dedicated to the conservation of pristine nature and representatives of fauna and flora.

But meanwhile, there are problems in particular, the region began to be actively used as year-round pastures for the cultivation of yaks, cattle and small cattle. Which in turn violates the food base of the snow leopard in some cases, geological exploration is underway in some areas, which in the future can also negatively affect these landscapes. There are also hunting grounds in the border territory of the Khan Tengri SNP (buffer zone), in some cases such hunting facilities are located inside the park itself, where the general silence of natural ecosystems is violated, etc.

In total 19 snow leopards are registered with camera traps and they are living in the Kaiyndy, Inylchek and Adyr-Tor sites of the Khan Tengri State Natural Park. Each of the designed 5x5 km grid cells have installed cameras and recorded animals and birds in their areal. 14 adult and 5 snow leopard cubs are detected and positioned in the digital map of study area with their further identification and registration. Whole area of the park will be investigated in the next research activities to record and monitor existing snow leopard's population.

The environmental conditions of the natural park as a whole are favorable for reproduction of animals, which creates the prerequisites for the reproductive core, and with the successful protection of high mountain ecosystems, even to increase its number in the indicated region. Use of Geoinformation systems and camera traps allows to continue research activities on detecting, identification and mapping of the snow leopards in the all territory of the Khan Tengri SNP as well as in the other wildlife reservation areas of the country.

This work was supported by the German Nature and Biodiversity Conservation Union (NABU) in Kyrgyzstan.

References

1. E. P. Koshkarev. *Snow leopards in Kirgizia (areal structure, ecology, protection)* [Снежный барс в Киргизии: Структура ареала, экология, охрана]. Frunze, Ilim, p.100 (in Russian) (1989)
2. E. P. Koshkarev, V. Vyrypaev. The snow leopard after the break-up of the Soviet Union [Что случилось со снежным барсом после распада СССР?]. Cat News, **32,** P.9-11 (in Russian) (2000)
3. A. Alamanov, F. Balbakova, O. Lipka. Climate change vulnerability assessment for Central Tian–Shan, Kyrgyzstan. Technical Report. DOI: 10.13140/RG.2.1.1696.5523 (2015)
4. V. I. Toropova, V.G. Radchenko, A. Rasakhun kyzy, S. Sagymbaev. Results of monitoring of snow leopard over a five year period (Kirgiz, Terskei and Kungei Ala-Too ranges) [Результаты монторинга снежного барса за пятилетний период (Кыргызский Терскей и Кунгей Ала-Тоо]. *Issledovaniya bioraznoobraziya na BT "Issyk-Kul"* **3**: P.85-86 (in Russian) (2005)
5. M. Sulaimanova. There are 330 snow leopards in Kyrgyzstan [В Кыргызстане имеется 330 снежных барсов]. Bishkek, Kabar News. Retrieved from https://www.meteoblue.com/en/weather/historyclimate/climatemodelled/inylchek-glacier_kyrgyzstan_1526997 (2020)
6. V. Tytar, A. DeKastle, M. Hammer. Mountain ghosts: protecting snow leopards and other animals of the Tien Shan mountains of Kyrgyzstan. Expedition Report, NABU. p. 88 (2016)
7. T. Mahmood, A. Younas, F. Akrim, S. Andleeb, A. Hamid, M.S. Nadeem. Range contraction of snow leopard (Panthera uncia). PLOS ONE **14(8)**: e0218460. https://doi.org/10.1371/journal.pone.0218460 (2019)

8. M. Y. Paltsyn, S.V. Spitsyn, A.N. Kuksin, S.V. Istomov. *Snow leopard conservation in Russia.* Krasnoyarsk: Gorod Publishing House. p.99 (2012)
9. G. A. Lazkov, A.T. Davletbakov, D.A. Milko, M.R. Ganybaeva. *Atlas of flora and fauna of the protected areas in Central Tien Shan (Kyrgyz Republic) [Атлас флоры и фауны особо охраняемых природных территорий Центрального Тянь-Шаня (Кыргызская Республика)].* – Bishkek, P. 298-320 (in Russian) (2016)
10. E. A. Kreuzberg-Mukhina, A.V. Esipov, E.A. Bykova, E.V. Vashetko, B. Aromov. Development of the National Action Plan for the conservation of Snow Leopard in Uzbekistan. Final Report on the ISLT project. Tashkent, Uzbek Zoological Society and State Committee of Nature Protection (2003)
11. *Red Book of the Kyrgyz Republic [Красная книга Кыргызской Республики].* State agency on environment protection and forestry under the Government of Kyrgyz Republic. Institute for Biology and Pedology of National Academy of Sciences of Kyrgyz Republic. Ecological Movement "Aleine" of Kyrgyzstan. – **2 ed.** – Bishkek. – p. 544 (in Russian) (2006)
12. F. Balbakova, A. Alamanov, O. Lipka, O. Pereladova. Climate Change Vulnerability Assessment in Snow Leopard Habitat Central Tian Shan Region of the Kyrgyz Republic. WWF Project Report (2017)
13. V. I. Toropova, T. Harder. Protection of the snow leopard in the Central Asian countries. Ecological Herald "Murok", Bishkek, No.1 (2002)
14. D. Mallon, M. Kulikov. *Aspects of trans-boundary snow leopard conservation in Central Asia.* Report of the FFI/CMS workshop, Bishkek, Kyrgyzstan (2014)
15. S. K. Alamanov, K.S. Sakiev et al. *Physical Geography of Kyrgyzstana [Физическая география Кыргызстана]* (Collective monograph: 589 pp.). Turar. Bishkek. P.140-141 (in Russian) (2013)
16. Atlas of the Kyrgyz Republic [Атлас Кыргызской Республики]. Moscow: Head Office of Geodesy and Cartography. p.157 (in Russian) (1987)
17. Meteoblue. Climate Inylchek Glacier. Retrieved from https://www.meteoblue.com/en/weather/historyclimate/climatemodelled/inylchek-glacier_kyrgyzstan_1526997 (2020)
18. R. M. Jackson, J.D. Roe, R. Wangchuk, D.O. Hunter. *Surveying Snow leopard Population with Emphasis on Camera Trapping: A Handbook.* The Snow Leopard Conservancy. Sonoma, California (2005)
19. M. Ancrenaz, A. Hearn, J. Ross, R. Sollmann, A. Wilting. *Handbook for wildlife monitoring using camera traps.* BBEC II Secretariat (2012)
20. A. S. Karnaukhov, M.P. Korablev, A.N. Kuksin, S.V. Malykh, A.D. Poyarkov, S.V. Spitsyn, M.D. Chistopolova, J.A. Hernandez-Blanco. *Snow Leopard Population Monitoring Guidebook.* – WWF. Krasnoyarsk. p.164 (2020)
21. A. A. Grachev, Yu.A. Grachev, S.K. Saparbaev, A.D. Djanyspaev. *Manual for monitoring snow leopards in Kazakhstan [Методическое руководство-пособие по мониторингу снежного барса в Казахстане].* Astana. p.32 (in Russian) (2016)
22. J. S. Alexander, A.M. Gopalaswamy, K. Shi, J. Hughes, P. Riordan. Patterns of Snow Leopard Site Use in an Increasingly Human-Dominated Landscape. PLOS ONE 11(5): e0155309. https://doi.org/10.1371/journal.pone.0155309 (2016)
23. R. M. Jackson, J.D. Roe, R. Wangchuk, D.O. Hunter. Estimating Snow Leopard Population Abundance Using Photography and Capture-Recapture Techniques. Wildlife Society Bulletin. **34.** P.772-781. 10.2193/0091-7648 (2006)

Basics of Atlas Mapping Optimization in the Fergana Valley

Kamolakhon Khakimova[1,*] , *Ilhomjon* Musaev [2], *Akbarjon* Khamraliev [2]

[1] FerPI, Department of Geodesy, Cartography and Cadastre, 150100 Fergana, Uzbekistan
[2] TIIAME, Department of Geodesy and Geoinformatics, 1000000 Tashkent, Uzbekistan

Abstract. Thematic maps reflecting the negative or positive impact of natural and anthropogenic environmental factors on human living conditions are of particular importance in modern world ecological and cartographic research. To optimize the ecological situation in the region, including the river regions and oases of Uzbekistan, such ecological mapping involves the creation of atlases of a systematic collection of analytical and synthetic maps. In the world, environmental mapping gives preference to synchronization methods where specific atlas maps are systematically created by interconnecting environmental characteristics, that is, several maps that interactively interact with their didactic texts in GIS software are deployed on a computer screen at the same time. These research programs ArcGIS, QGIS, and SWAT model software focus on database creation, electronic map visualization, modeling, and analytical mapping.
Keywords. Ecological Mapping, Digitization, Vectorization, Layers, Themes, Synthetic, Electronic Maps, Modelling.

1 Introduction

In the Fergana Valley, pollution of the environment, including air, water, land, by many industrial facilities, vehicles and various chemicals, causes great harm to the health of the population. Cities can be compared to active volcanoes. Urban industrial enterprises, vehicles, and household waste pollute air, water and soil [1].

Therefore, the main goal of creating an environmental atlas of the Fergana Valley is to provide a rich and comprehensive source of information and databases on the environmental situation in the Valley, as well as to develop a state environmental policy, optimize it and apply it in practice [2]. The ecological situation is unstable and constantly changes under the influence of the human factor. Consequently, practical ecological maps, nature conservation of territories and cartographic optimization of the natural environment are used as the basis for the implementation of the project and strategy, which is intended to strengthen environmental education, advocate for the general public and enhance environmental culture [3]. Maps of the ecological atlas allow you to visually determine the place and time of the ecological state of what happened in the past, at the present time and

* Corresponding author: kamolaxon@mail.ru

which, possibly, will occur in the future, depicting it in color, describing the possibility of counteracting this.

With the help of atlas maps, great opportunities have appeared for showing all the changes that occur in nature. Environmental situations are different, they have different effects on nature and people in the regions of the republic. Water, soil, atmosphere, pollution levels, health impacts and other environmental data cannot be plotted on a single map. Therefore, it is necessary to create a complex of maps, that is, an atlas covering all areas of ecology. Atlas is a system with specific sets of thematic maps that display human-nature interactions, social, natural and man-made elements of society. Therefore, the main purpose of mapping is to register and assess the state of the environment [6].

Environmental atlases include a system of maps dealing with interrelated and complementary environmental issues, and also consist of basic base maps of natural and socio-economic content. Industry and network maps describe and assess the ecological state of certain elements of the environment and represent the environmental problems of various industries (such as agriculture, forestry, water management, industry, etc.). The content of the ecological atlas maps covers various types of mapping, and also works with many objects and events, but among them there are several integrations and syntheses. Maps of different types of scientific and practical directions, their inventory, assessment, forecasting and combination in content, are shown in different directions, which are characterized by their distinctive features.

2 Methods

2.1. Study area

The Fergana Valley is a Valley in Central Asia spread across eastern Uzbekistan, southern Kyrgyzstan and northern Tajikistan. (Fig.1.)

Fig. 1. Map of the study area, Fergana Valley.

The Fergana Valley is an intermountain depression in Central Asia, between the mountain systems of the Tien-Shan in the north and the Gissar-Alai in the south. The Valley is approximately 300 kilometers long and up to 70 kilometers wide, forming an area covering 22,000 square kilometers. Its position makes it a separate geographic zone. The

Valley owes its fertility to two rivers, the Naryn and the Kara Darya, which unite in the Valley, near Namangan, to form the Syr Darya.

The central part of the geological depression that forms the Valley is characterized by block subsidence, originally to depths estimated at 6 to 7 kilometers, largely filled with sediments that range in age as far back as the Permian-Triassic boundary. Some of the sediments are marine carbonates and clays. The faults are up thrusts and over thrusts. Anticlines associated with these faults form traps for petroleum and natural gas, which has been discovered in 52 small fields. [1]

2.2. Methods

Methods of creating an electronic atlas and analysis using modern GIS - programs and methods for their creation with a hybrid type of data storage have the most promising opportunities. Maps of the ecological atlas allow to display the past, present and future, as well as to determine the place and time of the ecological state by depicting them in color, which makes it possible to develop measures to protect the components of nature.

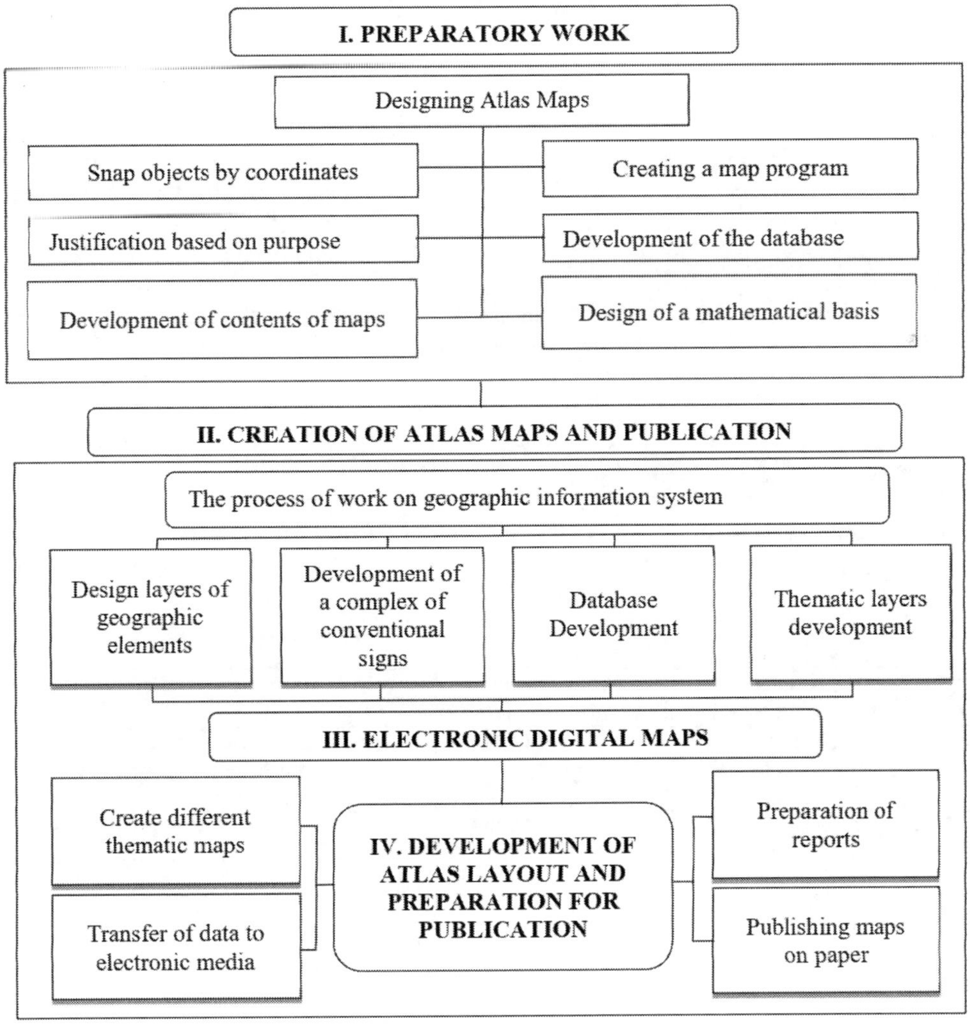

Fig. 2. Designing and mapping environmental atlas in a GIS program

The object of the study is the environmental atlas of the Fergana region. The subject of the study is the structure and content of atlas maps reflecting the environmental state of the Fergana region, developed on the basis of modern GIS technologies. The foundations of environmental atlas optimization are improved on the basis of maps of "degree of anthropogenic load on landscapes," demo ecological state "; Improved digital maps depicting the environmental condition of areas in need of protection in the Fergana region using TIN (Triangulated irregular network), SHAPE vector file and GRID raster files; Improved inventory maps using interactive maps as well as the DEM (Digital Elevation Model).

The cartographic method is most effective in assessing and forecasting environmental situations in the "Population-nature-society" system (Fig.2). At the same time, in the ecological atlas created for the first time, along with traditional indicators, a set of recommendations for the development of ecological culture and a set of measures to preserve the landscape were used.

The creation of an ecological atlas of the Fergana Valley will have a positive effect in the future in determining the state of geosystems changing under the intense influence of human economic activity, in developing a plan of action measures, as well as ensuring sustainability.

3 Results

The scientific significance of the results of the study is determined by the fact that in mapping the ecological condition of the Fergana Valley for the first time GIS-technologies with ArcGIS software, QGIS was used, the structure and content of the ecological atlas of the Territory were developed, as well as methodological approaches to their creation was improved.

The structure and draft of the ecological atlas of maps of the Fergana Valley were developed, on the basis of the geographic information system, moreover, the database "Ecology of Fergana" and its thematic layers also were developed. According to the content, the atlas maps are divided into 3 large sections: the reasons for the destruction of the ecological state; the consequences of environmental degradation; optimization of the ecological state of the Fergana Valley. Texts and legends were prepared on the maps, which facilitate the use of the atlas with modern GIS systems (table 1). Content and configuration changes have been made to reflect the new requirements. The rapidly growing trend in environmental cartography is highlighted based on new methodological approaches, assessment criteria and information sources.

Table 1. The structure of the ecological atlas

#	Name of map	Scale	The main content of the maps
Reasons for the destruction of the ecological state			
1	Map of anthropogenic factors affecting the environment	1:1 000 000	Anthropogenic factors are covered by the scope and measures to improve and protect land resources
2	Pollution map in industrial cities	1:1 000 000	Industrial centers, the amount of toxic gases emitted from industrial sites is given
Consequences of ecological degradation			
3	Air pollution level map	1:1 000 000	Air pollution level. Shown is the amount of toxic substances such as dust, gaseous sulfur, carbon monoxide, nitric oxide, ammonia, hydrogen sulfide compounds
4	Melioration map (irrigation	1:1 000 000	Newly developed regions, irrigation

	networks)		agrotechnical measures against deflation are highlighted. Areas of reclamation shown
5	Demo-ecological state map	1:1 000 000	The map is a database that provides anthropogenic pressure on the natural environment
6	Environmental stress assessment card	1:1 000 000	The degree of tension in the ecological state in the Fergana Valley, the regions of the ecological situation; stabilization
Optimization of the ecological state of the Fergana Valley			
7	Environmental maps	1:1 000 000	The horizontal range limit is indicated in the outlines. The distribution of natural resources that affect human health and the ecosystem is reflected in the distribution of territories
8	The degree of anthropogenic pressure on landscapes	1:1 000 000	The map lists two ecological spaces and economic groups that combine content. The degree of anthropogenic pressure on landscapes, agrodemographic pressure, anthropogenic pressure on agriculture, ecological gene groups in the land fund.
9	Environmental protection map	1:1 000 000	The degree of variability of natural complexes, economic and other human activities, anthropogenic impact on nature and its consequences
10	Ecological maps of the cities of Fergana, Margilan, Kuvasoi, Kokand, Andijan and Asaka	1:25 000	Shown are natural, demographic, technogenic, anthropogenic and other indicators of cities, reflecting the ecological state of tension

1. Data from electronic tachometers and GRS devices, imaging equipment, expedition materials, digital data, copyright originals, existing maps and other source materials were collected. The cartographic and raw materials were scanned and placed at the same scale as the raster images, and then placed in memory.

2. The thematic layers of the ecological atlas maps and their tables were compiled, they were analyzed using the Fergana Valley GIS programs, and a database was created.

3. Classifications of objects, existing tables (attributes) and text data were included in the computer memory.

4. A system of conventional symbols has been developed.

5. The thematic map layers were loaded and cartographic images were created, which were edited separately.

6. The layout of the map has been developed and refined for preparation for publication (Fig.3).

A cartographic base has been prepared for the creation, analysis of thematic layers of the models of the ecological atlas of the Fergana Valley and their drawings, as well as for the creation of a database and a map. Cartographic bases were developed from general geographic elements - the foundations of cartographic and aerospace materials from elements of mathematical foundations (relief, land users, soil cover, settlements, roads, borders) in accordance with the scale and projection of the map.

Creation and use of a multimedia ecological atlas of the Fergana Valley is intended to create a digital model of the valley through a raster image of the valley depression, create a three-dimensional 3D model of electronic ecological atlas maps, visualization and ways of using it in environmental protection. They were analyzed on the basis of GIS technology, methods of indexing space data and through the preparation of forecast and assessment maps, the monitoring methodology was improved.

ArcGIS geographic information mapping and QGIS-based software use the latest GIS database designs from raw data. The created system allows to solve the following tasks: collection and processing of databases; obtaining statistical information about the parameters of the content structure of any cartographic layer; search for objects by their attributes, plotting them on top of each other for comparative analysis; simultaneous viewing of the state of the territory in several layers using mathematical - cartographic modeling; the use of various imaging techniques, the technique of automatic mapping. (Fig.4.)

The latter method of using atlas maps can be shown with hypsometric color or coloring, which improves the readability of the shadows. Hypsometric staining is a set of colors used to provide additional information about the height of a shaded map. Thus, the ability to visualize the overall elevation height has been increased.

The above basic hill shade capabilities are available in almost all GIS software packages, but more advanced techniques are available in Adobe Photoshop.

Triangulated irregular network (TNT), SHAPE vector files, and GRID raster files are based on the vector data of this area and can often be used to describe concepts that can be referenced to a surface or generally oriented to a surface. There are many adjacent triangles, called TNT surfaces that do not collide with each other and they have created electronic maps of the environmental atlas from numerous points using the Delacney method.

Some 3D visualizations only allow one person to explore. For example, in some vertical real-time systems, the display may be viewed by a device and used by only one person. Other devices allow multiple people to view the display at the same time and directly lead a discussion (for example, CAVE software, which can accommodate multiple people.

The most important aspect of 3D visualization is the use of cartographers to create such visualizations that can be generalized using existing abstracts (for example, geometric shapes of objects) or high-level realities (for example, cities with mountains, trees, and streams filled with movement) (Fig.5)

The use of geoinformation technologies in the field of ecological and geographical research is the basis of the dissertation. The analysis of existing resources and capabilities of automated processing of environmental and geographical data and the results of creating digital electronic maps. The Environmental Atlas is a basic and convenient visual resource for assessing and predicting the state of the environment.

 https://doi.org/10.1051/e3sconf/202122702003

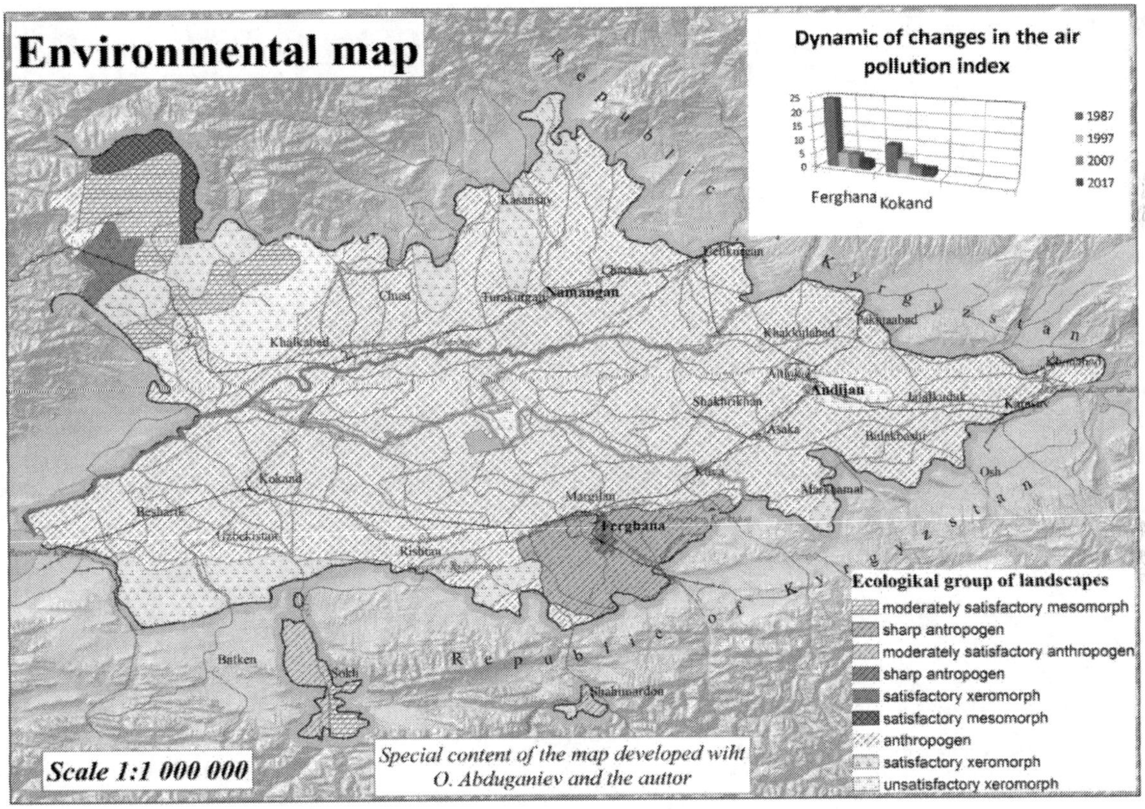

Fig. 3. Environmental map of the Fergana Valley

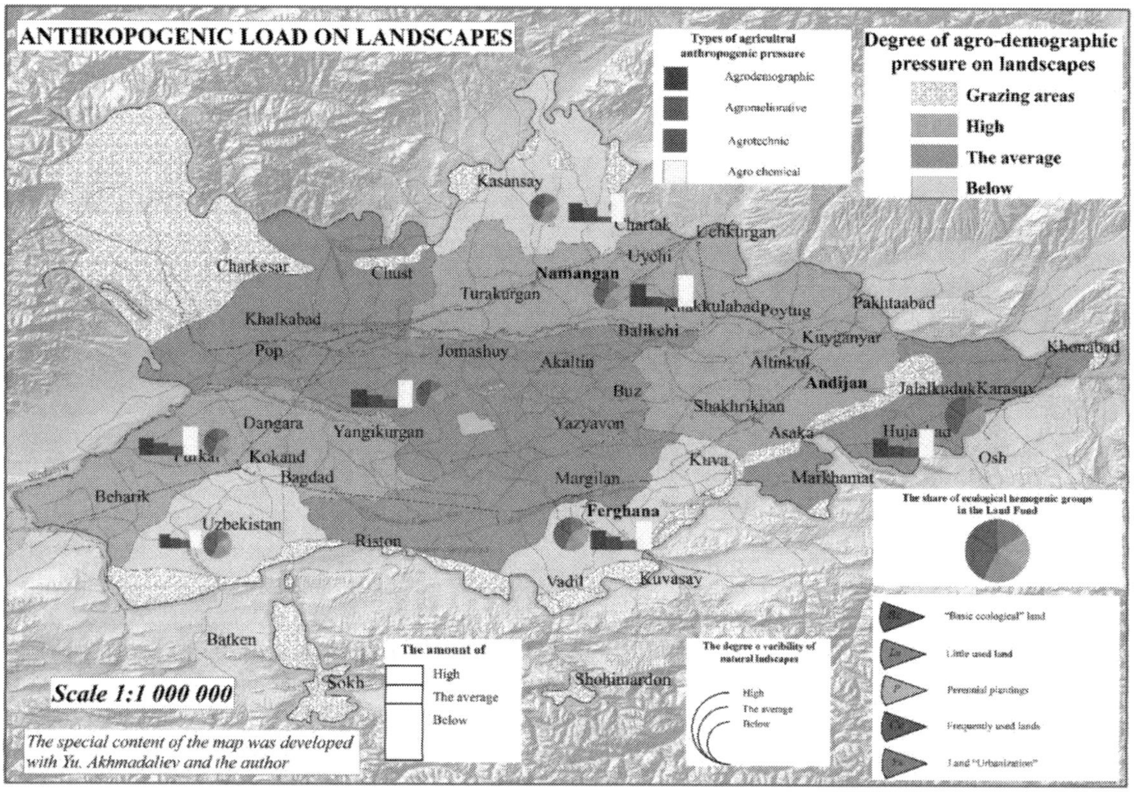

Fig. 4. Map of anthropogenic load on landscapes Fergana Valley

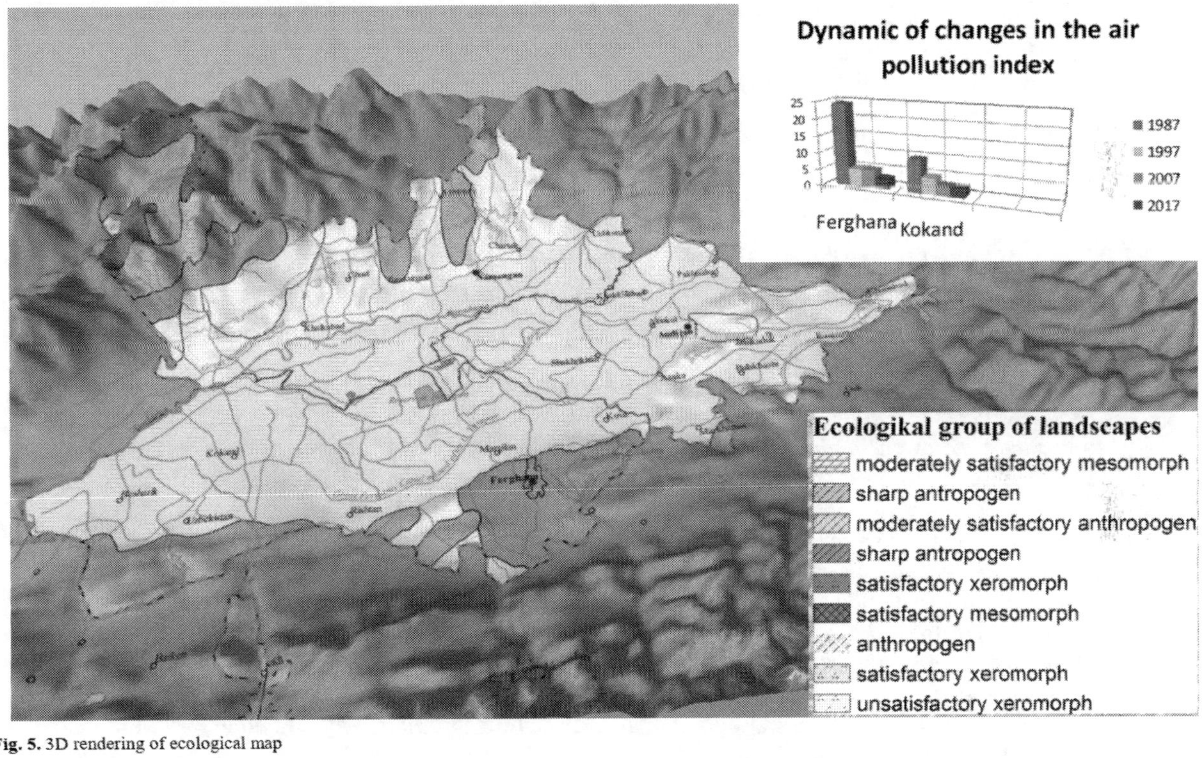

Fig. 5. 3D rendering of ecological map

4 Discussion and conclusion

The scientific significance of the research results is determined by the fact that when mapping the ecological state of the Fergana Valley, for the first time, GIS technologies with the ArcGIS and QGIS software were used, the structure and content of the ecological atlas of the territory were developed, and methodological approaches were improved in their creation.

The practical significance of the research results is determined primarily in the use of the above methods in the development of targeted government programs and practical measures to improve the environmental situation on the basis of conclusions and recommendations, thematic assessments and inventory maps.

Based on the study, the following main conclusions were drawn:

1. The experience of creating environmental atlases is improving every year. The methodological developments associated with the theory and practice of creating atlases are not so great. Modern ecological atlases, created at the present time, are characterized by the fact that they are based on a systemic principle for updating and processing geographic information, as well as on the necessary adaptive functions.
2. Proposals are given on the use of a systematic approach in mapping the nature and ecological state of the Fergana Valley, which made it possible to create a series of analytical and synthetic ecological maps.
3. Development of the composition and content of the ecological atlas using ArcGIS, QGIS and the SWAT model allowed the creation of an ecological database. It is proved that simplified and improved cartographic sources, consisting of several layers and obtained during the creation of maps of the ecological atlas, are the basis for the creation of other maps.
4. The use of geo-innovative technologies in atlas mapping will lead to a decrease in labor costs and time, and an increase in the accuracy of the work performed. At the same time, proposals for software, mathematical basis, layout layout, and projection, generalization of visualized maps of the ecological atlas using ArcGIS, QGIS and SWAT modeling programs based on data characterizing the ecological state of the Fergana Valley were developed and given.
5. The use of imitations of the inventory of space data based on the objectivity and practical orientation of information with the interpretation of images and aerospace images of the atlas, made it possible to create interactive maps.
6. Of particular importance is an operational database of information on geodesy, cartography and cadastres in the shortest possible time with monitoring of the environmental situation and operational information in emergency situations, as well as providing timely access to updated information on the environmental state of the region.
7. When modeling in a GIS suite of commonly used map servers using internal or additional tools, the raster and vector data model made it possible to transform the field in which the spatial distribution of events in the Fergana Valley changes. This database serves as an important source for obtaining relevant scientific conclusions, assessing situation, monitoring, and identification of priority areas for the future.
8. The use of GIS technologies in the field of ecological-geographical research and the use of three-dimensional 3D automated methods for processing existing resources and ecological-geographical data in the development of content on this basis of ecological atlas maps, synchronization of maps is proposed.
9. In order to develop assessment and recommendation maps with cartometric indicators of the ecological state of protected areas in the Fergana Valley, proposals and

recommendations on the trend of planning, assessing and taking into account the environmental impact are given.

10. Ecological atlas of the Fergana Valley and its practical thematic ecological maps will serve as a basis for assessing the ecological situation, nature conservation and rational use of natural resources, as well as optimization of the ecological situation.

The authors wish to thank the "Uzdavyerloyiha" Scientific-Design Institute of Fergana branch and the Joint Research Center of GeoInformatics at TIIAME for constant encouragement and necessary support as well as data used in this research and within the framework of the research plan of the applied project I-8-20 "Creation of the National Atlas of the Republic of Uzbekistan". We also thank the anonymous reviewers for their valuable comments and suggestions.

References

1. S. Musayev, J. Mellor, I. Musaev, M. Nuretdinova. Impact of climate uncertainties on agriculture in Fergana Valley of Uzbekistan. Intl J. Innovative research in a multidisciplinary field. **86.87**, 85-91 (2019)
2. S. Musayev, E. Burgess, J. Mellor. A global performance assessment of rainwater harvesting under climate change. Conservation, and Recycling. **132**, 62-70 (2018)
3. A. Marupov, A. Abdurahmanov, B. Ahmedov. Main Ways to Improve the Efficiency of Agricultural Land Use in the Fergana Valley Sample. Intl J. Advanced Research in Science, Engineering and Technology. October 6-10, 11211-11215 (2019)
4. K. Khakimova, A. Marupov, G. Mirzakarimova. Maintaining Cadastral Valuation for the Effective Use of Agricultural Lands of the Fergana Region. Intl J. Advanced Research in Science, Engineering and Technology. October 6-10, 11022-11026 (2019)
5. D. Eshnazarov, M. Nuretdinova, S. Ibrokhimova, M. Abdukadirova. Ways and bargaining methods of geodescents in regional systems of the statecadastre of the Republic of Uzbekistan. Intl J. Research Culture Society. October 3-10, 41-44 (2019)
6. K. Khakimova. Special issues of correction of the map of nature protection of Fergana region *(Farg'ona viloyati tabiatini muxofaza kilish kartasini tuzatishning alohida masalalari)*. Inf. Geo. Soc. Uzb., **6,** 199-202 (2010)
7. A. Egamberdiev, K. Khakimova. Atlases are unique cartographic encyclopedias (Atlaslar - uziga xos kartografik entsiklopediyalar). Inf. Geo. Soc. Uzb., 6, 226-227 (2015)
8. A. Egamberdiev, Sh. Prenov, K. Khakimova. History, modern state and current problems of cartography development in Uzbekistan (История, современное состояние и актуальные проблемы развития картографии в Узбекистане). Inf. Geo. Soc. Uzb., 6, 219-222 (2017)
9. M. Egamberdieva, K. Khakimova, R. Maxamadaliev. Changes in the structure of cities of Uzbekistan during the years of independence (Mustaqillik yillarida O'zbekiston shaharlari tarkibiidagi o'zgarishlar). News of the NUU, 7, 389-391 (2017)
10. K. Khakimova, E. Safarov. The Main Theoretical and Methodological Provisions of the Educational Geoinformation System. Eur. Sci. Rev. Austria.Vienna, 7, 70-71 (2018)
11. K. Khakimova, R. Maxamadaliev, Kh. Manopov. Development of maps for studying nature conservation and rational use of natural resources in the Fergana Valley (Разработка карт для изучения охраны природы и рационального

природопользования в Ферганской долине). Inf. Geo. Soc. Uzb. Special issue 6, 152-154 (2018)

12. A. Hamidov. Geoecological significance of natural geographical research in the Fergana Valley. Modern geography: theory and practice (Int. Sci. Pra. Conf., Tashkent. 92-95, 2006)
13. K. Khakimova. Some issues of creating an ecological atlas of Fergana region. The role of women scientists in the development of science and technology. Materials of the Republican scientific-practical conference. 448-449 (2006)
14. K. Khakimova, M. Egamberdieva, D. Eshnazarov. The use of GIS - technologies for creating electronic ecological maps (Использование ГИС – технологий для создания электронных экологических карт). Inf. Geo. Soc. Uzb. Special issue 6, 158-160 (2018)
15. A. H. Robinson, J. L. Morrison, P. C. Muehrcke. Elements of Cartography. John Willey & Sons, USA, 674 (1995)
16. J. Bertin. Semiology of Graphic: Diagrams, Networks, Maps. University of Wisconsin Press, Madison (1983)
17. P. Wiegand. Learning and Teaching with Maps, Taylor & Francis, London: Routledge, 180 (2006)
18. P. Hurst, P. Clough. Will We Be Lost without Paper Maps in the Digital Age? J. Inf. Sci. **39**, 48–60 (2013)
19. W. Johnson, H. Jellinek, L. Klotz Jr, R. Rao, and S. K. Card. Bridging the Paper and Electronic Worlds: The Paper User Interface." *INTERACT'93* and CHI'93 Conference on Human Factors in Computing Systems, Amsterdam, April 24–29 (1993)
20. K. Ooms, P. Maeyer, L. Dupont, N.V. Veken, N.V. Weghe, S. Verplaetse. Education in cartography: what is the status of young people's map-reading skills? Car. GIS, **43:2**, 134-153 (2015)

E3S Web of Conferences **227**, 02004 (2021) https://doi.org/10.1051/e3sconf/202122702004
GI 2021

Climate indicators for forest landing and evaluation of forest shelterbelts

Maftuna Begimova[1,*]

[1]Tashkent Institute of Irrigation and Agricultural Mechanization Engineers, Faculty of Land Resources Management, 39, Qori Niyazi Street, Tashkent 700000, Uzbekistan

Abstract. Forest shelter belts are one of the effective approaches in the fight against wind erosion, improving the microclimate of fields, protecting roads and reservoirs. The problem of assessing forest plantations and prospects for their creation is relevant for many countries of the world, due to the problem of desertification, water and wind erosion, as well as the global food crisis. The carried out analysis of scientific literature has shown that the existing assessments of forest shelter belts with insufficient degree are being taken into account the modern methods of remote sensing.Therefore, the problem of assessing forest shelterbelts from satellite imagery is relevant.The main part and research results. During the summer period of 2019, in the territory of the protective forest plantations of Fergana Valley of the Republic of Uzbekistan there were laid 170 testing plots. Then 20 test sites, which were used as reference for thematic mapping, there were laid with the collection of additional information: location coordinates and bookmark location.The purpose of the study is to assess the area of forest shelter belts zones of the Republic of Uzbekistan using remote sensing.The object of the study is forest shelter belts of Fergana Valley of the Republic of Uzbekistan.

1 Introduction

Forest shelterbelts and desert fodder plants are one of the effective approaches in the fight against with wind erosion for livestock feed, improving the microclimate of fields, protecting roads and reservoirs.

The problem of assessing protective forest stands and perspectives of their creation is relevant for many countries of the world, due to the problem of desertification, water and wind erosion, as well as the global food crisis.

The carried out analysis of scientific literature has shown [1,2,3,4,5] that existing assessments of forest shelterbelts and desert fodders are not sufficiently taken into account the modern methods of distance sensing. In particular, the possibilities of remote monitoring of such plots haven't been studied enough. Therefore, the problem of assessing forest shelterbelts from satellite imagery is actual.

* Corresponding author: e-mail: Begimova-1993@mail.ru

Analysis of the current state of the problem. The climate of Central Asia, including Uzbekistan is determined by the situation in this region, the territory of which is located at a considerable distance from the main sources of moisture - oceans.

Air masses, passing the distance from the Atlantic Ocean to Central Asia, gradually reduce moisture content. Deserts play a certain role in climate formation. In contact with the hot surface of sand, air masses entering to Central Asia are even more dried out. Small reservoirs of the Caspian and Aral seas at very high summer air temperatures do not affect the humidification regime. The flat part of Uzbekistan, unprotected by the mountains and open from the side of the desert is the most arid. Uzbekistan belongs to the Turan climatic province, whose characteristic features are a sharp continental climate, high temperatures and dry air. Large temperature fluctuations are observed both during the year and in the diurnal course. The annual amplitude is 29-30 ° C in the south and 38-40 ° C in the north, daily in the southern regions 18-20 °, sometimes much more.[7,8]

The determining factor of climate formation in Central Asia is two centers of atmospheric action: the winter Siberian anticyclone and summer thermal depression (69).

The distribution of precipitation over the territory of the republic depends on the direction of the moist air masses and the surface features of the area. Most of the desert zone is the poorest in precipitation. The average annual amount here is in the range of 80-120 mm with fluctuations from 40 to 150 mm. As you move to the foothills, the amount of rainfall increases to 400-550 mm, which allows you to grow crops in these areas due to rainfall and the accumulation of moisture in the soil (rainfed farming).

In the regime of precipitation throughout the republic, seasonality of their distribution is pronounced. The greatest amount falls in the spring, slightly less in the fall and winter. March and April are especially rainy, when up to 80 mm and more can fall in a day. Almost no precipitation occurs in the flat part of the summer.

For Uzbekistan it is characteristic low air humidity: the average annual absolute humidity is 6-8 mb. in winter 11-12 mb. Relative humidity in winter months reaches 70–80%, in summer months it drops to 25–30%. On some days, it does not exceed 10%.

The plains of Uzbekistan are characterized by high air temperatures: average annual 15-17 °. The average air temperatures of the coldest month – January, in most areas range from -: 0.6 to -6.9 °. So, in the lower reaches of the Amu Darya, the average air temperature in January is -6.9 ° C (Nukus). The sum of negative temperatures is 300-550 °. On the Ustyurt plateau –860 °, the absolute minimum is -38 °. The average January temperatures in Termez and Sherabad are + 2.8 ° and + 3.6 °, the sum of negative temperatures is only -17 °. The average temperature of the warmest month — July — ranges from 27.1 (Nukus) to 30.1 ° (Sherabad).In foothill areas, the average annual air temperature is 2–2.5 ° lower than in the lowlands. A characteristic feature of the thermal regime here is the frequency of frosts during transition periods, especially in March.[9,10]

Beyond the republic, the absolute maximum and minimum air temperatures are relatively high. The absolute minimum in Nukus, drops to -32 °, in Termez -21°. In the foothill zone, the air temperature drops to -35 °. In summer, it continuously increases and in July reaches its maximum values. The absolute maximum in Termez and Sherabad is 44–48 °. The surface of the soil heats up to + 70 °, sand up to + 90 ° (167). Such high air temperatures in combination with low humidity and winds cause the occurrence of dry wind phenomena.

Dry climate is also characterized by high evaporation: up to 700 mm in the north and 1200-1500 mm in the south of Kyzyl-Kum. Evaporation is an indicator of soil drought (8).

The vegetative period with temperatures above 5 ° C lasts 264 days in Tashkent, up to 321 days in the southern regions (Sherabad), with temperatures above 10 ° C respectively 214—227 days.

The sum of positive temperatures in the north-eastern part of the republic ranges from +4400 to + 4500 °, in the southern (foothills of Surkhandarya) —from -1-4750 to + 5100 °, in the desert - to + 6000 °.

On the territory of Uzbekistan, especially in the flat part, strong wind activity is developed with a predominance of northern, eastern and adjacent wind directions. The occurrence of large storm winds in conditions of narrow gorges and narrow exits from valleys and hollows is due to southern cyclones. Wind activity is especially active in Golodniy Steppe, Kokand group of regions of Ferghana Valley, Bukhara oasis, Karshi steppe and Surkhan-Sherabad valley.[11,12]

In the cold period, north and northeast winds prevail over the territory of Central Asia. Western winds during this period are observed only during the passage of cyclones. The average wind speed ranges from 1.5-2.0 m / s in protected oases to 3.5-5.0 m / s in the plains of deserts and semi-deserts. In the open areas of the Aral Sea, winds reach 9-10 m / s.

In the southern deserts and deep valleys, wind activity is activated in the summer, due to the increased atmospheric turbulence.

Mountain elevations, deep depressions and vast valleys cause a peculiar local circulation of the atmosphere. High-speed winds develop mainly at the exits from gorges and pits. A typical example is "Bekabad" wind breaking out of Ferghana Valley. Its maximum speed is 34 m / s, the highest repeatability is noted in the spring.

In a cold season, as a result of various atmospheric circulations between the northern and southern regions of Central Asia, stormy winds of cold front arise. The orographic anticyclones of the alpine zone also contribute to their formation. Storm winds are widespread in the western part of Ferghana Valley, where the number of days with strong winds exceeds 40-50 per year. Western rumba prevail at a speed of 25-30 m / s.

The local ones also include strong dry winds that occur in the southeastern part of Central Asia, carrying a huge amount of sand and dust, called Afghans. These are westerly winds, reaching 17–20 m / s and lasting up to 5 days.

Dry and hot winds are widespread in Central Asia, blowing on the mountain slopes and in the foothill plains of the southern mountain systems- Garmsili are usually of the eastern and southeastern directions, developing speeds of up to 10-15 m / s and more. Garmsili, burning plants, cause great damage to agriculture.

In Bukhara oasis, the northern and northwestern winds blowing at a speed of 10-15 m / s are most active. In summer, there are usually southerly winds, accompanied by high temperatures. There are more than 30 dry days per year in Bukhara region, in Karshinekiy steppe and Surkhandarya valley- 2 times more.

Analysis of climatic data shows that there are many unfavorable factors in the republic (dry air, high evaporation, strong and hot winds, low water) for the successful cultivation of crops. Years of research and production experience show that these obstacles can be greatly reduced by creating systems of protective forest plantations in the agricultural territory, improving its aesthetic appearance.[13,14,15,16]

2 The main part and research results

During summer period of 2018, 170 test plots were laid in the territory of the protective forest plantations of the Ferghana Valley of the Republic of Uzbekistan. 20 test plots, which were then used as reference for thematic mapping, were laid with the collection of additional information: location coordinates and bookmark location. Bookmarking was carried out in forest shelter belts along roads and cultivated agricultural land (Fig. 1). The remaining test sites were collected using high-resolution satellite imagery, cadastral and forest inventory data, and map services.[6]

Fig. 1. Caption of the Figure 1. Below the figure.

The coordinates of each test site were recorded using a GPS receiver. The information received was entered to the account sheet a (Table 1).

Table 1. Fragment of the accounting statement.

№ points	Location	Latitude, des. hail.	Longitude, des.hail.
1	along the road	71,0451	40,6945
2	along the field	71,2825	41,0525
3	along the field	71,3955	40,8939
4	along the road	71,4406	40,8639
5	along the road	71,3184	40,6582
6	along the road	70,996	40,6531
7	along the field	70,8867	40,5117
8	along the field	70,6782	40,3831
9	along the field	71,1462	40,4098
10	along the road	71,2454	40,4399
11	along the road	70,7457	40,5851
12	along the field	71,4376	40,5597
13	along the road	71,3184	40,6582
14	along the road	71,4406	40,8639
15	along the road	71,3357	40,3365
16	along the road	70,6782	40,3831
17	along the road	70,6541	40,5678
18		70,7457	40,5851

https://doi.org/10.1051/e3sconf/202122702004

3 Preparation of satellite data

For the work, scenes of the high-resolution satellite Sentinel-2 were selected, which covered the study area (Fig. 2).

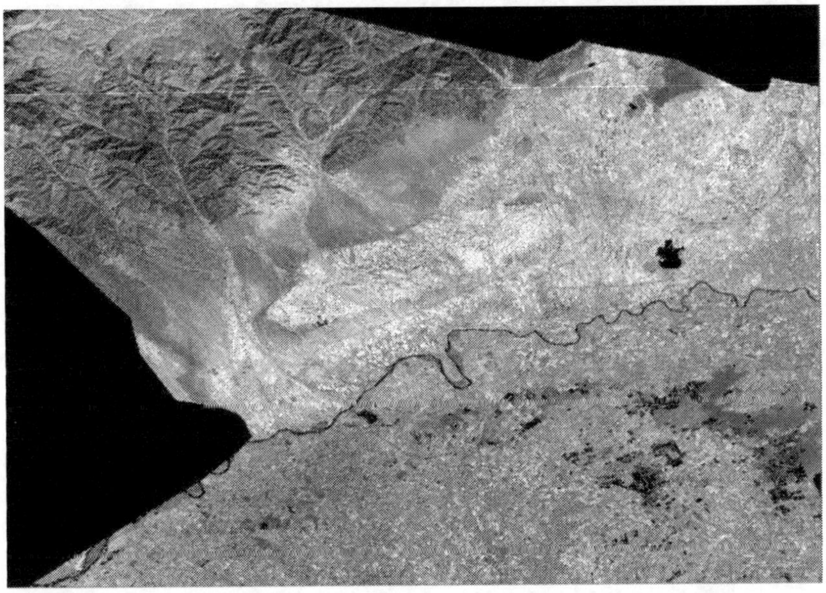

Fig. 2. Scene fragment of the Sentinel-2 satellite used in the study (date of filming 08/31/2018)

Images selected for the study were prepared for thematic mapping by atmospheric and linear correction, as well as geometric alignment (Fig. 3).

Fig. 3. Sentinel-2 snapshot fragment (8th spectral channel) before (left) and after (right) linear alignment

Selected images were cropped to the mask of Ferghana Valley and sewn into a seamless mosaic for further work (Fig. 4).

Fig. 4. Seamless mosaic of fragments of satellite images to the study area.

The pledged test sites were highlighted and tied to matched satellite images.

4 Conclusion and recommendations

In this article, the author's research results, which are aimed at filling the gaps in the study of monitoring and evaluation of protective forest belts of the Republic of Uzbekistan according to remote sensing data.

The research work includes the results of the author's research, which are aimed at filling the gaps in the study of the tasks of monitoring and assessing the protective forest belts of the Republic of Uzbekistan according to remote sensing data.

The use of the method of separability of vegetation cover classes in high-resolution satellite images of Sentinel-2 is justified by the high accuracy of thematic mapping.

When constructing a thematic map of protective forest plantations of Fergana Valley of the Republic of Uzbekistan, their total area for 2019 was determined. The total area of protective forest bands is just over 27 thousand hectares, which is about 5% of the territory. This figure is generally consistent with official data, according to which the total area of forest shelter belts as of 2007 was 40 thousand ha.

Reference:

1. Z. Yu, T. Qin, D. Yan, M. Yan, The impact on the ecosystem services value of the ecological shelter zone reconstruction in the upper reaches the basin of the Yangtze river in China. International Journal of Environmental Research and Public Health. p.15. (2018).

2. Y. Zhang, Z. Yang, S. Guo, Q. Wang, K. Zhan, Ecological changes in the Minqin oasis belt over the past 20 years. Acta Prataculturae Sinica. P.14-24. (2018).
3. J. Zhengzhong, W. Yingju, L. Jiaqiang, Influence of plantation of a shelter-belt on component of organic carbon in the Taklimakan desert over last decade. E3S Web of Conferences. p.53. (2018)
4. Z. Zuo, Z. Pan, A. Zhang, D. Yu, Z. Zuo, Spatial wind speed and surface wind erosion characteristics of farm-shelter forest nework in arid sandy area. Transactions of the Chinese Society of Agricultural Engineering. P.135-141. (2018).
5. V. Kachova, G. Hinkov, E. Popov, L. Trichkov. Agroforestry in Bulgaria: history, presence status and prospects./ V // Agroforestry Systems P.655-665 (2018).
6. G. Lantzanakis, Z. Mitraka, N. Chrysoulakis, Comparison of Physically and Image Based Atmospheric Correction Methods for Sentinel-2 Satellite Imagery. Perspectives on Atmospheric Sciences P. 255-261. (2016).
7. P. Marcela, P. Marco, S. Andreas, S. Harald, Relationships between forest variables and remote sensing data in a Nothofaguspumilio forest. Journal Geocarto International. P.25-43. (2009)
8. G. Mateo, J. Tomislav, The effect of fusing Sentinel-2 bands on land-cover classification. Johnson International Journal of Remote Sensing. p. 39. (2017)
9. A. Mitchell, B. Mora, Current remote sensing approaches to monitoring forest degradation in suport of countries measurement, reporting and verification (MRV) systems for REDD. Carbon Balance Management P. 9-10. (2017)
10. K. Miyamoto, Decreased Rice Production as the Result of Windbreak-cutting. Journal of Agricultural Meteorology. P.145-148. (1961)
11. R. Nisa, Evaluation of roadside green belt trees damaged by mistletoes parasite plant in Mean Merdeka Road, Central Jakarta, Indonesia. IOP Conference Series: Earthand Environmental Science p.203. (2018)
12. H. Pengyu, L. Fabian, B. Chandrashekhar. Pengyu, Annual Cropland Mapping Using Reference Landsat Time Series A Case Study in Central Asia. Distance Sensing P.10-12. (2018)
13. E.P. Da, M. Fleckenstein, P. Leinenkugel, A. Parker, N. Oppelt, C. Kuenzer, Tropical forest cover dynamics for Latin America using Earth observation data: a review covering the continental, regional, and local scale. International Journal of Distance Sensing. Published online. P.3196-3242 (2015).
14. J.W Roberts, S. Tesfamichael, M. Gebreslasie, J. Aardt , Forest structural assessment using remote sensing technologies: an overview of the current state of the art. Southern Hemisphere Forestry Journal. p.69. (2010)
15. E. Romijn, C. B. Lantican, M. Herold, Assessing change in national forest monitoring capacities of 99 tropical contries. Forest Ecology and Management **Vol. 352**. p. 109-123. (2015)
16. N. Yakimov, The value of afforestation afforestation to combat erosion processes on agricultural land. Land of Belarus – (2014). p. 3-7.A.

Land cover-adjusted index for the former Aral Sea using Landsat images

Ilhomjon Aslanov[1*], *Sayidjakhon* Khasanov[1,3], *Yakhshimurad* Khudaybergenov[2], *Michael* Groll[4], *Christian* Opp Ch[4], *F.* Li[3], and *Ramirez* Del-Valle E[5]

[1]Tashkent Institute of Irrigation and Agricultural Mechanization Engineers, Tashkent, Uzbekistan
[2]Department of Geodesy, cartography and natural resources, Karakalpak State University, Nukus, Karakalpakstan, Uzbekistan
[3]Institute of Geographical Sciences and Natural Resources Research, Chinese Academy of Sciences, Beijing, China
[4]Faculty of Geography, Philipps-Universität Marburg, Germany
[5]Institute of Biology, National Autonomous University of Mexico, Mexico

Abstract. The Aral Sea was the fourth largest inland lake on the globe until 1960, with a surface area of about 68,000 km2. Mainly, the huge irrigation projects in many parts of its transboundary catchment were responsible for the catastrophic desiccation and ecological crises of the Aral Sea after second part of 20th century. Ecological crisis surrounding the Aral Sea (lake) regions is one of the critical environmental problems of Central Asia. As a result, monitoring of desertification processes and determining the aerosol concentration in the atmosphere are highly relevant for any attempts to mitigate environmental changes in the Aral Sea basin. Remote sensing is the most appropriate method for studying desertification and dust storms as it easily covers large areas with a high spatial and temporal resolution. Satellite images provide detailed multispectral information about the earth's surface features, which proves invaluable for the characterization of vegetation, soil, water, and landforms at different scales. Vegetation cover, biomass, and soil properties were analyzed with remote sensing methods (NDVI, SDVI). It is emphasized that vegetation indices have little sensitivity at low leaf area which is common to all desert ecosystems.
Keywords: Aral Sea, Water shrinkage, Land-Cover, Remote Sensing, GIS, Central Asia, NDVI

1 Introduction

The immediate cause of the lake water disappearance of the Aral Sea is the fact that its two major rivers, the Amu Darya and Syr Darya, only contributed an average of 5 km³ of water to the lake's annual water balance instead of the former quantity of 56 km³. Besides that, in many years, the river water did not even reach the lake at all [12, 47]. Since there was no sufficient water to ensure the lake water balance, the evaporation rate from shallow

*Corresponding Author: ilhomaslanov@tiiame.uz, ilhomaslanov@gmail.com

depths of the lake has been dramatically increasing, thus urging to actively develop potential remediate measures to combat with this human-induced environmental issue [30, 33, 18, 14]. Detailed analysis of land cover using remote sensing data in large area provides a complete profile for better analyzis and mapping. Instead the lake surface a new desert has been covering the Aral Sea region during the last six decades. Sand and dust storms often do have its source area from this bare sandy area of the Aral Sea accelerating desertification processes [45, 3]. They are often reflecting an early warning about the ongoing local and regional environmental changes around the Aral Sea [46, 5]. Once they progress from slight to serious and severe categories they contribute to the spread of desertification through the transport and deposition of sediments that can destroy local ecosystem and infrastructure, and render areas uninhabitable [43, 30]. During sand transportation, airborne dust is involved in many physicochemical processes of the atmosphere, hydrosphere, and geosphere. Atmospheric sand deposition composes Aeolian mineral particles which strongly contribute to the composition of the sediments, both on the continents and in oceans worldwide [11, 39]. Conducted an in-depth and systematic analysis of long-term space and time variations in of both area and water volume of the Aral Sea, and their potential drivers, which was lacking prior to these experiments[47, 42, 28, 9, 26]. Firstly, they extracted the lake surface area of the Aral Sea using one of the Remote Sensing (RS) indices, Normalized Difference Vegetation Index (NDVI), based on the dataset captured by MODIS during 2000-2018. This extraction enabled to quantify and distinguish the inter-annual variations of lake surface area and water volume of the Aral Sea during 1960–2018 based on long-term hydrological data generated by the MODIS sensor. The results of these experiments revealed that from 1960 to 2018 the area of the Aral Sea shrank dramatically and the average water shrinkage rate per annum accounted for around 1,000 km^2. They claim that this shrinkage rate has decelerated over recent years. Besides that, during the experiments, marginal expansion of the Aral Sea lake surface area was observed. As these scientists assert that in 1986, the Aral Sea split into two lakes. Since then, these two lakes have encountered opposing trends: the much bigger South Aral Sea located in Uzbekistan has shrunk dramatically despite having some trend oscillations, while the North Aral Sea located in Kazakhstan has enlarged obscurely [1].

Considering the above mentioned results, this paper aims to provide an improved insight to spatio-temporal land-cover change, aligning with water shrinkage, from 1960 in the Aral Sea was change water balance and start shrinking but however in the last two decades and to determine a water shrinkage rate per annum based on Landsat sensor. Besides that, as an objective of this paper, once a water shrinkage rate has been ascertained, performing a simplified correlation analysis between land-cover change of the Aral Sea and climate data is contemplated. Based on the expected results some conclusions should be given regarding the water shrinkage effects both on the land use changes and on the formation of sand and dust storms.

2 Study area

Central Asia is the largest landlocked region in the world. The majority part of its landscapes is desert, semi-desert, dry steppes, and high mountains [32]. Due to the fact that the Aral Sea region became after the desiccation disaster of the Aral Sea a very vulnerable ecological zone, where desertification processes have been accelerated during the last four - six decades [24], this study site was selected for further investigations. The Aral Sea is located in the central part of Turan lowland in Central Asia and surrounded by several deserts, and is sandwiched between two deserts, the Karakum and the Kyzylkum. [41, 15] (Fig 1). The Aral Sea is nowadays considered as a salty lake bestriding the Uzbek-Kazakh border, covering an area of approximately 68,478 km^2 with a total volume of 1,064 km^3, a maximum depth of 69 m, an average depth of 16 m and a shoreline stretching for more than 4,430 km in 1960 [28].

Since the Aral Sea is situated in a dry region with high summer temperatures, where the average annual rainfall is below 250 mm, there is immense evaporation. However, over the past eight decades, annual precipitation in the Aral Sea region has increased, with winter suffering the largest growth trend with 0.7 mm [47]. The Aral Sea region has a distinct continental climate that features exceptionally improved solar radiation and extremely low humidity, negligible rainfall, uncertain time of precipitation, and huge differences between seasonal and annual average temperatures. In the lower locations of this region, average air temperatures in July are around 30 °C while the highest air temperature often reaches 45 °C and 50 °C, respectively; Average air temperatures in the winter season range from −8 °C to 0 °C, with the lowest temperature hovering around −38 °C [4].

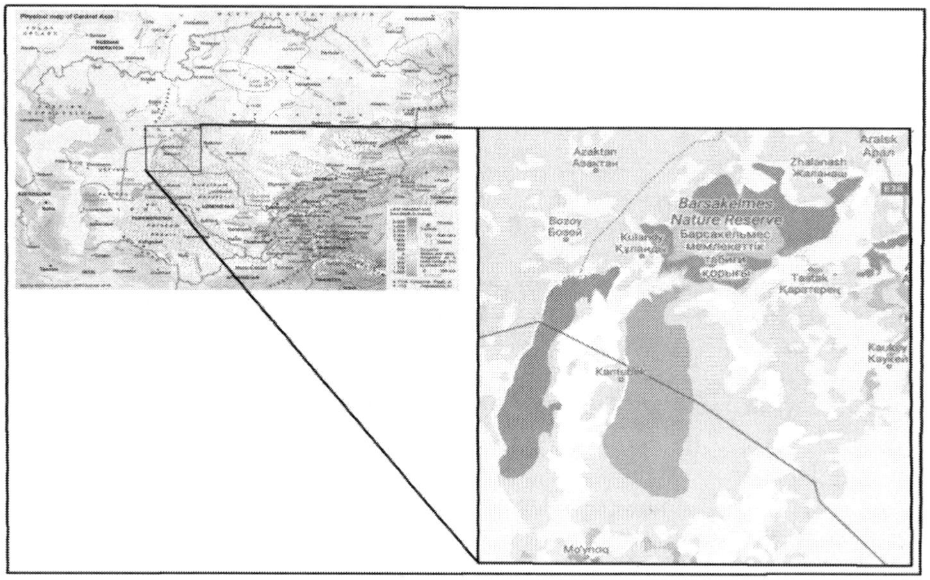

Fig.1. Location of the Aral Sea (Source: GRID-Arendal)

3 Methods

Application of remote sensing tools and techniques are important to instantly assess the dynamics of land-cover change at the wider spatial resolution, rather than spending more time on the in-situ research. These tools have a great potential to monitor and evaluate drought processes and water shrinkage including the Aral Sea disaster [17, 41] first started to locate, assess and monitor the land-cover in drylands with geo-information systems (GIS) and RS techniques. They claimed that GIS and RS-based data provide a permanent record of the land-cover changing rate in such a format that records measurable changes. [17] considered that the RS approach towards the land cover analysis plays an important and key role as one of the major sources of state-of-the-art and physically-based information, whereas GIS provides the toolbox that improves the ability of data integration, data analysis and data extraction from the source. According to [2, 7, 25, 36] it is possible to understand the desertification rate as a result of land-cover changes. The amounts of land-cover changes can be assessed by using RS indices like NDVI and Soil-Adjusted Vegetation Index (SAVI). These indices might function as a basis for an early warning of desertification. NDVI and SAVI, calculated using satellite images, have revealed the capability of RS for systematic, reliable and spatially extensive monitoring of drought after water shrinkage [7, 29, 37, 38]. Recent studies have also shown that with the help of the integrated use of GIS and RS analyses of both temporal and spatial dynamics of land-cover changes caused by drying surface water bodies become possible, [46] analyzing potential changes occurring among land cover features and develop the baseline maps of desertification or land-cover [6, 16, 44]. Concerning these studies, RS implies either satellite images or aerial photography to perform a trend analysis for future scenarios and create trend maps showing the possible changes in land cover condition over a certain period [27].

4 Data

The Physical map of Central Asia from 1960 was taken as a source to extract the former extent of the Aral Sea, which was 68,000 km^2, to gain a deep insight how land-cover of this region has spatially and temporally changed in combination with water shrinkage. However, the base year of the present analysis is set in 1999, because the nationwide assessment of the Aral Sea shrinkage effects was commenced, and the analysis was conducted till 2019. This Physical Map was digitized in order to extract the borders of extent of the Aral Sea for further analysis.

For secondary data satellite images taken especially in the end of each year (1999, 2009, and 2019) were downloaded from the open source databases Earth Explorer and Glovis (www.earthexplorer.usgs.gov; www.glovis.usgs.gov). This period was used to learn similar patterns at the end of the years. For this reason, the last cloud-free days of December are highly appropriate to observe the dynamics of land-cover in the Aral Sea region. Regarding the sensors of satellite images, Landsat TM 5, Landsat ETM+ 7 and Landsat 8 OLI (for further information on Landsat sensors, see CRISP) were used. The temporal resolution of the selected sensors is 16 days, whereas, the spatial resolution of images is 30 meters, which means one pixel of the image covers 30 m x 30 m of land surface.

4.1. Data processing and analysis

The analysis of land-cover in the shrinking Aral Sea region required several steps to interpret the results. Firstly, initial preprocessing steps and operations on satellite images (e.g. atmospheric and geometric corrections, outlier removal and mosaicking) were undertaken which directly improved the quality and the accuracy of the remote sensing maps [38, 25, 23, 21, 8]. Then the further analysis was conducted by using ArcGIS and Erdas Imagine software packages. Then, a GIS tool called "Resample" was used in order to convert the actual pixel size of the satellite images to 150 meters x 150 meters to expedite analyzing processes, return an enhanced visualization, and reduce the size of the remote sensing images to ensure proper data storage. Afterwards, the Normalized Difference Vegetation Index (NDVI) was applied to detect the canopy cover from the satellite images. The NDVI ranges from -1 to 1 and assesses whether the target being analyzed contains photosynthetically active vegetation or not by using Equation 1 [2]:

$$NDVI = \frac{NIR-RED}{NIR+RED} \tag{1}$$

Where: NIR is the Near InfraRed band of Landsat sensor (band 4 for Landsat TM 5 and Landsat ETM+ 7; band 5 for Landsat 8 OLI); and RED is the red band of Landsat sensor (band 3 for Landsat TM 5 and Landsat ETM+ 7; band 4 for Landsat 8 OLI).

Once the NDVI was calculated, vegetation cover above 0.3 NDVI was removed from each satellite image [22, 35] in order to improve the accuracy of the Soil-Adjusted Vegetation Index (SAVI) values ranging from -1.5 to 1.5. The SAVI enables the sufficient description of the soil-vegetation system and soil type classification which is essential for estimating the consequences of land-cover change by using Equation 2 [19, 2].

$$SAVI = \frac{NIR - RED}{NIR + RED + L} \times (1 + L) \tag{2}$$

Where: L is a soil adjustment factor (according to Huete (1988), L = 0.5)

After all, the NDVI and SAVI-based maps were stacked to create final land-cover maps. When a potential RS index, SAVI, was found to describe the soil system and reflection, the footprint (to verify whether sandy soil or vegetation cover is dominantly occupying the Aral Sea region) of water shrinkage in the Aral Sea was evaluated for all years (1999-2019).

When outcome maps were created for all years, the dynamics of the salt-sand-canopy cover close to the study area were showing whether the actual area of the lake was preoccupied by canopy or bare (salty-)sandy soil. As far as the actuality was identified, the above-mentioned step-by-step methods were enough to tell the average annual water shrinkage rate of the Aral Sea.

A simple correlation analysis was performed using R software to identify how water shrinkage correlated with wind speed and air temperature. Data on wind speed and air temperature were recorded in two different meteo-stations, 'Aralskoe More' and 'Uyaly', located close to the Aral Sea and were retrieved from an open source "NASA Earth Observatory". Additionally, for statistically proving the correlation between wind speed and water shrinkage, a simple linear regression analysis was undertaken to check whether the p-values of all variables were statistical significant ($p > 0.05$ – statistically insignificant; $p < 0.05$ – statistically significant).

5 Results

As a consequence of the drying up of the Aral Sea for the last 20 years, the landscape of the dried-up part of the Aral Sea was completely transformed, thus a new born desert – Aral-Kum desert appeared **(Fig 2)**. New map of Aral-Kum illustrates the land-cover map of the study area for the last two decades.

According to the land-cover maps **(Fig. 2)** created by the combination of NDVI and SAVI indices, already visually, the lake surface area of the Aral Sea has experienced a critical shrinkage. On the other hand, aligning with sandy areas, salty sand areas have dramatically extended to the eastern side and, over decades, due to natural phenomena, salty sand has also migrated from the east to the central part. It is interesting to note that the of mud and vegetation areas were still not preoccupied by hazardous sandy or salty layers. Thereafter, in order to derive quantitative values on the distribution of each land-cover class, the "Reclassify" operation was performed in ArcGIS software to transfer the qualitative data into quantitative. The result of this operation is displayed in **(Fig. 3)**.

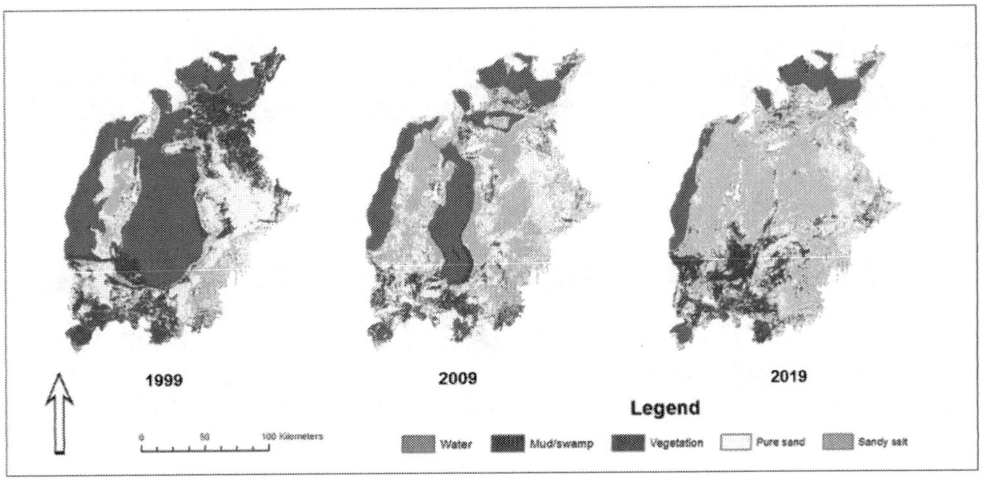

Fig. 2. Dynamics of land-cover change in the Aral Sea from 1999 to 2019.

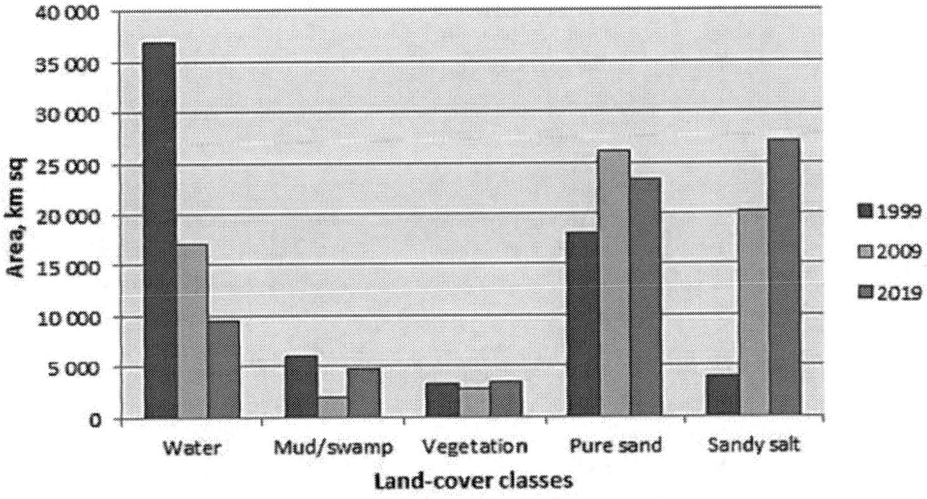

Fig. 3. Breakdown of land cover classes from 1999 to 2019

As can be seen in Fig. 3, there were evidentially upward trends for the extension of the actual sandy salt and sandy areas, despite experiencing fluctuations. Sandy areas occupied approximately 18,000 km^2 in 1999, whereas sandy salt areas accounted for below 5,000 km^2. Both land-cover classes have invaded another roughly 5,000 km^2 and 20,000 km^2 of the Aral Sea territory, respectively, by the end of 2019. The lake area sharply has been shrunk by approximately 25,000 km^2 over 20 years. However, the other two land-cover classes, mud and vegetation cover, almost remained stable in the entire period of the analysis, where vegetation cover in the Aral Sea region was rather certain than muddy area.

Referring to Figure 3 it is possible to calculate the average shrinkage value. Compiling all of the enlarged and decreased land-cover areas together, the annual water shrinkage rate per annum of 1,372.65 km^2 was determined.

After having calculated the shrinkage rate, a simple correlation analysis was carried out for water and all land-cover classes to understand how the changes in each class interconnected each other (Table 1).

Table 1. Correlation analysis of land cover classes

	Water	Mud/Swamp	Vegetation	Pure sand	Sandy salt
Water	1	0.66	0.59	-0.99	-0.99
Mud/Swamp	0.66	1	0.74	-0.91	-0.65
Vegetation	0.59	0.74	1	-0.46	0.01
Pure sand	-0.99	-0.91	-0.46	1	0.80
Sandy salt	-0.99	-0.65	0.01	0.80	1

According to the interesting aspect of the correlation table (Tab. 1), water was positively correlated with mud (0.66) and vegetation (0.59). More specifically this means, when lake surface area increases, mud and vegetation will stay proportional as a response to the increase of lake surface area. However, the lake surface area was totally and negatively correlated with pure sand (-0.99) and sandy salt (-0.99) causing an increase when lake area shrinks.

For the continuation of the statistical analysis, ANOVA was applied for water areas to check its dynamics and the shrinkage whether it was significant or not. As a result, R^2 was equal to 0.83 in three observations, whereas F and F critical were 3.532 and 3.478, respectively, according to the results of ANOVA (Table 2). Last but not least, the p-value was calculated to test against statistical significance of water shrinkage rate in relation with other land-cover classes. The p-value here is the main character of the table, indicating if the relation between variables is statistically significant ($p > 0.05$ – insignificant; $p < 0.05$ – significant).

Table 2. ANOVA output

Source of variation	SS*	df**	MS***	F ratio	p-value	F critical
Between groups	1.04E+09	4	2.59E+08	3.532	0.0028	3.478
Within groups	7.34E+08	10	7.34E+07			
Total	1.77E+09	14				

*-sum-of-square – sum of the square of variation;
**-degree of freedom – the maximum number of logically independent values (in our case, the water shrinkage rate of the Aral Sea) that have the freedom to vary;
***-mean squares – an estimate of population variance;

Once the correlation between classes was revealed, the correlation between such land-cover classes (lake, pure sand and sandy salt) and climate factors was founded to determine the role of wind speed and air temperature for water shrinkage and for the formation of sand and dust storms.

As emphasized in the methodology, climate data like wind speed and average seasonal air temperature in summer were additionally included in the correlation analysis for such land-cover classes (Table 3).

Table 3. Correlation coefficients between climate factors and land-cover classes

		Water	Pure sand	Sandy salt
'Aralskoe More' meteostation	avg. T, °C	-0.91	0.52	0.93
	WS, m/s	-0.86	0.99	0.84
'Uyaly' meteostation	avg. T, °C	-0.94	0.96	0.93
	WS, m/s	-0.98	0.90	0.98

According to the data presented in Table 3, in general, there is a huge impact of climate data on water shrinkage and these two climate factors boost the actual evaporation rate. Herein, a decrease of lake area in the Aral Sea region was exceptionally and significantly correlated with wind speed and seasonal average air temperature registered in two different meteostations close to the study area. As the main primary or natural drivers of ecosystem destruction and dust and sand storms formation in the region, these climate factors were seriously correlated with pure sand and sandy salt which can migrate through wind due to their minor particle size.

Referring to the correlation tables, it is assumed that the impact of climate factors was mainly detected and significantly relevant to the change in lake surface area. While an increase in air temperature leads to an adverse change in lake areas as a result of evaporation, it can be assumed that the current policies for regional climate action need to be sustainably amended.

6 Conclusions

This paper shows that as a result of the drying up of the Aral Sea over the last 20 years, a new salty desert, namely Aral-Kum desert, has formed on an extensive area in its dried part. As the Aral Sea shrinking water is replaced by newly formed salty desert, known as Aral-Kum, the potential for the mobilization of salts and heavy metals included in the exposed lake sediments raises considerable concerns for the southern Aral Sea region, which is mainly affected by dust from the Aral-Kum. Most of the Aral-Kum sediments show negligible heavy metal concentrations, indicating that the Aral Kum might not be the major source of these pollutants, but that local geological and anthropogenic factors are prevalent. The high variability of both spatial and temporal dust deposition intensity, the grain size of the chemical composition of the dust [32] leads to the conclusion that continuous on-site monitoring is needed for a competent understanding of the dust translocation processes in the densely populated areas of the southern Aral Sea region, their impact on the human health, and their effects to the intensely used land. This research is very cost-effective and easy to maintain so that a continuation of this long-term ecological monitoring is feasible, even during times of financial restrictions.

References

1. N. V. Aladin. W. T. W. Potts. Changes in the Aral Sea ecosystem during the period 1960-1990. Hydrobiologia. (1992)
2. A. Bannari. D. Morin. F. Bonn. A.R. Huete. A review of vegetation indices. Remote Sensing Reviews, **13**(1-2), 95-120. (1995)
3. A. Burton. Northern Aral Sea Filling up Fast. Frontiers in Ecology and the Environment, Vol. **4**, No. **5**. http://www.jstor.org/stable/3868774 (2006)
4. CAWATERinfo. The Aral Sea Basin. Available online from http://www.cawater-info.net/aral/index_e.htm/Accessed 13 August (2018)
5. E.J. Crighton. Barwinl. I. Small. R. Upshur. What have we learned? A review of the literature on children's health and the environment in the Aral Sea area. Int J Public Health (2011)
6. S. I. Deliry. Z. Y.Avdan. N. T.Do. U Avdan. Assessment of human-induced environmental disaster in the Aral Sea using Landsat satellite images. Environmental Earth Sciences. (2020)

7. A. Diouf. E.F. Lambin. Monitoring land-cover changes in semi-arid regions: remote sensing data and field observations in the Ferlo, Senegal. Journal of Arid Environments, **48(2)**, (2001)
8. D. V. Fedorov, L. M. G. Fonseca, C. Kenney, B. S. Manjunath. Automatic registration and mosaicking system for remotely sensed imagery. In Image and Signal Processing for Remote Sensing VIII Vol. **4885**. (2003)
9. B. Gaybullaev, S.C Chen, D. Gaybullaev. Changes in water volume of the Aral Sea after 1960. Appl. Water. Sci. **2 (4)**. https://doi.org/10.1007/s13201-012- 0048-z (2012)
10. M. H. Glantz Aral Sea Basin: A Sea Dies, a Sea Also Rises. Ambio, Vol. 36, No. **4** http://www.jstor.org/stable/4315834 (2007)
11. L. Gomes and A. Gillette A comparison of characteristics of aerosol from dust storms in Central Asia with soil-derived dust from other regions. Atmospheric Environment Vol. **27A**, No. 16. (1992)
12. B. Grobety et al. Airborne particular in the urban Environment. *Elements.* Vol. No **6**. (2010)
13. M. Groll et al. Long term analysis of Aeolian dust in Central Asia – Results from the CALTER-Project. In: Marburg International Dust & Sand Storm (DSS) Symposium "DSS and Desertification". (2009)
14. M. Groll. Chr. Opp. I. Aslanov. Spatial and Temporal distribution of the dust deposition in Central Asia – results from a long term monitoring program. Aeolian Research, Elsevier, DOI: 10.1016/j.aeolia.2012.08.002 (2012)
15. M. Groll. Ch. Opp. G. Issanova. N. Vereshagina,O. Semenov. Physical and Chemical Characterization of Dust Deposited in the Turan Lowland (Central Asia). Central Asian Dust Con. Vol. **99**. DOI: https://doi.org/10.1051/e3sconf/20199903005 (2009)
16. T. Higginbottom., E Symeonakis. Assessing land degradation and desertification using vegetation index data: Current frameworks and future directions. *Remote Sensing*, **6(10)**, (2014)
17. P.Hostert. A. Roder. T. Jarmer. T. Udelhoven. J. Hill. The potential of remote sensing and GIS for desertification monitoring and assessment. *Annals of Arid Zone,* **40(2)**, (2001)
18. X. Huang et al. Dust deposition in the Aral Sea: implication for the changes in atmospheric circulation in central Asia during the past 2000 years. In: *Quaternary Science* reviews. (2011)
19. A. R. Huete, A. R. A soil-adjusted vegetation index (SAVI). *Remote sensing of environment*, **25(3)** (1988)
20. R. Indoitu, L. Orlovsky, N. Orlovsky. Dust storms in Central Asia: Spatial and temporal variations. *Journal of Arid Environments,* **85** (2012)
21. R. R. Irish, J. L. Barker, S. N. Goward, T Arvidson,. Characterization of the Landsat- 7 ETM+ automated cloud-cover assessment (ACCA) algorithm. *Photogrammetric engineering & remote sensing,* **72(10)** (2006)
22. K. Ivushkin, H. Bartholomeus, A. K. Bregt, A. Pulatov. Satellite thermography for soil salinity assessment of cropped areas in Uzbekistan. *Land degradation & development,* **28(3)** (2017)
23. N. G. Kardoulas, A. C. Bird, A. I. Lawan, Geometric correction of SPOT and Landsat imagery: a comparison of map-and GPS-derived control points. *Photogrammetric Engineering and Remote Sensing,* **62(10)** (1996)

24. K. L. Kiessling Conference on the Aral Sea: Women, Children, *Health and Environment. Ambio*, Vol. **27**, http://www.jstor.org/stable/4314791(1998)
25. S. Liang, H. Fang, M. Chen. Atmospheric correction of Landsat ETM+ land surface imagery. I. Methods. *IEEE Transactions on geoscience and remote sensing*, **39(11)**, (2001)
26. P. N. Makkaveev, V. V. Gordeev, P. O. Zav'yalov, A. A. Polukhin, P. V. Khlebopashev A. I. Kochenkova. Hydrochemical characteristics of the Aral Sea in 2012–2013. *Water resources*, **45(2)**, (2018)
27. M. Masoudi, P. Jokar, B. Pradhan. A new approach for land degradation and desertification assessment using geospatial techniques. *Natural Hazards and Earth System Sciences*, **18(4)**, (2018)
28. P. Micklin. The future Aral Sea: hope and despair. *Environmental Earth Sciences*, **75(9)**, (2016)
29. S. E. Nicholson, C. J. Tucker, M. B Ba. Desertification, drought, and surface vegetation: An example from the West African Sahel. *Bulletin of the American Meteorological Society*, **79(5)** (1998)
30. Chr Opp. Desertification in Uzbekistan. Geographische Rundschau International Edition Vol. **1, 2**: (2005)
31. Chr.Opp, Vom Aralsee zur Aralkum: Ursachen, Wirkungen und Folgen des Aralseesyndroms. Asien (Reihe Planet Erde). (2007)
32. Chr.Opp, M. Groll, I. Aslanov, T. Lotz, N.Vereshagina. Aeolian dust deposition in the Southern Aral Sea region (Uzbekistan) ground base monitoring results of LUCA project. Quaternary International **429**, (2017)
33. Ch.Opp, M. Groll, O. Semenov, N. Vereshagina,A. Khamzina. Impact of the Aral Sea Syndrome – the Aralkum as a Man-Made Dust Source. E3S Web of Conferences **99** https://doi.org/10.1051/e3sconf/20199903003 (2019)
34. L. Orlovsky, G. Tolkacheva, N. Orlovsky, B. Mamedov. Dust storms as a factor of atmospheric air pollution in the Aral Sea basin. *WIT Transactions on Ecology and the Environment*, **74**. (2004).
35. A. Platonov, A. Karimov, S. Prathapar. Using satellite images for multi-annual soil salinity mapping in the irrigated areas of Syrdarya province. Journal of Arid Land Studies, **25-3** (2015)
36. S Prince. Mapping desertification in southern Africa. In: Land Change Science: Observing, Monitoring, and Understanding Trajectories of Change on the Earth's Surface (Gutman G, Janetos A, Justice CO, Moran EF, Mustard JF, Rindfuss RR, Skole D, Turner II BL, eds), pp. 163-184. Kluwer, Dordrecht, NL. (2004).
37. S. D. Prince, C. O. Justice. Coarse resolution remote sensing of the Sahelian environment. Ecology and Society, **13(7)** (1991)
38. R. Richter. A fast atmospheric correction algorithm applied to Landsat TM images. Title Remote Sensing, **11(1)** (1990)
39. K. Schepanski. Transport of mineral dust and its impact on climate. Geosciences **8, 151** doi:10.3390/geosciences8050151(2018)
40. I. V. Severskiy Water-Related Problems of Central Asia: Some Results of the (GIWA) International Water Assessment Program. *Ambio*, Vol. **33**, No. 1/2, Transboundary Issues in Shared Waters http://www.jstor.org/stable/4315455 (2004)
41. A. S. Walker, C. J. Robinove, Annotated bibliography of remote sensing methods for monitoring desertification (No. 851). US Geological Survey. (1981)

42. X. Wang. Y. Chen, Z. Li, G. Fang, F.Wang, H. Liu, The impact of climate change and human activities on the Aral Sea Basin over the past 50 years. Atmospheric Research, **245**, (2020)
43. G. F. S.Wiggs, S. L.O'hara, Wegerdt, J. Van Der Meer, I. Small, R. Hubbard. The Dynamics and Characteristics of Aeolian Dust in Dryland Central Asia: Possible Impacts on Human Exposure and Respiratory Health in the Aral Sea Basin. The Geographical Journal, Vol. **169**, No. 2 (2003) http://www.jstor.org/stable/3451395.
44. T. Wu, S. Sang, S. Wang, Y. Yang, M. Li, Remote sensing assessment and spatiotemporal variations analysis of ecological carrying capacity in the Aral Sea Basin. Science of The Total Environment, (2020).
45. I. Ying et al. Characterization of Asian dust storm and non-Asian dust storm PM2.5 aerosol in southern Taiwan. Atmospheric Environment **40** (2006)
46. H. Yongxiang et al. Dust storm in Asia continent and its bio-environmental effects in the North Pacific: A case study of the strongest dust event in April, 2001 in central Asia. *Chinese Science Bulletin* Vol. **51** (2006)
47. X. Yang, N.Wang, J.He, T.Hua, Y. Qie. Changes in area and water volume of the Aral Sea in the arid Central Asia over the period of 1960–2018 and their causes. CATENA, **191** (2020)

Evaluation of desertified delta plant communities using spectral indexes and landscape transformation models

Mamanbek Reimov[1]*, *Viktor* Statov[2], *Polat* Reymov[2], *Nizamatdin* Mamutov[2], *Salavat* Abdireymov[2], *Yakhshimurad* Khudaybergenov[2], *Shynar* Matchanova[2] *Aziz* Orazbaev[2]

[1] Tashkent Institute of Irrigation and Agricultural Mechanization Engineers, Tashkent, Kary Niyazi, 39, Uzbekistan
[2] Karakalpak State University, Nukus, Ch. Abdirov 1, Uzbekistan

Abstract. The work concerns the application of spectral indexes (NDVI, SAVI and other types of spectral indexes) for monitoring and evaluation of the plant communities of desertified Amudarya river delta plant communities. The described approach includes the multi-level spatial model of the landscape transformation with desertification, soil erosion, soil salinity increasing, climatic changes to improve the adequacy of the remote sensing methods for complicated and non-stable post-deltaic ecosystems of the Aral Sea region. The key role of the geomorphological and pedological analysis for the spatial model implementation had been shown. Some geostatistical techniques had been used to link general data on vegetation cover calculated from remote sensing information with the type of the plant community, with consequent clarification of the vegetation map using both succession-based model approaches and field data.
Keywords. Environment, Plant Communities, Remote Sensing, Aral Disaster.

1 Introduction

Aral Sea crisis and geoecological monitoring. The Aral environmental disaster is one of the largest man-made natural crisis. Desiccation of the Sea and the lack of waters causes a lot of negative eco-logical trends in the Aral Sea region. There was a great impact on the delta of Amudarya River with the decay of unique ecosystem, with the lose of biodiversity as well as soil erosion and soil salinization. Transformation of the seashore and riparian landscapes of the low coarse of Amudarya river was a subject of intensive studies of outstanding geographers, botanists and soil scientists during a time of the Aral Sea devastation, and now we have some general view onto the process of the deltaic landscape

*Corresponding Author: nukusgiscenter@gmail.com

transformation during desertification. But development of modern remote sensing methods, especially last results in processing of multi-spectral images and implementation of landscape trans-formation model allow to extend traditional method of change detection and give us new spatial representation of the delta desertification process [1, 2].

2 Methods

2.1 Optical indexes as change detection tools.

We used high-resolution space imaging from LANDSAT-7 and SETNTINEL satellites, SRTM-based DEM products and some retrospective information on Amudarya delta landscape transformation, including field data. All the space images after proper radiometric correction had been used to calculate of set of optical indexes to mirror current state of terrestrial ecosystem.
We calculate NDVI, SAVI, vegetation indexes, BI, GI, COSRI salinity indexes and some badland indexes (NBLI, EBBI) and wetness index (MNDVI). All these spatial referenced grids are combined into 3-dimensional array. Also we used MODIS-derived annual NDVI sequences to discriminate type of flora [3-7].

2.2 Landscape change detection and plant communities. Landscape change model.

V.A. Popov proposed an extinctive landscape transformation model for Amudarya delta using limited number of landscape classes, based on the water regime of the ecosystem presumably with some geomorphological corrections. His classification includes hydromorphic, mesomorphic, halomorphic, xeromorphic and anthropic landscapes. Transition coefficients from one landscape class (state of the ecosystem) had been derived from empirical data [8].

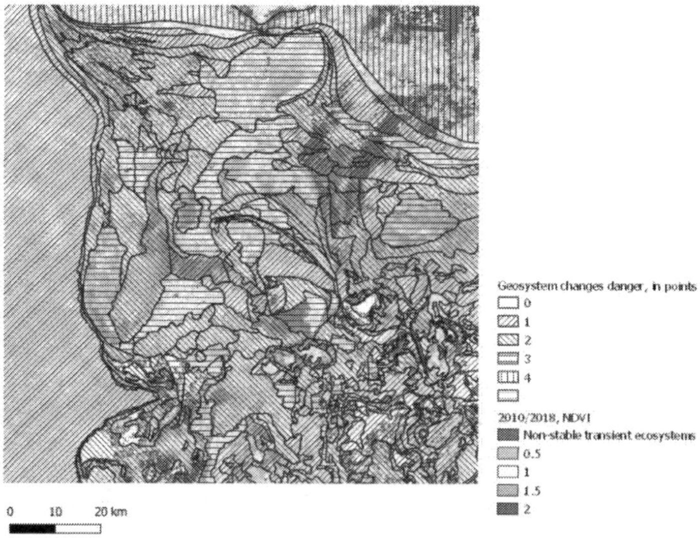

Fig1. Landscape changes map of Amudarya delta with ratio of inter-annual vegetation indexes.

Also there is other plant community transformation model for Amu Darya delta based on the groundwater table and surface water accessibility [9, 10].
We suppose that primary landscape units changes as a whole system, under-going some external influence - climatic, edaphic, water stress, water table lowering and, thus we can simulate model landscape transformation as a pat-tern reconstruction [11]. Also, as any state in transformation sequence of the particular landscape unit is connected with specific plant communities and plant succession is a natural phenomenon confirmed with a lot of geobotanic indication exploration, we can extrapolate accumulated information about local flora development for the modern and foretasted stages of landscape transformation.
Notwithstanding reaction of the every landscape unit on the disturbances has some individual features, geoecological and botanical laws support some constant trends during desertification process.
Landscape transformation in deltaic ecosystem is a path-dependency phenomena, as current state of the landscape unit depends on local geomorphology, soils, site position in hydrology network and, in that way, we got additional crucial prerequisites to evaluate further transformation details, treating reference data collected during past decades by numerous researchers of Amudaria delta [12-18].
This empirical-based transition matrix can be used to determine possible landscape transformation detected through changes of spectral indexes, and, using known consistent pattern of the plant community's digression, we can make some prognoses on vegetation characteristics and geomorphological properties of the parcel, connected with soil properties (fig.2) [17].

<table>
<tr><td></td><td>Geosystems</td><td>Changes in optical indexes and landscape picture of the Earth Surface under investigation</td></tr>
<tr><td rowspan="5">↓</td><td colspan="2">initial deltaic hydromorphic geosystems - wetlands, Tugay forest and pastures
Populus spp,, Salix spp., Eleagnus turkomanica,
Halimodendron halodendron, Climatis orientalis</td></tr>
<tr><td colspan="2">↓</td></tr>
<tr><td>mesomorphic geosystems.
Phreatophitic communities and some species with good adaptation mode *(Tamarix Sp., Alhagi)*</td><td>NDVI values lowering, some increase of EBBI index, retention of key components of spatial structure of landscape.
Lowering of MNDWI index</td></tr>
<tr><td colspan="2">↓</td></tr>
<tr><td>halomorphic geosystems
Communities of halophitic plants</td><td>NDVI values lowering, slight growth of EBBI index value, changes of key components of spatial structure of landscape and growth of soil-salinization sensible indexes - BI, and COSRI</td></tr>
<tr><td colspan="3">↓</td></tr>
<tr><td colspan="2">xeromorphic geosystems (zonal landscapes)</td><td>significant NDVI lowering, trend to seasonal desert-like fluctuation of SAVI, slow growth and stabilization of NBLI and EBBI indexes, badlands formation, slight lowering of BI and SI indexes</td></tr>
</table>

Fig.2. Diagram of landscape transformation with remote sensing indication signs.

3 Results

We found that integration of landscape transformation model with optical change detection based on a spatial distribution and long-term dynamics of vegetation can be effective tool for estimation and prognosis of plant community's development and soil erosion. Comparison of the simulated results of landscape transformation with field da-ta demonstates good correlation of the predicted stages with detected plant communities, especially for halophitic and phreatophitic communities. Results for the psammophitic communities needs more complex interpretation due to combination of plant succession and intensive sandy soil erosion, sometime these sites demonstrates meta-stabilization.

4 Discussion and conclusion

In the current studies we want to discuss how integration of very sensitive methods of space images processing with model-oriented interpretation could lead to help in landscape mapping of transformed ecosystems and reveal some additional geoecological phenomena. There are some appropriate rules for the landscape transformation transition matrix derived from known plant succession studied before; using a set of spectral indexes to estimate deepness and speed of the transformation we can predict plant community compound.
Application of multiple optical indexes with landscape transformation model to describe current state and dynamics of desertified delta is effective instrument for geoecological studies of aridisation processes of former wetlands.
Also the phenomenon of landscape interference, when some non-stable state of local ecosystems (if few landscape-forming forces are compatible in power) could be meta-stable during decades, found a new confirmation for some parcel between xeromorphic and halomorphic states.
Complex approach based on optical index interpretation based on landscape transformation model could be used of operative geoecological mapping of non-stable landscapes as a tool for land use cartography and environment changes prognosis.

References

1. P. Reimov, D. Fayzieva. The Present State of the South Aral Sea Area in "The Aral Sea The Devastation and Partial Rehabilitation of a Great Lake" eds. Micklin P. et al., Springer-Verlag Berlin Heidelberg 171-206 (2014).
2. A.A. Rafikov, Sh.E. Ergashev, E. Khayadaorov. South Aral Sea basin desertification processes. Tashkent (1997)
3. A. Bannari, D. Morin, F. Bonn, R. Huete. A review of vegetation indices//Remote Sensing Reviews, 13**(1-2)**, 95-120 (1995)
4. Y. Xie. Remote sensing imagery in vegetation mapping: a review//Journal of Plant Ecology, 1**(1)**, 9–23 (2008)
5. M. Govender, K. Chetty, H. Bulcock. A review of hyperspectral remote sensing and its application in vegetation and water resource studies//Waters SA, 33**(2)**, 145-151 (2007)
6. J. Xue, B. Su. Significant Remote Sensing Vegetation Indices: A Review of Developments and Applications//Journal of sensors. 1-17 (2017)
7. M.V. Konyuskova. Digital soil mapping of solonetz regions in the Northern Caspian lowland. Moscow. KMK Scientific press. (2014)

8. V.A. Popov. Aral Problems and Amudaria delta landscapes. Tashkent, Fan, (1990)
9. A.A. Chibilve. Modern dynamics of ecosystem components in the deser-steppe regions of Russia Eds. et al. Moscow (2001)
10. V.S. Zaletaev. Ecosystem of rivers' floodplains: structure, dynamics, resource potential. Moscow, IWP RAS. (1997)
11. M.G. Turner, R.H. Gardner, R.V. O'Neill. Landscape ecology in theory and practice. Pattern and Process. Springer. 401 (2001)
12. A.B. Bakhiev, S.E. Treshkin, N.K. Mamutov. Geobotanic aspects of degradation and productivity enhancement for arid zone plant communities. Vestnik KKO AN RUz. №**4.** 11 (2010)
13. Z.V. Kuz'mina, S.E. Treshkin, N.K. Mamutov. An Influence of Climatic Changes on Formation of Vegetation Modelled in Dry Area of the Aral Sea //Arid Ecosystem. 10 **(21),** 82-94 (2004)
14. N.M. Novikova, Zh.V. Kuz'mina, T.V. Dilkareva, N.U. Trofimova. Preservation of the tugai bio-complex diversity within Amudarya and Syrdarya river deltas in aridization conditions // Ecological research and monitoring of the Aral Sea deltas. A basis for restoration. UNESCO Aral Sea Project. 1997-2000 Final Scientific reports. – Paris. (2001)
15. N.K. Mamutov. New approach and methods of nature protection activity management in Aral Sea region. Proceeding of the VIII INTECOL International Congress of Ecology/ Seoul. Korea. 176 (2002)
16. N.K. Mamutov, N.M. Novikova. Productivity of conditions of artificial watering at the Amudarya river delta. African Journal of Range & Forage Science. 20**(2),** 147-152 (2003)
17. S.E. Treshkin, N.K. Mamutov. The modern state of vegetation cover of the Amudarya Delta and problems of biodiversity conservation. XVII International Botanical Congress/ Vienna, Austria 17-27 July, 578-579 (2005)
18. N.K. Mamutov, P.R. Reymov, V.A. Statov, S.J. Abdreymov Change of main type of flora of the south of the Aral Sea region as result of environment degradation. XVII International Botanical Congress Vienna, Austria, Europe 17-23 July, 617 (2005)
19. F.I. Khakimov. Soil and ameliorative conditions of desertified deltas. Puschino. (1989)

Application of GIS and RS in real time crop monitoring and yield forecasting: a case study of cotton fields in low and high productive farmlands

Zokhid Mamatkulov [1,*], *Eshkobil* Safarov [2], *Rustam* Oymatov [1], *Ilhom* Abdurahmanov[1] and *Maksud* Rajapbaev[1]

[1] Tashkent Institute of Irrigation and Agricultural Mechanization Engineers (TIIAME), Koriy Niyaziy str., 39, 100000, Tashkent, Uzbekistan
[2] National University of Uzbekistan named after Mirzo Ulugbek (NUUz), University str., 4, 100174, Tashkent, Uzbekistan

Abstract. Badland reclamation and low productive farmlands always have been one of the most detrimental effects on the national economy, typically in agricultural sector of Uzbekistan. Nonetheless, such kind of lands has been used extensively for major crops like cotton and winter wheat. However, it is difficult to assessing real productivity of them. Advanced technologies as GIS and RS are vital tool for geospatially analysing and making decisions on this type of fields. This research was carried out for real-time crop monitoring and yield forecasting in case of low productive (3.5 ha) and high productive (8.3 ha) cotton areas of Jarkurgan district (Surkhandarya region, Uzbekistan) based on geospatial analyses of multi-temporal satellite images, condition of groundwater, soil salinity, and ground truth data. For monitoring vegetation phenology of cotton and forecasting its harvest, False Colour, NDVI (Normalized Difference Vegetation Index) and SI (Salinity Index) analyses of areas were carried out by using 6 temporal windows of multi-temporal Sentinel 2 from April through August 2019. Besides, groundwater condition data which was obtained from observation wells these located in massives consists of both cotton fields was analysed by IDW (Inverse Distance Weighting) interpolation algorithm method to determine groundwater's effect to vegetation development and yield.
Keywords: GIS, remote sensing, crop monitoring, yield forecasting, NDVI, soil index, Sentinel 2.

1 Introduction

Crop development monitoring and crop production estimate data is very important for all country in case of decision making on the food supply of the day-by-day growing population. Traditionally, crop yields forecasting throughout the growing season include

* Corresponding author: zohid3095@gmail.com

models that assimilate climate, soils and other ancillary data as response functions to describe development, photosynthesis, evapotranspiration and yield for a specific crop based on strong physiological and physical concepts, these models are poor predictors when spatial variability in soils, stresses or management practices are present [1].

However, remote sensing of crop canopies has been promoted as a potentially valuable tool for agricultural monitoring because of its temporal coverage and capability of obtaining images in many spectral wavelengths [2]. All major crop models including wheat, cotton, rice, maize etc., were calibrated for yield forecasting during initial to peak growing season and estimated near harvest time [3]. As well the quality and condition of irrigated lands are the main criteria affecting expected yields in the agricultural industry [4]. About 45% of the total irrigated agricultural lands of Uzbekistan are saline in various levels [5] and in fact, this indicator has a very high negative impact on the agricultural sector of the country.

Assessment of the quality of irrigated lands and monitoring of crop development on low-yield lands are counted as a crucial issue [4]. Geo-Information Systems (GIS) and Remote Sensing (RS) technologies can be useful tools for addressing such complex issues in the agricultural sector. The main reason for this is the advances techniques like GIS and Remote Sensing can greatly contribute to managing timely, accurate, cost-efficient and repetitive information about agriculture [6]. For instance, by applying GIS and RS technologies can be of great use for monitoring the development process of cotton, identifying affected areas, determining soil quality condition, yield forecasting, visualizing real topography of crop fields and other spatial analyses in modern agriculture [7].

The scientists [8-10, 22] have revealed major application of remote sensing in agriculture including estimation of crop yield and vegetation vigour, assessment of water demand and development of vegetation, crop mapping, monitoring of land cover/use changes as well as in precision agriculture and irrigation management (Fig. 1).

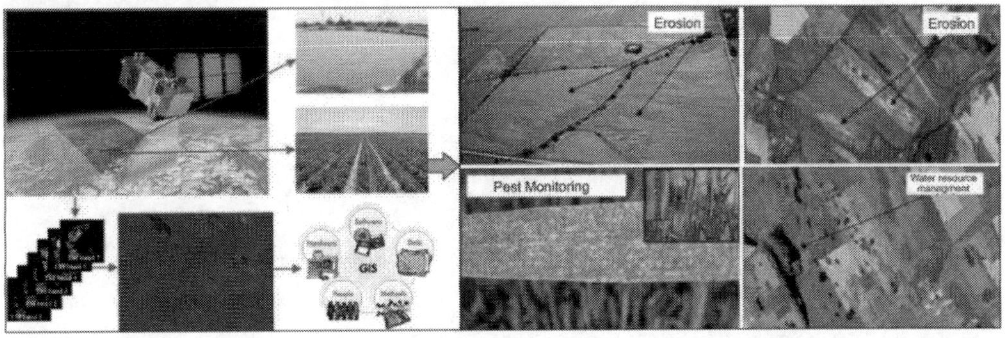

Fig. 1. Application of RS and GIS in agriculture (https://regi.tankonyvtar.hu/hu/tartalom/tamop425/0027_FHT7/ch01s07.html)

These technologies also have the great potential to make rapid decisions on complex issues and positions [6]. Remote sensing gives easy access to get important expensive information about large agricultural fields. With a change of type and status of vegetation, spectral reflectance of vegetation will change.

This research is conducted on monitoring development process of cotton and predicting its harvest by using GIS and RS. Producing cotton plays an important role in the economy of Uzbekistan with widely contribution in gross domestic product (GDP), national exports and rural employment. The total cotton area covers about 40% of entire irrigated land in the country [11]. Predicting harvest earlier, helps to planning economical activities earlier. In this term particular RS methods favour to getting very important but cheap data. Many researchers as [12-14] have got positive result on cotton yield forecasting and monitoring cotton development by GIS and RS with particular approaches. Most of them used NDVI based on different resolution of satellite images in their research. In this research was performed based on multi data approach method in terms of monitoring cotton development and forecasting yield.

2 Study area and methodology

2.1 Study area

For this research, we selected two types: high-yielding (8.3 ha) and low-yielding (3.5 ha) cotton fields located in the Jarkurgan district of Surkhandarya region, Southern part of Uzbekistan (Fig. 2).

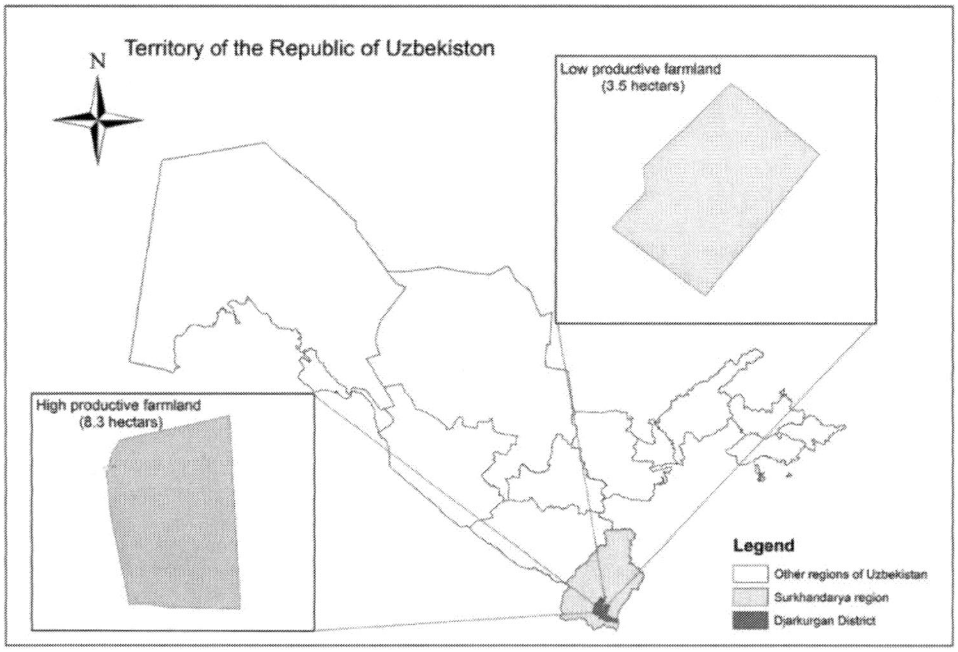

Fig.2. Locations of research areas

The overall area of the Jarkurgan district of the Surkhandarya region is 81423.0 hectares, 58878 hectares of the total area are agricultural land types and 19474.9 hectares of agricultural land types were which the irrigated agricultural lands on January 1, 2019 [15]. Especially gardening, fruits and vegetables, melons, root-fruit crops, cotton and wheat crops are considered leading main technical crops for on irrigated lands [16]. In 2019, the total irrigated agricultural land area was allocated for the cultivation of 7065 hectares of cotton and 8180 hectares of wheat, while productivity was planned to reach 23000 and 22100 tons, respectively [17]. Since the district is located in the arid region, mainly early ripening and medium-ripening varieties of cotton are grown, such as Bukhara-102, Namangan-77, Sultan, Beshkahramon and Jarkurgan. The average growth period of these cotton varieties changes between 100 and 125 days. As for the wheat varieties, which are resistant to heat and drought varieties as Yaksart, Hazrati Bashir, Babur, Asr as well as in the regions with good water supply, Grom, Krasnodar-99 and Tanya varieties are planted [18].

In the territory of the district, several different types of soils are distributed, depending on the natural geographic conditions and the influence of anthropogenic factors, as well as alluvial – prolyuvial, delyuvial - prolyuvial and according to alluvial deposits of the Amudarya, soil types: irrigated grey clay, brown, brown-meadow, desert - meadow, meadow-desert and desert sandy soils are distributed.

Research conducted 8.3 hectares of the cotton field with high productivity, where the mechanical content of soils, light and medium, as well as medium and none- saline soils, are widely spread. The middle level of groundwater is up to 2-3 meters. At the same time, in the area with a low fertility 3.5 hectare of the cotton field, light coarse and fine-grained, sandy soils with a low, medium, partially strongly saline and non-saline soils are common. Groundwater levels range from an average of 1.5-2.5 meters [18].

In 2013, as a result of the research carried out by the State Research Institute of Soil Science and Agro-chemistry of the State Committee of "Yergeodezkadastr", 1:10000 scaled map was created on the topic of assessing the quality of irrigated lands by the bonitet levels (from 0 to 100) of the soil of per unit fields. According to the map, bonitet levels of low and high productive areas, 38 and 69 respectively.

2.2 Methodology

To monitor the real-time crop development processes in research fields, 6 temporal windows of Sentinel 2 satellite images obtained on April 10, May 5, June 24, July 9, July 29 and August 23 of the year 2019 which consist of the planting of seeds until the maximum development period of the cotton. The main superiority aspects of Sentinel 2 having ability to deliver updated images of the area every 5 days interval with medium 10 meter resolution of bands: blue (2), green (3), red (4) and near-infrared (8) (Fig.s 3A, 3B and 3C). With the help of combinations of bands, false colour composition is defined as the use of near-infrared, red and green bands to form map with a 3-band image instead of red, green and blue bands. False colour composition helped to monitor the growth of cotton in per field unit. Along with this, green indicators were identified through analyses of NDVI (Normalized Difference Vegetation Index) in both research cotton areas (Fig.s 3A, 3B and 3C).

The NDVI is a widely used remote sensing indicator for crop growth monitoring, farmland management and crop production prediction [19]. For NDVI analyse, fourth (red, R) and eighth (near-infrared, NIR) bands of Sentinel 2 were used and calculated as:

$$NDVI = (NIR-R) / (NIR+R) \tag{1}$$

Crop growth monitoring and yield forecasting programs typically use the NDVI as a crucial indicator of the crop condition.

Indices of NDVI changes between -1.0 and 1.0. It means, if vegetation development well accrued in the field, most of the pixel values of the image indicate 1.0 or near to 1.0.

https://doi.org/10.1051/e3sconf/202122703001

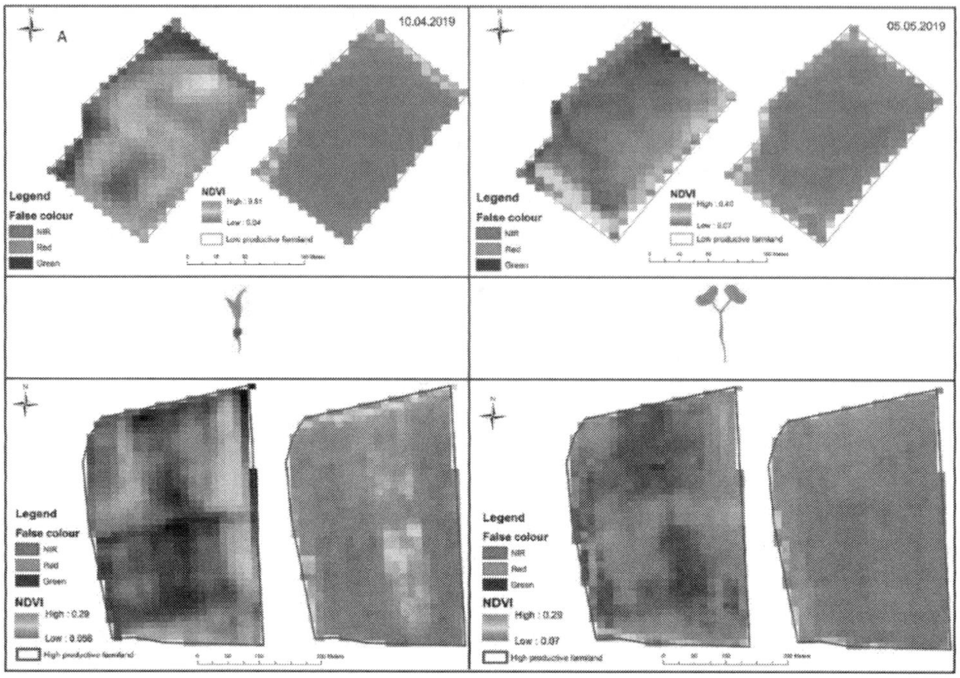

Fig. 3A. False colour composition and NDVI analyses of cotton fields with crop phenology on April 10 and May 5, 2019

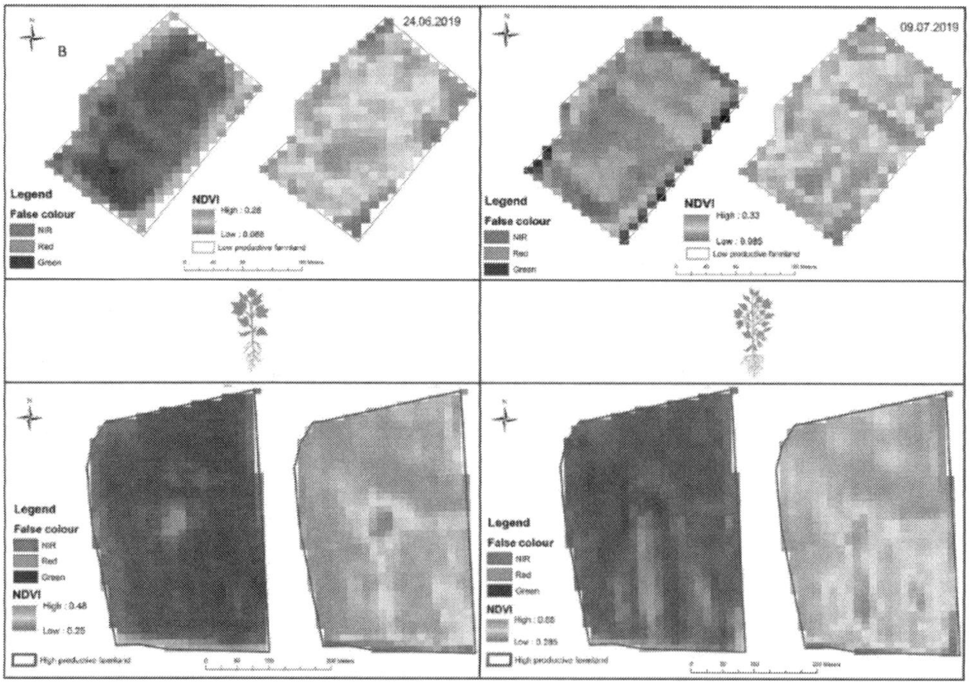

Fig. 3B. False colour composition and NDVI analyses of cotton fields with crop phenology on June24 and July 9, 2019

https://doi.org/10.1051/e3sconf/202122703001

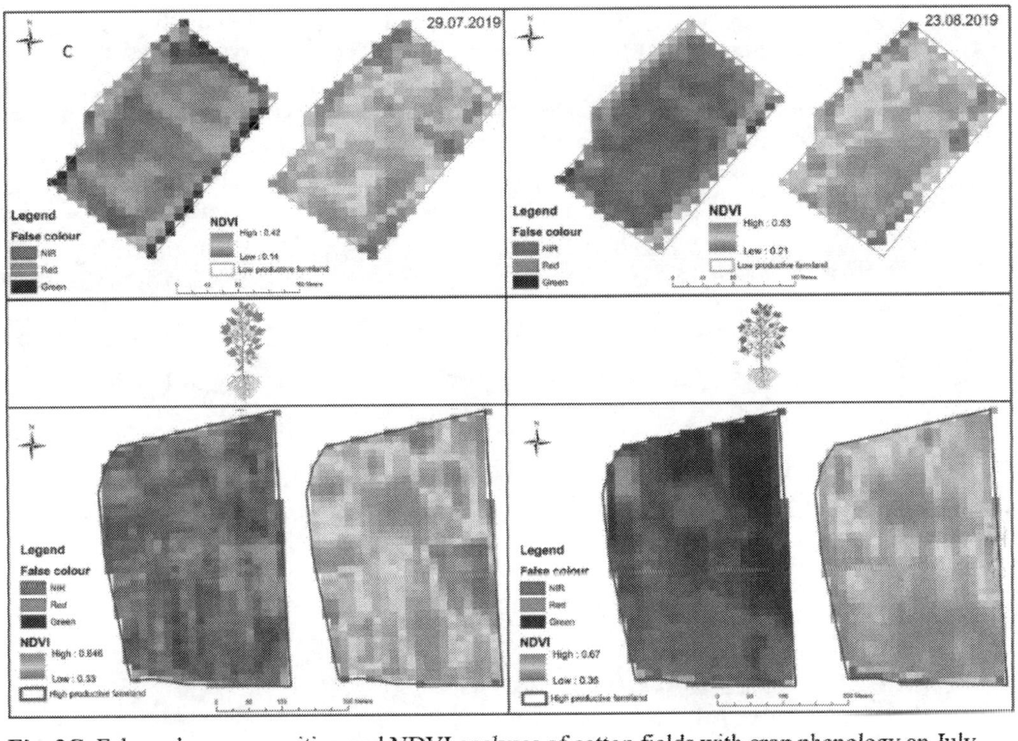

Fig. 3C. False colour composition and NDVI analyses of cotton fields with crop phenology on July 29 and August 23, 2019

The main factors affecting the deterioration of the productivity state of agricultural lands and normal development of the crop groundwater level and mineralization [20]. Therefore, data obtained from existing 9 observation wells in N.Baymurodov massive where high fertility cotton field is located and 11 observation wells in the Mirishkor massive where cotton field with low productivity is located were analysed by the IDW (Inverse Distance weighing) interpolation algorithm during the period of the high irrigation regime, July 2019 (Fig. 4).

IDW basically depends on two assumptions: (1) the unknown value of a point is influenced more by nearby control points than by those farther away, and (2) the degree of influence (weight) of points on each other is directly proportional to the inverse of the distance between the points raised to a power6and can berepresented by the following equation [21]:

$$Z = \frac{\Sigma_{i=1}^{n} w_i Z_i}{\Sigma_{i=1}^{n} w_i} \tag{2}$$

where, Z is the interpolated value for a point with an unknown observed value, wi the weighting function that determines the relative importance of each individual control point Zi in the interpolation procedure, Zi the observed value at control point i (i=1,…, n), which is in the closest neighborhood of the interpolated point, and n the total number of such points.

The key indicator contributing to the deterioration of soil reclamation and reduced productivity is soil salinity. Remote sensing and GIS integrated approaches have been applied by numerous research methods in attempts to analyses model soil salinity in an effectively efficient way. In case of analyses, SI (Salinity Index) was carried out in order to determine the degree of salinity of soil in a cotton field with low fertility (3.5 ha) by remote sensing (Fig. 4). For determining the coverage of salinity of the soil, chosen image which was taken before planting cotton in the field (in April 2019). Otherwise, vegetation may restrict for obtaining reliable result.

Generally, there are 14 most used calculation of spectral indices of soil salinity. In this research, we used the method to calculate as:

$$SI = (R \times NIR)/G \tag{3}$$

where, G (Green)-second, R (Red) – fourth and NIR (Near infrared) eighth bands of Sentinel 2 images.

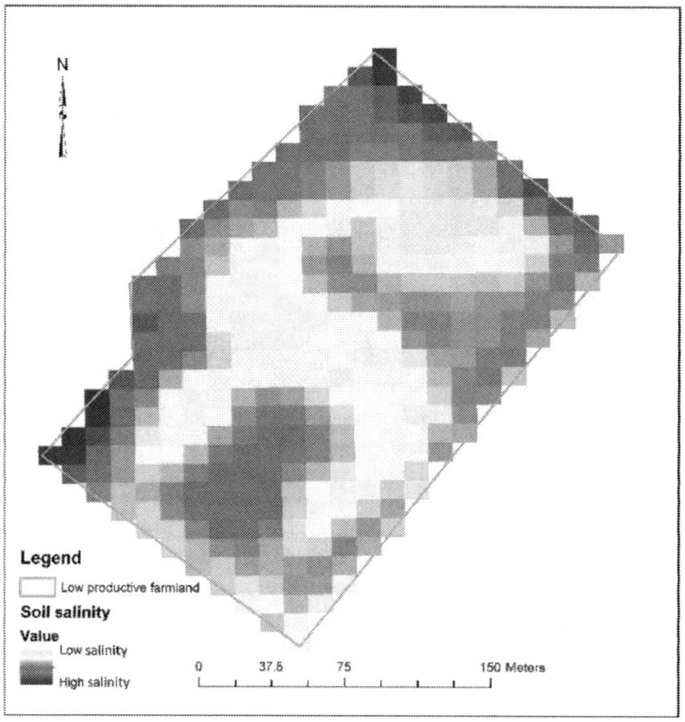

Fig. 4. Salinity index map of low productive farmland

In order to forecasting yield for both fields, a time period with the highest NDVI indicator in the cotton field with low productivity during the observation, 23.08.2019 was selected. Ground truth data were obtained from the field and linear correlation was analysed in order to compare the reliability of the results obtained and how close they are to the reality (Fig. 5).

3 Results and discussion

In terms of analysing crop monitoring, research reveals that (Figures 3A, 3B, 3C), the vegetation process was not smoothly developed in a 3.5 ha cotton field with low productivity. According to False Colour and NDVI, the normal development of the crop was occurred only in the central parts of the field, far from the centre of the field, it can be seen a significant sparsely and not well development of vegetation. The maximum value of NDVI analysis of images taken in July-August when the time of highest vegetation period, the low productive cotton field was considered as 0.53 while this indicator showed 0.67 for the highly productive cotton field (Figure 3C). As the main factors influencing to the development of the plant, it can be possible to include poor agricultural activities, inadequate irrigation, low consumption of mineral fertilizers, etc. Also, the level of groundwater and mineralization (salinity) in the area have a great influence on the land reclamation state as well as on the normal growth of the plant. If the groundwater level and mineralization are high, the crop might be damaged [4]. From the models developed through the IDW interpolation algorithm (Figures 3A and 3B), it can be seen that in the research conducted low productive cotton field is located in 1.57-1.80 meters of groundwater levels and 1.40-1.60 g/l of groundwater mineralization (salinity) part of massive. Taking into account the fact that the root of the cotton is the arrowroot and penetrates to a depth of 1.5-2.0 meters, the groundwater level is higher than the accepted norm and it was assumed that GW has a significant negative effect to cotton development. The cotton field with high fertility is located in an area with groundwater levels at a depth of 3-5 meters, which does not show a negative result in the development of the crop. Results showed that mineralization of groundwater is normal for both cotton fields.

However, when determining the degree of soil salinity with SI analysis, (Figure 5) showed that the presence of salinity in the field with low fertility of almost half of the field (43%) was saline.

Using the NDVI, GPS points obtained on the way to the locations and the comments given to their condition, harvest forecasting map of low productive farmland developed was developed (Figure 6). The minimum value of NDVI under the condition of 23.08.2019 of the low yielded research field was 0.21 while the highest value was 0.53. It was concluded that when comparing the results obtained by in situ data and performing a correlation analysis, the yield could be available if the minimum value of NDVI indicates higher than 0.35 in cotton fields. When the NDVI indicators which higher than 0.35 extracted, only 2.27 (64 %) ha of the overall area with well-developed and yielded cotton. The main reason that the field with high productive farmland was not conducted, that in the case of 23.08.2019 the minimum value of NDVI from this field was higher than 0.35 (Figure 3C).

The yield of the major crops by soil bonitet level is calculated as follows:

$$Y = A \times B \times C, sentner \tag{4}$$

where, A is the vegetation-covered area of the field (2.27 ha a part from 3.5 ha of the low productive area covered with vegetation), B is soil quality, which ranges from 0 to 100. In our case, soil quality is equal to 38 point. C=0.4, which is a special coefficient for cotton yield calculation.

If we calculate the yield of low productive farmland field in future based on the above formula, harvest forecasted as:

$$Y=2.27\times0.4\times38=34.5\ sentner=3.45\ tons \tag{5}$$

In fact, for this area expected harvest was 5.3 tons before planting. In this case, the damage that may occur near to 1.8 tons. The formula which was used for harvest calculation applied for high productive farmland (8.3 hectares) and expected harvest was equal to 23.2 tons. However, real harvest was 22.5 tons and accuracy is 96%. For low productive farmland, yield forecasting accuracy is some lower (85%) than other. The real harvest was 2.93 tons.

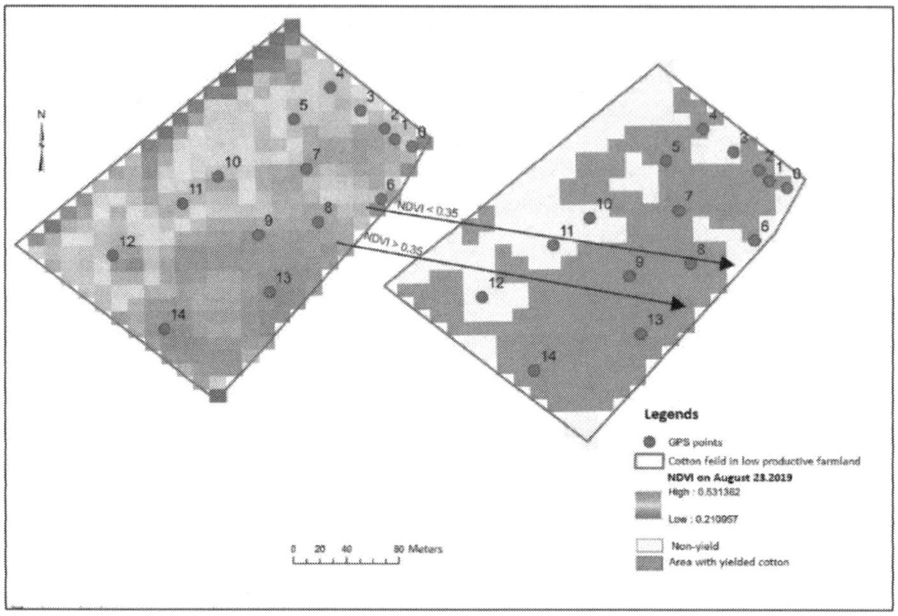

Fig. 5. Extracting yielded cotton area from NDVI

4 Conclusion and recommendation

In the case of making a decision through remote sensing and GIS-based Multiple Data Approach (MDA) may give a higher accuracy for conducted research. Remote sensing and GIS-based agriculture monitoring is an important component of the food security information system which provides reliable and timely crop area estimates and crop production forecasts at national, regional and global scales. Formulated indices (NDVI, SI) from spatial-temporal Sentinel 2 images as well as other ancillary environmental data like groundwater condition, soil quality and expert knowledge may give more reliable model on crop development and predicting yield. For utilizing these kinds of advanced technologies, require a high level of ability to use software devices and computer technologies as well as the high level of theoretical and practical expert knowledge in the application sphere.

References

1. C. L. Wiegand. The value of direct observations of crop canopies for indicating growing conditions and yield. The 18th International Symposium on Remote Sensing of Environment, **18**, 1551 (Paris, 1984).
2. N. A. Quarmby, M. Milnes, T.L. Hindle and Silleos. The use of multi-temporal NDVI measurements from AVHRR data for crop yield estimation and prediction. Inter. J. Remote Sens., **14,** 199 (1993)
3. I. Ahmad, D. A. Awan, M. Bhatti, I. H. Akhtar, M. Ibrahim. Satellite Remote Sensing and GIS based Crop Forecasting System in Pakistan. Crop monitoring for improvement food security, 95 (FAO, 2014)
4. T. Sh. Beisenboyev, N.F. Bespalov, *Salinization dynamics of irrigated soils and cotton productivity*. Tashkent (1993)
5. Annual report of Ministry of Water Resources of the Republic of Uzbekistan (2018)

6. P. Kingra, D. Majumder. Application of Remote Sensing and GIS in Agriculture and Natural Resource Management Under Changing Climatic Conditions. Agri. Research J., **53(3)**, 295 (2016)
7. Z. J. Mamatkulov, E.Yu. Safarov, R.K. Oymatov, I.I. Abdurahmanov. Application of GIS and Remote Sensing in crop monitoring and yield forecasting in case of low-yielded farmlands. J. Problems Architecture. Const., Samarkand, Spatial Volume, **130** (2019)
8. C. Atzberger. Advances in remote sensing of agriculture: context description, existing operational monitoring systems and major information needs. J. Remote Sens, **5**, 949 (2013)
9. T. Bernardes, M. A. Meriera, M. Adami, A. Giarolle and B. F. Rudorff. Monitoring biennial bearing effect on coffee yield using MODIS remote sensing imagery. Journal of Remote Sens, **4**, 2492 (2012)
10. T. Sakamoto, M. Yokozawa, H. Toritani, M. Shibayama, N. Ishitsuka. A crop phenology detection method using time series MODIS data. Journal of Remote Sens Environ, **96**, 366 (2005)
11. S. Macdonald, *United States Department of Agriculture.* Retrieved from www.usda.gov/oce/forum (2018)
12. Y. Huang, S. Thomson, *Remote Sensing for Cotton Farming* (2015)
13. N. Ramarao, *Conformity Analysis of Cotton Crop using Remote Sensing and GIS*, (Geospatial World, 2009)
14. D. Atakhanov, *Cotton yield forecasting in Tashkent province by using Remote Sensing techniques*.https://www.wur.nl/en/activity/Cotton-yield-forecasting-in-Tashkent-province-by-using-Remote-Sensing-techniques.htm (2013)
15. Land Fund of the Republic of Uzbekistan (2019)
16. Report of the scientific-research institute of Soil Science and agro-chemistry on the creating soil maps of irrigated land areas and soil evaluation of massives of Jarkurgan District of the Surkhandarya region (Tashkent, 2013).
17. Annual report of the Ministry of Agriculture of the Republic of Uzbekistan (2018)
18. Reports of scientists of the centre of scientific -industrial agriculture of Uzbekistan, (2015)
19. L. Congcong, L. Hongjun, L. Jiazhen, L. Yuping, L. Chunqiang, K. Manevski, Y. Shen. Using NDVI percentiles to monitor real-time crop growth. Journal of Computer and Electronics Agri., **162**, 357 (2019)
20. K. Isabaev, M. Khamidov, D. Alieva, *Crop irrigation and productivity*. (Tashkent, 1991)
21. P. M. Bartier, C.P Keller. Multivariate interpolation to incorporate thematic surface data using inverse distance weighting (IDW). Journal of Computers & Geosciences, 22, 795 (1996)
22. M. Verőné Wojtaszek, Földhasználati tervezés és monitoring (precíziós mezőgazdaság). Land use planning and monitoring (precision agriculture) https://regi.tankonyvtar.hu/hu/tartalom/tamop425/0027_FHT7/ch01s07.html (2010)

Creating large-scale maps for agriculture using remote sensing

Gayrat Yakubov[1]*, *Khamid* Mubarakov[1], *Ilkhomjon* Abdullaev[1], *and Azizjon* Ruziyev[1]

[1]National University of Uzbekistan, Faculty of Geography and Natural Resources, Department of Geodesy and Geoinformatics, 100174, Tashkent, Uzbekistan

Abstract. Reliable information on the real state of agricultural lands will be required to the development of appropriate measures for the rational use of agricultural lands. To obtain such information, it is necessary to keep permanent and systematic records and inventories of land resources. Large-scale special plans and maps will be required for accounting, inventory and classification of agricultural land. Currently in Uzbekistan such cartographic materials are being created on the scale 1: 10 000 and 1: 25 000 by administrative and territorial units, farms or individual land plots. The article considers the issues of creation of special maps of agricultural land in scale 1:10000 on the example of Sharof Rashidov district of Jizzakh region using remote sensing data with very high spatial resolution KOMPSAT-3.

1 Introduction

Agriculture is one of the most important sectors in the economy of the Republic of Uzbekistan. According to the latest data provided by The State Committee on Land Resources, Geodesy, Cartography and State Cadastre, 46.25% of the total territory of the country is used for agricultural purposes [1]. For rational, effective and targeted use, as well as for the protection of land resources, it is necessary to carry out accounting, analysis and evaluation by regularly and systematically obtaining information about their state. Large-scale agricultural maps are important sources that provide the ability to quickly obtain information about the state of the land areas of specific administrative, territorial units. In order, cartographic, statistical data or results of terrestrial measurements and remote sensing materials are used to create agricultural maps [2]. However, remote sensing materials can be considered essential data in terms of their spatial, spectral and temporal resolution capabilities.

Nowadays, the data obtained from Earth Observing Systems such as Sentine-2, Landsat-7, Landsat-8, EOS AM-1 can be used for agricultural purposes [3], they provide multispectral images with spatial resolutions from 10 meters to 250 meters. However, the images obtained from these spacecraft do not allow the creation of large-scale maps (1: 25000 and 1: 10000) of relatively small land plots (for example, farms) [4, 5]. Studies have shown the feasibility of using satellite images of very high spatial resolution (≥0.5 m) for

* Corresponding author: yakubov1203@gmail.com

the creating and updating of large-scale maps, in terms of accuracy, reliability and efficiency [6, 7].

In recent years Uzbekistan has paid special attention to the introduction of modern technologies in agriculture. In particular, the Decree of the President of the Republic of Uzbekistan from May 31, 2017 (PU-5065) directed to the use of remote sensing materials for monitoring and mapping of land resources was adopted [8]. In addition, Presidential Decree No. PP-4709 was adopted on 11 May 2020 [9]. The Decree defines measures to specialise certain regions of the Republic of Uzbekistan, in particular the districts of Jizzak region, in the cultivation of agricultural products. Based on the above, the article deals with the issues of creation of large-scale (1: 10000) maps of agricultural land based on geographic information systems using global navigation satellite technology (GNSS) and remote sensing (RS) materials on the example of Sharof Rashidov district of Jizzakh region. (Figure 1).

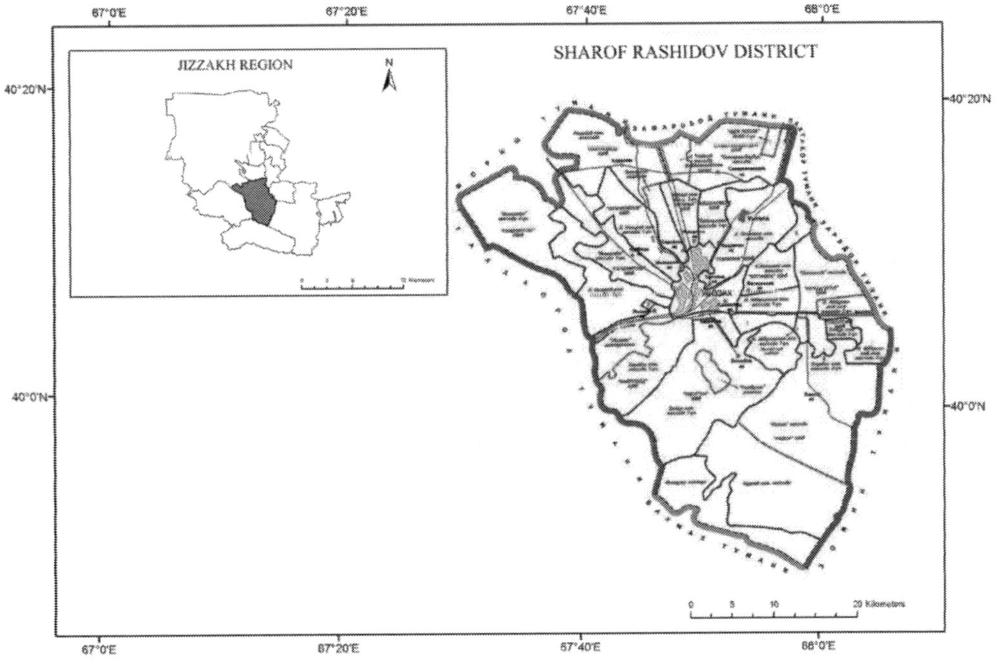

Fig. 1. Area of study – Jizzakh region, Sharof Rashidov district.

2 Materials and methods

The State Committee for Land Resources, Geodesy, Cartography and State Cadastre of the Republic of Uzbekistan is the main competent authority for the creation of cartographic products of various scales in the country. Currently, the Committee uses images with very high spatial resolution of the KOMPSAT (Korean Multi-Purpose Satellite) spacecraft for the rational, targeted use and protection of land resources. Currently, the KOMPSAT-3 spacecraft provides images over five spectral channels (panchromatic, blue, green, red and near infrared) with very high spatial resolution (2.8 m in multispectral mode and 0.7 m in panchromatic mode).

Using satellite images of very high spatial resolution KOMPSAT-3, were created agricultural maps of 1: 10000 scale of Tashkent, Bukhara, Andijan, Ferghana and other regions by the State Unitary Enterprise Geoinformkadastr under the the Committee.

To creation map of agricultural land users for the area of study, satellite images of KOMPSAT-3 with very high spatial resolution with an ortho-ready level were used. Orthophotoplan generation from the satellite images was performed in the following order:

• reconnaissance of the area of study and determination of the location of ground control points (GCP);

• determine the coordinates of the GCP;

• photogrammetric processing of satellite images in a special software product.

To georeferencing satellite images, 6 triangulation points and 21 additionally defined control points were used (Figure 2). Geodetic measurements were performed with a Stonex S9 III PLUS GNSS receiver in RTK mode.

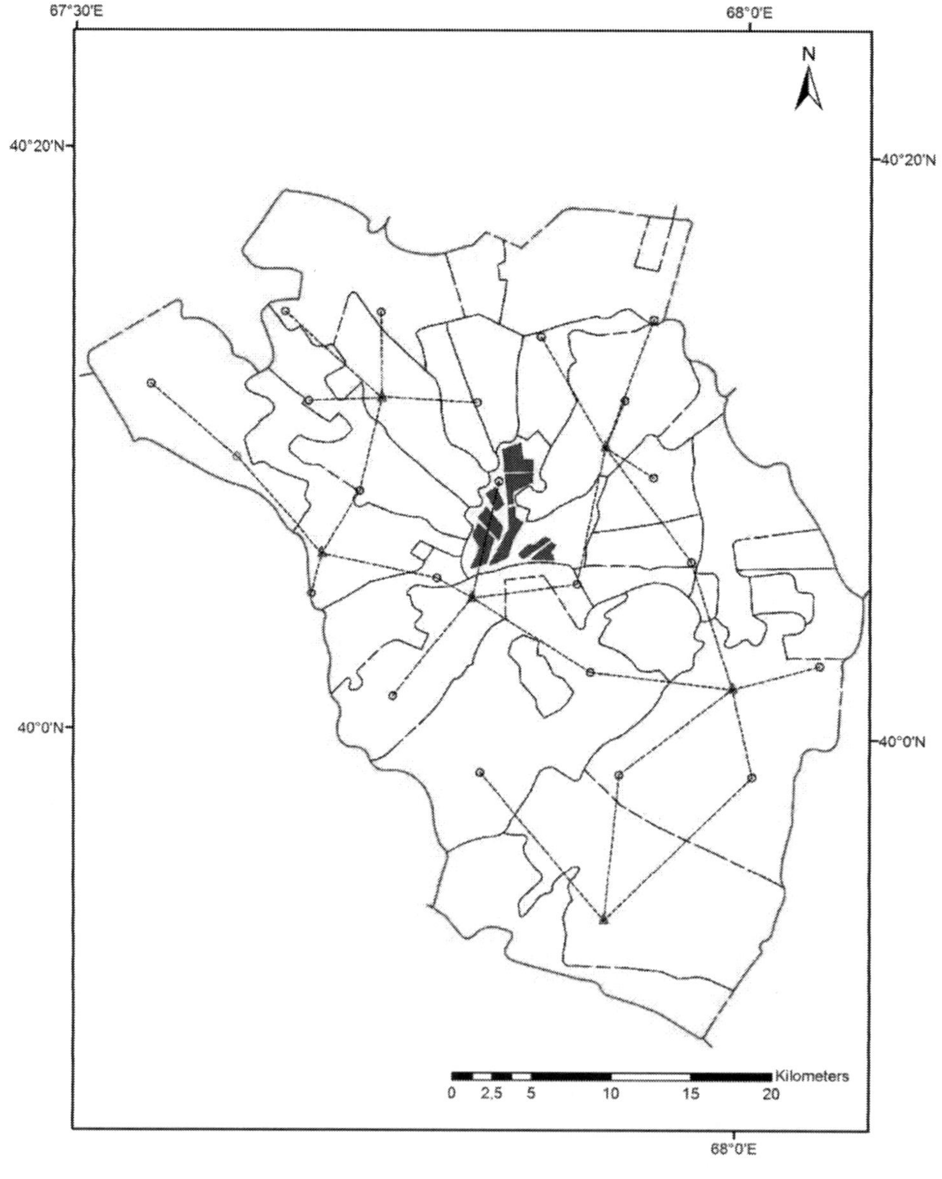

Fig. 2. Scheme of location of the GCP.

The generation orthophotoplan using the coordinates of ground control points and the satellite images with the ortho-ready level was performed by the PHOTOMOD digital photogrammetric station in PhotoMOD Lite 6. 2 software package in the following order:

- determination of the location of control points in satellite images;
- photo triangulation;
- satellite image orientation;
- orthomosaic generation;
- orthophotoplan generation.

The last and complex step in creating the large scale map of the study area is the interpretation of satellite images. Interpretation of satellite images was carried out strictly in accordance with special instructions [10]. In the process of interpreting remote sensing materials, terrain objects were identified and marked with appropriate symbols. Interpretation or analysis of remote sensing data was carried out by visual method. At the same time, terrain objects, the reliability and correspondence of which are beyond doubt, were interpreted from space images using the Panorama 11 program of the Racurs company, and terrain objects that could not be identified by space objects were identified directly in the field. It should be noted that one of the most important objects of interpretation is the boundaries of land use and land ownership, settlements and lands of the state reserve. Boundaries, in terms of interpretation, refer to special objects.

3 Results and discussions

As a result of interpretations of satellite imagery, the location of agricultural lands, boundaries of land users, their areas and other natural and artificial objects was determined (Figure 3) and a map of land users in agriculture was created (Figure 4).

Fig 3. Fragment of the 1:10000 scale agricultural map.

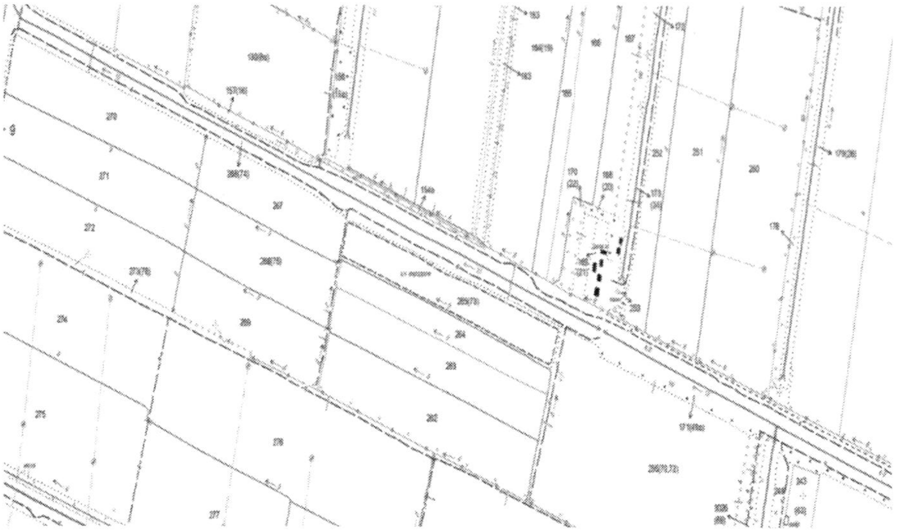

Fig 4. Fragment of the interprated orthophoto.

Such large-scale maps, which graphically represent information about the boundaries of land users and land uses (land tenure) and other real estate on land, as well as data on the actual use of land, are usually included in the group of cadastral maps and serve as the spatial basis for the formation of the cadastral database. They can serve to solve the following tasks:

- demonstration of natural features and the current state of land;
- displaying the structure of land holdings and their reclamation status;
- display of factors of negative anthropogenic, including technogenic, impact on the state of land;
- improving the accuracy of determining the boundaries and calculating the area of land;
- transition to new technologies and methods for studying and mapping lands;
- provision of detailed information in order to ensure the rational use and protection of land, the efficiency of land management [11].

Such maps serve as a topographic basis for creating maps of special content (soil, geobotanical, hydrogeological, underground communications, etc.).

4 Conclusions

The technology for creating land user maps in agriculture using very high-resolution satellite imagery was considered on the example of the study area - Sharof Rashidov district, Jizzakh region. Thanks to the use of ortho-ready satellite imagery of the KOMPSAT-3 spacecraft with a very high spatial resolution, it significantly reduced the time for photogrammetric processing, and the main attention was paid to the work on image interpretation. A large scale map of land users provides real information on the state of land used in agriculture, which subsequently serves as the basis for creation soil salinization maps, identifying factors affecting soil fertility, and developing important measures to address them.

References

1. *"National report on the state of land resources of the Republic of Uzbekistan"*, "Davyergeodezkadastr", Tashkent, Uzbekistan, (2019)
2. N. Stupen, M.Stupen, O.Stupen, *"Electronic agricultural maps formation on the basis of GIS and earth remote sensing"*, Sci. Pap. Ser. Man. Eco.Eng.Agr. Rur. Dev, vol. **18**, no.4, pp. 347-352, (2018)
3. I.Rodin, *"Analysis and preparation of RS data from open sources for precision farming systems"*, Sci. Tech. J. Geod. Cartog. Nav "Geoprofi", Vol. **6**, pp. 16-19, (2018)
4. V. Lebourgeois, S. Dupuy, É. Vintrou, M. Ameline, S. Butler and A. Bégué, *"A Combined Random Forest and OBIA Classification Scheme for Mapping Smallholder Agriculture at Different Nomenclature Levels Using Multisource Data (Simulated Sentinel-2 Time Series, VHRS and DEM)"*, Rem. Sen. Op. Acc. J, Vol. **9**, no.3, pp. 1-20, (2017)
5. E. Bello, N. Chigbu, I. Agbaje, *"Large scale mapping: an empirical comparison of pixel-based and object-based classifications of remotely sensed data"*, South Afr. J. Geo, Vol. **6**, no.3, pp. 277-294, (2017)
6. M. Gianinetto, *"Updating Large Scale Topographic Databases in Italian Urban Areas with Submeter QuickBird Images"*, Int. J. Nav. Obs. Vol. **2008**, pp. 1-9, (2008)
7. W. Teijun, *"The capability of IKONOS images for 1:10000 mapping informatsion China"*, Master Thesis, (2005)
8. *Presidential Decree No. UP-5065 "On measures for strengthening control over protection and rational management of land, improvement of geodetic and cartographic activity, regulating keeping state cadastre and state registers"*, (Date of text: 31 May 2017)
9. *Presidential Decree No. PP-4709 "On additional measures for specialization of the regions in the production of agricultural commodities"*, (Date of text: 11 May 2020)
10. *(8) Instructions for interpretation aerial images and photoplans at scales of 1: 10,000 and 1: 25,000 for the purposes of land management, state land registration and land cadastre,* (1978)
11. E. Pokhorova, *"Socio-economic maps"*, (2018)

Experience in creating a soil-reclamation map of the Zarafshan river valley based on the system analysis of lithodynamic flow structures

Naila Sabitova[1], *Oykhumor* Ruzikulova[2*], *Ilhomjon* Aslanov[2*]

[1]National University of Uzbekistan, 2 University str, Tashkent, Uzbekitan.
[2]Tashkent Institute of and Agricultural Mechanization Engineers, 39 Qori-Niyoziy str, Tashkent, Uzbekitan

Abstract. Analysis of lithodynamic flows and soil-reclamation studies, based on the plastic relief map, allowed for a systematic analysis of the main components of the natural environment. The ways of migration of chemical elements in the soil are established. For the first time, causal relationships of spatial differentiation of the degree and chemical composition of saline soils in the Zarafshan river valley were revealed. The Zarafshan river basin has a clearly defined border, as we can judge from the map of plastic relief made up in 1: 100 000 scale. Here three cascade systems are well distinguished, which is very important for the formation of the reclamation state of the territory. In this area, that is, the subsystem, on the proluvial-alluvial deposits is the main volume of groundwater, which is closely related to the surface. This subsystem is part of the catchment area of the Zarafshan river basin, which is the main source of removal from the territory of a huge mass of dissolved and solid substances.
Keywords. Plastic relief, lithodynamic flow structures, chemical composition and salinization of soils, system analysis, soil reclamation map.

1 Introduction

Irrigated agriculture, in particular, the cotton growing zone occupies a special place in the Middle and lower Zarafshan valley. (Fig.1). Over the past 15-20 years, under the influence of the current water management situation, there has been an increase in the process of salinization and erosion, deterioration of water quality, which led to the degradation of irrigated soils. Every year, the area of near lying mineralized groundwater increases. This situation is typical not only for the Zarafshan river valley, but also for the entire Republic. For example, of the water resources used in the national economy in the amount of 60.6 km3, 54.8 km3 or 90.4% of water is spent on irrigation of crops. [9]. According to the «Waterproject» Association, only 40% of the water taken from water sources reaches the plants, and the rest of it is lost in the irrigation network (-40%) and

*Corresponding authors: oyhumor.ruzikulova@gmail.com, ilhomaslanov@tiiame.uz

during irrigation (-20%). More than 60% of the total water loss forms a return flow. The area of irrigated land in need of drainage is 3.3 million hectares in Uzbekistan, of which 2.4 million hectares are provided with horizontal wells and 0.4 million hectares with vertical wells. Closed horizontal drainage is built on an area of 580 thousand hectares - this is less than 18% of the total area in need of drainage [1]. Large investments in the reconstruction and construction of the collector-drainage network, water costs for annual washing of saline lands in the amount of 12-13 km3 are ineffective. Studies have shown that the area of medium and highly saline land is continuously increasing.

Analysis of domestic and foreign experience of land irrigation has shown that the effectiveness of water reclamation is largely determined by the completeness of accounting for all the main natural factors that determine the degree of salinity. The optimal water-salt regime of saline soils is achieved by carefully taking into account the natural features of the territory: the degree of its natural drainage, salinity of soils in the aeration zone, mineralization and pressure of groundwater, mechanical composition and lithological structure of soils in the saturation zone. Not unimportant is also the correct understanding of the structure of the surface of the territory. Under the influence of endogenous and egzogenous processes, the formation of relief forms of convex and concave (increases and decreases) occurs. These forms of relief in the form of lithodynamic flows are reflected on the map of plastic relief. These forms determine the direction and speed of surface and subsurface, as well as biogeochemical runoff-the result of the influence of the gravitational field. The organization of irrigation, taking into account the above provisions, creates an optimal-rational water-salt regime that provides maximum salinity of saline soils and desalination of groundwater with minimal labor costs and scarce irrigation water. To achieve this goal, it is necessary to conduct comprehensive research based on a systematic analysis of the study area, which is relevant.

2 Research area

The object of study is the soil geosystems of the Zarafshan valley. The Zarafshan valley is characterized by a high natural discharge dynamic [5]. in the mountainous upper part located Tajikistan. Middle and lower part located Uzbekistan (Fig 1).
The research objective is to study the spatial differentiation of salts in soils within the geosystems of the Zarafshan river valley.

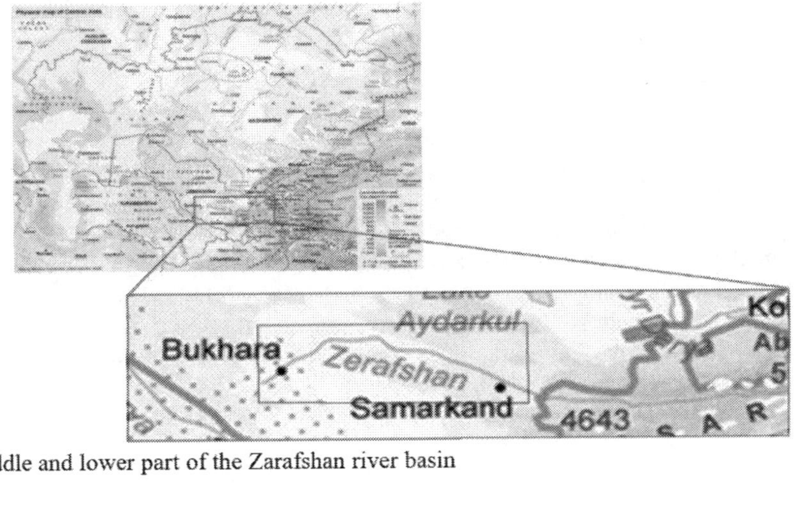

Fig. 1. Middle and lower part of the Zarafshan river basin

3 Data and Methods

In the study of soil salinization [12] system-structural, comparative-geographical, soil-geographical, soil-geochemical and laboratory-analytical methods, interpretation of aerial photographs, methods of key sites and expert-analytical assessments of the current reclamation state of irrigated soils were used [16]. The soil-reclamation map is based on the map of plastic relief [21].

Stock and published materials, as well as the results of field research, were used as factual material in the preparation of the soil-reclamation map. Full-scale observations were carried out to identify the dynamics of changes in the depth and chemical composition of groundwater, as well as the degree and type of salinization, the content of water-soluble and toxic salts in irrigated soils. To study the process of salt migration and identify their spatial distribution, the objects in the basin of the river were selected [22]. Zarafshan consists of 10 reference (key) farms, including 5 in the second (middle part) (Samarkand basin) and 5 in the third segment (Bukhara and Navoi-Kanimeh). On the reference farms, full-scale studies were carried out using the salt survey method (key sites) on a scale of 1: 10000. In total, 69 main soil sections were laid on the study area and in addition, several auxiliary intermediate pits of the second order (up to 1 m). about 350 soil samples, 32 ground samples and about 30 irrigation and collector-drainage water samples were taken from them, which were examined in the analytical center of the Institute. In the main reference sections, the description of the morphological structure of soils on genetic horizons was performed, and in 15 of the most typical ones, the determination of the volume weight of soil-soils by the cylinder method, in three - five times repetitions, and in some workings water-physical properties were studied [23].

4 Results

Analyses of water extract of type I and II were carried out with the determination of the entire composition of easily soluble salts-the dense residue (the sum of salts), CO_3, HCO_3, CI, SO4, Ca, Mg, which allowed us to obtain the most detailed characteristic of saline soils. In the selected soil samples, the quantitative content and qualitative composition of salts, the degree and type of salinization were determined, the reserves of water-soluble salts were determined by genetic horizons and by layers 0-1, 1-2 m and 0-2 m. Laboratory studies of saline soils were conducted in the analytical center of the research Institute of soil Science and Agrochemistry (Tashkent, Uzbekistan).

The obtained data are used in the preparation of the soil-reclamation map (fig) valley of the river Zarafshan. The morphological basis of the soil-reclamation map, when conducting field research, as well as analysis of the actual material was a map of plastic relief (landforms). The map is made up by establishing the most clearly defined boundaries between the systems of convex and concave forms of relief. This method made it possible to display flow structures on maps, which are essentially geochemical flows that carry information about the reclamation state of the studied territory. The flow structures selected on the map made it possible to determine the areas of formation, transit and discharge of salts, since the main organizational unit of the entire system of contours of the relief plastic map is a catchment basin of any order. It is known that the basins of the geosystems there is an absolute predominance of not strictly horizontal, but inclined movements of water and solid masses from higher to lower levels. As a result, the basin territory is dynamically interfaced with horizontal migration flows. The studied territory, the Zarafshan river basin has clearly defined borders, where three cascade geosystems are well distinguished, which is very important for the reclamation assessment of the studied territory.

The Zarafshan river basin has a clearly defined border, as we can judge from the map of plastic relief made up in 1: 100 000 scale. Here three cascade systems are well distinguished, which is very important for the formation of the reclamation state of the territory. The upper part of the studied territory is occupied by the basins of the upper arms of the Zarafshan river, which is formed at the fork of the spurs of the Turkestan, Zarfshan and Hissar ridges. This is the inner mountain part of the valley the width of which in the upper part does not exceed 2 km and in the area of the Panjiken corridor reaches up to 12km. In this area, that is, the subsystem, on the proluvial-alluvial deposits is the main volume of groundwater, which is closely related to the surface. This subsystem is part of the catchment area of the Zarafshan river basin, which is the main source of removal from the territory of a huge mass of dissolved and solid substances.

The second, that is, the middle part of the basin-the Samarkand basin, lies sub-latitude between the Panjikent intermountain corridor in the East and Karmaninsky in the West and is surrounded by mountain systems. The valley of the Zarafshan river has, as can be seen on the map, a single stream structure in the form of "bulbs". The valley itself is composed of numerous "bulbs", that is, subsystems that are closely Packed together. Such a complex and natural structure of stream structures is associated with the morphology of the relief and it is a pronounced form of abstract geological formations, which form the soil cover with their inherent reclamation features. The overall slope of the average subsystem of the Zarafshan river basin is insignificant-0.004-0.008 m. the main area of irrigated land is concentrated here. In the Meridian city of Juma, the strip of irrigated land reaches the greatest width and gradually moves in the direction of the Karmaninsky corridor. On the studied subsystem, ground and surface waters are closely interconnected and directed to the riverbed. In the Central part, they coincide with the direction of irrigation water flows. Ground water is fresh, hydrocarbonate-sulphate. The relatively deep occurrence and rather low mineralization of groundwater in irrigated soils is explained by the fact that the relief and geomorphological conditions here contribute to a better outflow of groundwater. Dominated by the process of natural desalinization of soils.

The lower, third section of the Zarafshan basin is a huge Delta plain. As can be seen from the map, starting from the Karmaninsky corridor, the flows fan out to the West and South-West, and in the extreme West, wide flow structures come out to the modern AmuDarya valley. The flows of matter coming from above accumulate within it, thus the subsystem is located in the zone of accumulation of water-soluble salts.

The analysis of the actual field material showed that the heterogeneity of geosystems-basins is manifested in the lithological-facies composition, geofiltration and geophysical properties, the intensity of water exchange in the aeration zone, which leads to salinization of soils and increased mineralization of groundwater. On the map (Fig.2.) each area is heterogeneous and reflects the vector movement of the soil-geological mass under the influence of the earth's gravity field.

Flow structures on the map show that against the background of the General slope of the relief and the composing geological bodies, there is a concentrated movement of matter and energy of different Genesis. It is known that basins are characterized by an absolute predominance of not strictly horizontal, but inclined movements of water and solid masses from higher to lower levels. As a result, the basin territory is dynamically interfaced with horizontal migration flows. We know that in the tops of the removal cone and its periphery, the soils constantly become heavy and saline, the depth of the ground water level decreases and their consumption for surface evaporation increases, which leads to intensive salt accumulation in the soils.

https://doi.org/10.1051/e3sconf/202122703003

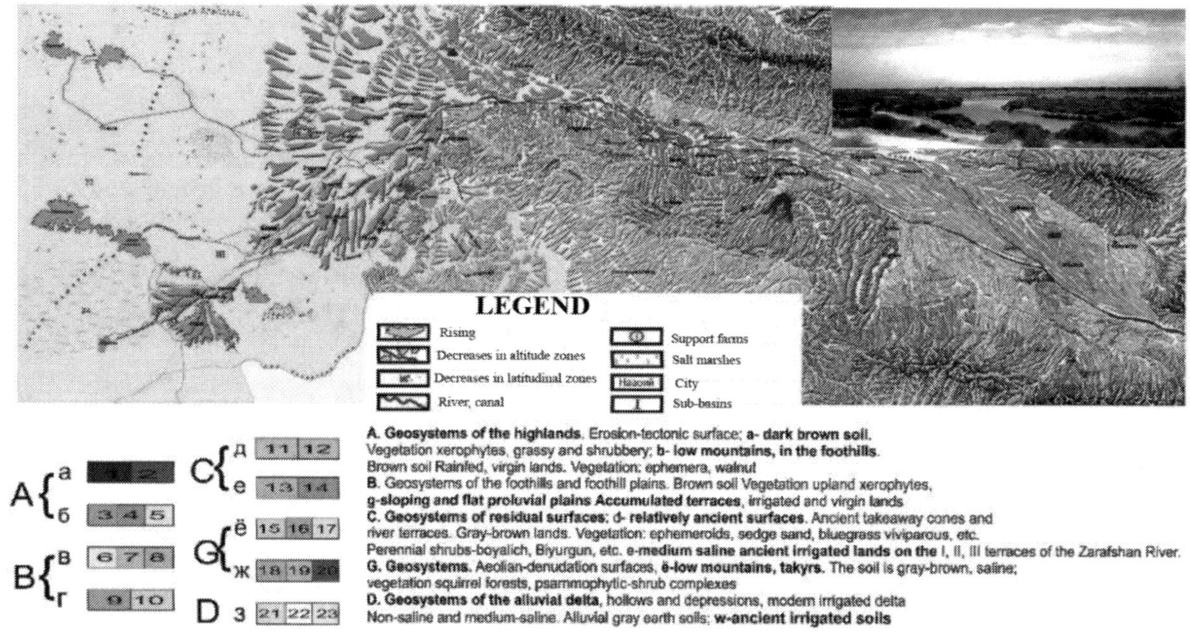

Fig. 2. Soil-reclamation map of the middle and lower part of the Zarafshan river basin

According to the geometric shape of flow structures, it is possible to distinguish the boundaries of geosystems within which there is a sharp change in the reclamation state of soils.

The chemistry of salinization of ground water in the majority of cases is chloride-sulphate and sulphate, magnesium-sodium, magnesium-calcium and calcium-magnesium. The qualitative composition of the salts is dominated by sodium sulfates (Na2S04) magnesium (MgS04), a fairly high proportion of calcium sulfate (CaS04), less than calcium bicarbonate (Ca(HCO3)2). The accumulation of sulfates at low levels of mineralization of groundwater is quite intense and its content increases in a straight line depending on the increase in mineralization to about 5-6 g/l. Further, the rate of increase in the concentration of sulfates slows down. The formation of gypsum (CaS04*2H20) and sodium sulfates (Na2SO) of magnesium (MgSO4) is observed.

In the area of Navoi, the valley expands somewhat (Navoi-Kanimeh oasis). The studied soils on the key site are old-irrigated, gray-earth-meadow, meadow, alluvial-saz, mainly medium-loamy, to varying degrees' saline, composed of proluvial-alluvial and alluvial deposits. Observations have shown that the level of ground water on the studied irrigated massifs in recent years has risen almost 1.0-1.5 m. the Seasonal amplitude ranges from 0.8-1.2 m. the Mineralization of groundwater 1-3 and 3.0-3.5 g/l. A natural relationship between the salt composition of soils and the degree of mineralization of groundwater was revealed. The content of chlorine, sodium and magnesium in the studied groundwater increases in parallel with the increase in mineralization. Accumulation of Na and CI occurs mainly in the form of NaCI. The third lower sub-basin after about 50 km between Kesilapanku Automatischem and plateau moves in the Bukhara oasis. The key site in the Bukhara oasis, located on a low-flow area, is highlighted on the map in the form of a triangular shape. Soils old and new-irrigated, meadow and takyr, desert, medium-and light-loamy, varying degrees of salinity. They are composed of sand-loam and clay alluvial deposits and agroirrigation sediments with a capacity of up to 5-10 m. they are Developed on the ancient seaboard (ancient cone of the Zarafshan river) plain and on the P-floodplain terrace. Here begins the ancient Bukhara cone of removal, the width of which in the area of Bukhara reaches 60-65 km. Field observations have shown that groundwater, depending on the terrain, lies at various depths from 56 to 320 cm. They have a generally established character with seasonal highs at depths of 0.52.0 m, and lows of 2.5-3.2 m, the amplitude of oscillations is on average 1.2-1.5 m. The degree of mineralization of groundwater varies greatly. In Navoi-Kanimeh oasis, it ranges from 2 to 10 g/l, in Bukhara-from 1.7 to 11.2 and Karakul - from 3 to 12 g/l.

West of the railway station Yakka-tut the valley is compressed by the exits of tertiary rocks, where it passes into the Karakul oasis-the fourth subbasein, which occupies the lower position in relation to the upper floors of the river basin. The streams of matter coming from above are deposited within it, thus the subsystem is predisposed to the accumulation of water-soluble salts. Here, the mineralization of groundwater in fallow and empty peripheral areas increases to 20-50 g/l or more. As can be seen from the presented map, starting from the Karmaninsky corridor, the flows fan-like expand to the West and South-West, and in the extreme West, wide flow structures go out to the modern Amu Darya valley. The difference in the amplitude of altitudes within the plain part of the valley from Karmadonskom of the corridor to the bed of the Amudarya is 190-200 m. the Terrain is flat and low-lying parts of the subsystems local pool is broken inside the basin depressions Agitminski, Karakirski, Dengizkulsky basins etc. They are used to collect collector and drainage water diverted from the irrigated area. Soils are old - irrigated meadow alluvial desert, medium-loamy, medium-and highly saline. They are composed of loamy-clay-sand alluvial and agroirrigation deposits, mainly sand and pebble layers.

In irrigated fields, spotty salinization is widespread, and in non-irrigated fields, there is a sudden salinization. Convex and concave forms on the compiled soil-reclamation map explain the pattern of distribution of spotted salinization. They are located on the concave structures of the selected streams. Depending on the shape of the terrain, the degree of artificial and natural drainage of the territory, the type of cultivated crops, groundwater lies at different depths. During the growing season, ground water lies from 0.50 to 2.10 m and from 2.50 - 3.20 m - in non-vegetative, and their mineralization ranges from poorly mineralized to highly mineralized (10-12 g/l).

The chemistry of salinization of ground water in the pre-fertile majority of cases is sulphate and chloride-sulphate, rarely sulphate-chloride. The qualitative composition of the salts is dominated by sodium sulfates (NaSO4) and magnesium (MgS04), a high proportion of calcium sulfate (CaS04). As the salt concentration increases, the content of chloride salts (NaCl) increases. CaC12 and MgCl2 salts were detected in separate groundwater samples. Moderate hydromorphic humidification from the ground water lying above the "critical depth", high temperature and dryness of the air at a relatively high wind activity of the territory leads to a wide development of the saline process.

5 Conclusion

The upper part of the studied territory is occupied by the basins of the upper arms of the Zarafshan river, which is formed at the fork of the spurs of the Turkestan, Zarfshan and Hissar ridges. This is the inner mountain part of the valley the width of which in the upper part does not exceed 2 km and in the area of the Panjiken corridor reaches up to 12 km. System analysis of lithodynamic flow structures and conducted full-scale studies of saline soils of the studied territory, allowed to identify the causes and patterns of formation, regional features of salt accumulation, as well as to establish the dependence of spatial differentiation of soil salinity on geological and geomorphological conditions. The dependence of soil salinization and their chemical composition on the movement of stream structures is established, which is reflected on the soil reclamation map. There is an alternation of unsalted (washed) soils (areas) with weak medium and strong, and sometimes very saline (salt marshes). It is revealed that the depth of the salt horizon (salt maximum), its power and the degree of salinity of the soil of objects represent a great variety, determined mainly by lithological-geomorphological, soil-reclamation and anthropogenic-economic conditions. In saline (salt is found in a layer of 0-30 cm). There are soils highly saline (30-50 cm), saline (50-100 cm), deep saline (150-200 cm) and unsalted (deeper than 200 cm) differences. The number of salts in the profile of saline soils is different, often with several second-order maxima in layers of heavy mechanical composition. According to the content of water-soluble salts in the arable layer, the studied soils are mainly medium and highly saline, there are soils with a weak degree of salinity, and not infrequently there are saline soils with a salt content of 2-3 to 8-9% with a chlorine content of 1.2-2.8%. The total reserves of soil salts of the studied territories are extremely variegated and in the upper meter layer of soil their content ranges from 47-58 to 297-462 t / ha.

The study was funded by the Uzbekistan Foundation of Basic Research, № 11.1.15. (System research on the genesis of irrigated saline soils in Uzbekistan, development of a set of measures (technologies) aimed at optimization and management of water-salt regime of the regions taking into account regional peculiarities.

References

1. I. Abdullaev, S. Rakhmatullaev. Transformation of water management in Central Asia: From State-centric, hydraulic mission to socio-political control. Environ. Earth Sci. **[73]** (2015)
2. V. I. Antonov Water resources of Uzbekistan as part of the total water resources of the Aral Sea basin and their use in modern conditions and in the future. Water resources, problems of the Aral Sea and the environment. Collection of proceedings. (2000)
3. O. A. Borsuk Lithodynamic flows in river basins. Ecology of river basins: Proceedings of the [8] (2016).
4. E. A. Dmitriev Concepts of plastic relief and soil science soil science. **[3]** (1998).
5. M. Groll, Chr. Opp, R. Kulmatov, M. Ikramova I. Normatov, Water quality, potential conflicts and solutions—an upstream–downstream analysis of the transnational Zarafshan River (Tajikistan, Uzbekistan). Environ Earth Sci.(2013) DOI 10.1007/s12665-013-2988-5
6. K. W. F. Howard, K.K. Howard, The new "Silk Road Economic Belt" as a threat to the sustainable management of Central Asia's transboundary water resources. (2016)
7. V. M. Korsunov Methodology of soil ecological and geographical research and soil cartography //V.M.Korsunov, E.N.Kraseha, B.B.Raldin-Ulan-Ude: Publishing house of BNC SB RAS. (2002).
8. R. Kulmatov, M Groll, Ch. Opp, M. R. Rosen and D. Kulmatova, "Spatial and Temporal Changes of Transboundary Pollutants in the Zarafshan River, Uzbekistan," Water, Air and Soil Pollution (2012).
9. R. Kulmatov, Ch. Opp M. Groll, M. Kulmatova, Assessment of Water Quality of the Trans-Boundary Zarafshan River in the Territory of Uzbekistan. Journal of Water Resource and Protection (2013) doi:10.4236/jwarp.2013.51003
10. Method of plastic relief in thematic mapping. Pushchino: ONTI NCBI, 1987.
11. I. K. Nazarov, Abiogenic flows in arid geosystems: optimization of nature management (based on materials from the lower part of the Zarafshan river basin). (1992)
12. O. Ruzikulova Ways to improve methods of land-reclamation map of Zarafshan oasis. GOSNIIPA (2005).
13. N. I. Sabitova, Application of the method of relief plasticity in drawing up a map of groundwater mineralization. Method of plastic relief in thematic mapping. NCBI of the USSR Academy of Sciences (1987)
14. N. I. Sabitova, U.K. Abdunazarov, O.Sh. Ruzikulova, The role of topographic maps in identifying the formation of flow geosystems of the earth's surface of the Zarafshan river valley //Actual problems of geography: materials of the Republican scientific and practical conference. (2006).
15. N. I. Sabitova, A.U. Akhmedov, G.T. Parpiev, O.Sh. Ruzikulova, Geometric feature of morphological structure and water-salt regime of the middle and lower part of the Zarafshan river valley //New technologies that increase soil fertility: international scientific conference materials. - Tashkent: GOSNIIPA, 2004. – p. 85-88.
16. N. I. Sabitova, A.U. Akhmedov, O.Sh. Ruzikulova, System-structural analysis and mapping of soil-reclamation state of irrigated lands of intermountain valleys. Ways to improve the efficiency of irrigated agriculture (2016).

17. N. I. Sabitova, O.U. Akhmedov, System analysis in the study of the reclamation state of soils (on the example of the Zarafshan river basin). Materials of the IV Congress of soil science and agrochemists of Uzbekistan. Tashkent GosNIIPA, 2005, -p. 94-104.
18. N. I. Sabitova, A.A. Neskhodimov, P.B. Neiman, Application of the method of relief plasticity to geomorphological zoning on the example of the Eastern part of the Zarafshan river valley //Uzbek geological journal. **[4]** (1990).
19. P. A. Shary, Topographical method of second derivatives. Collection of articles "Geometry of structures of the earth's surface". **[1]** (1991).
20. V. B. Sochava, Introduction to the doctrine of geosystems. - Novosibirsk: Science, 1978.
21. I. N. Stepanov, Space and time in the Sciences of soils. - Moscow: Science, (2003).
22. I. N. Stepanov, Theory of plastic relief and new thematic maps. Science (2006).
23. I. N. Stepanov, N.A. Loshakova, A.I. Satalkin, M.I. Andronova, Composing soil maps using the system cartographic method-plastic relief. Method of plastic relief in thematic mapping. Pushchino: NCBI of the USSR Academy of Sciences. (1987)
24. H. A. Toychiev, N.I. Sabitova, Plasticity of modern relief in solving geoecological problems on the example of the Eastern part of the Zarafshan river valley. Geoecology-problems and prospects. (2000)

The role of GIS technology in determining irrigated geosystems

Oykhumor Ruzikulova[1,*], *Naila* Sabitova[2] and *Gulbakhor* Kholdorova[3]

[1]Tashkent Institute of Irrigation and Agricultural Mechanization Engineers, 39 Kari Niyazov str., Tashkent Uzbekistan
[2]National University of Uzbekistan named after Mirzo Ulugbek, 4 University str., Tashkent Uzbekistan
[3]Jizzakh state pedagogical Institute named after A. Qodiriy, Sharaf Rashidov prospect 4, Jizzakh city, Uzbekistan

Abstract. Irrigated lands of Bukhara, Samarkand and Navoi regions are located in the Zarafshan river oasis and are called geosystems. Assessment of the reclamation state of geosystems of the region has been carried out in the area of classification and mapping for the effective use of land. When classifying a research object, it is much easier to identify, evaluate and apply reclamation measures. In the classification of landscapes, they are grouped in a certain order and there is an opportunity to apply such reclamation measures. So far, the basin-level classification of the Zarafshan river oasis and geosystems has not been carried out. In the scientific paper, landscapes were analysed as a form of geosystem. The geosystems of the region were grouped and summarized for the performance of the reclamation geographical assessment based on water and salt parameters. The classification of the object of study using GIS is of great scientific and practical importance. An analysis of the origin, structure, development and other features of an object when it is classified will determine the scientific significance of the case. Geosystems were isolated using the forms depicted on the relief plastic card in the scientific paper. Measures have been developed to improve the reclamation state of existing geosystems. They reached scientifically sound conclusions.
Keywords: Plastic method of relief, flow patterns lithodynamics, melioration, salinization, landscape, groundwater and surface water, assessment of melioration, mapping.

1 Introduction

Actuality of the topic. State reforms implemented in Uzbekistan in recent years cover all sectors of agriculture, among other areas. It is known that the amount of irrigated land and inland water in Uzbekistan is limited. Therefore, the use of GIS in classifying and evaluating the ameliorative status of geosystems of the oasis is considered as a solution to existing problems. Object of research: Irrigated lands of Samarkand, Navoi and Bukhara regions, geosystems of the oasis. In the context of natural population growth in Uzbekistan,

* Corresponding author: oyhumor.ruzikulova@gmail.com

the issue of rational use of resources is still relevant. This research work is based on the Decree of the President of the Republic of Uzbekistan dated October 23, 2019 No PF-5853 "On approval of the Strategy of agricultural development of the Republic of Uzbekistan for 2020-2030", as well as land and to some extent serve the objectives set out in other regulations on the efficient use of water resources. Moreover, the classification of any geosystems on the surface using morphometric analysis of horizontal lines on topographic maps ensures efficient use of the land.

Geosystems are defined as their stability, their resistance to various external influences. Geosystems differ from the second geosystem by mechanical composition, basic physico-chemical properties of soil, water and other parameters. The hydrographic net is well represented in these areas. Therefore, the boundaries of geosystems are important for their formation, stratification and development. In turn, separation and classification of boundaries of geosystems will help to define separate meliorative-geographical measures in each element. At present, there is no consensus on the geosystem boundaries on the regional and typological orientation maps. Even on maps drawn directly in the field, the boundary is crossed, depending on the knowledge and skills of the landscape scientist.

From the literature analysis, it was found that the most convenient way to limit geosystems was mapping techniques relief plastic. Relay plastic is a description of the sequence of altitudes and lows in nature. One of the key issues in land reclamation is the disclosure of reasons for spatial development of processes within the boundaries, quality and quantity changes. In the determination and evaluation of the reclamation state of small geosystems, the upper, middle and lower parts of the geosystems were allocated. In geosystems, areas of salinity occurrence, movement and accumulation of salts are highlighted. It is important to put these places clearly on large-scale maps.

One of the key issues in land reclamation is the disclosure of reasons for spatial development of processes within boundaries, quality and quantity changes. In the determination and evaluation of the reclamation state of small geosystems, the upper, middle and lower parts of the geosystems were presented. In geosystems, areas of salinity occurrence, movement and accumulation of salts are highlighted. It is important to put these places clearly on large-scale maps.

These geomorphological forms represent the locations of formation, movement and accumulation of salts in soils. For example, the elongated forms on the relief plastic card indicate the locations of the salts, and the triangular shaped forms indicate the locations of the salts.

Regional geosystem at work; natural boundary - subsystem; facies-studied as elements of the system. The accumulation of salts in the soil is mainly dependent on the following local natural conditions: relief and geological structure of the area, climate, soil-water permeability properties and so on. Depending on the rate of flow of the Zarafshan river and the mechanical composition of the river bed, the rock brings sandy, sandy, sandy and other deposits. In the selection of streams, large boulders, gravels, and then smaller sandy and sandy rocks, then sludge and sediments are selected.

According to [1] saline soils are the soils with a total content of easily soluble salts, accumulating more than 0.25% of the soil weight and preventing or cultivating cultural plants. Dynamic transformation of salts in soils of the region's geosystems is also influenced by the principles of stratification of geosystems. Of these salts the specificity of salts plays an important role in land reclamation.

Leading soil scientist and land reclamation [24, 3] show that salt flowing in water is based on the following law.

Carbonate salts are more water-soluble than sulphate and chloride salts. Chlorine salts, on the other hand, are difficult to dissolve in water. Surface water (spring, stream, small and large rivers) in the oasis and the chemical composition of the oceans provide information on the chemical composition of the water-soluble substances of the soils. Many scientists have already noted that the chemical condition of the soil cover of irrigated lands can be determined by the rivers and lakes flowing through it. One of the causes of salt accumulation and secondary salinization in the soils is also determined by the composition of the rocks in the river bed.

2 Study area

In particular, according to the geomorphological condition of the valley of Zarafshan (Uzbekistan) (Fig.1) owing to the geomorphological features of AN Hodzhimatov. The oasis is divided into delta plains, alluvial terraces, proluvial plains, lowlands, cones, slopes, plateaus and hills. Delta Plains: head, middle and periphery; alluvial terraces, respectively: to the lower (I and II) and to the upper terraces (III and IV); the middle and main parts of the proluvial plains and surrounding (peripheral) parts; lowlands; cone spreads on the head, middle and surrounding parts; hills to slopes; plateau and hills are subdivided into lowlands [32, 23]. Studying the chemical composition of water entering lakes and reservoirs in the Zarafshan river basin will help to identify and re-climate the reclamation state of the soils in the basin. All river basins have areas of formation, movement and accumulation of matter and energy. Part of the formation is caused by atmospheric precipitation and constant flow of surface and groundwater from rocks. There is a transformation of surface and groundwater at the places of movement. At the site of accumulation, salts come from solid and organic matter and accumulate in the bottom and in the sediments of geosystems. This causes soil salinity. Soil and geochemical conditions of the river basin require the use of agrotechnical measures for land reclamation. The extensive involvement of the region's geosystems in irrigated agriculture has led to profound changes.

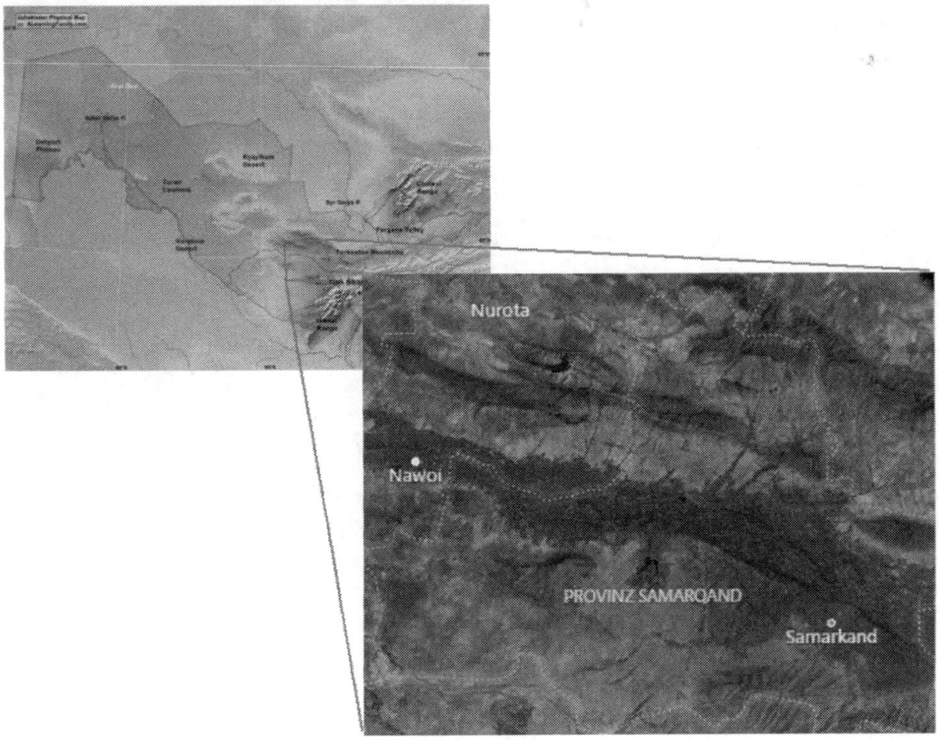

Fig.1. Middle Zarafshan river (Zarafshan valley) (Source: https://www.mapsofworld.com/uzbekistan/ and bing map)

3 Methods

Sochava in his scientific work developed [6] the scientific basis for the stabilization and efficient use of the natural environment [27, 31, 9]. Geosystems are subdivided into planetary (geographical), regional (natural geographic zones, large countries, provinces and landscapes), and topological (structural, morphological parts of landscapes: place, urochishche, facies).

The concept of geosystems contributes to the development of modern areas of natural geography. It has been a ground for new approaches in science that have made it possible to apply the results.

The classification indicators developed by [10] are remarkable for their scientific accuracy and convenience. Therefore, the classification was used in scientific work [10, 8, 14]. In classification of geosystems in the middle and lower reaches of the river, [6, 15, 23]. It was based on landscapes isolated [9].

The classification of geosystems used a method of relief plasticity. This method is based on scientific research [16, 17, 22] and others have done research. This method was first used for the reclamation of the northern Mughan and southeastern Shirvan plains (1939-1940) [7].

The method of relief plastics was used by [34] in his work. The top and bottom of the aquifers, which are the main elements of relief, are depicted as a line along the relief slope [31]. The relief plastic card clearly separates the boundaries of the on-site reservoirs. The advantages of using this method in meliorative-geographical research have been described in the work [29, 8]. The water-salt parameters of river basin geosystems change with the occurrence of natural effects of melioration in the river basin: hydrological, hydrogeological, lithological, soil, etc.

The quality of river water in the oasis is subdivided into 2 classes:

1) Closed geochemical, inactive, that is, salts move through low water flow (low irrigation).

2) Open geochemical, active - movement of water-soluble salts, rapid migration (excessive irrigation, leaching, mainly drainage activities). The basin method should also take into account the accumulation of salt in the soil and its spatial distribution.

The methods used to create a relief plastic card were used. The altitude and altitude were determined by baseline equations of the earth surface based on a 1: 100,000 scale topographic map of the study [16, 25, 13]. On this map, geosystems represent the basins formed by surface and groundwater flows. Their geomorphological forms served as an important criterion for delineation and classification of boundaries (Fig 2). The maps of this scale show their location on the map [13-15].

4 Results

In the classification of geosystems, their morphotectonic features were taken into account, and the classification of mountain and plane geosystems was separated. The main difference between them is based on the fact that natural zones are represented in two types (plains - horizontal, mountains vertically) (table). Subclasses are distinguished according to the geosystems development. Mountains - to moderate and low mountains; Plains - divided into high, moderate and low plains.

Relative altitudes up to 700–800 m are precipitated, mountains below 1000–1500 m, and uplands with 2000–2500 m are referred to as moderate mountains (up to 4,000 m in wet tropics) [10, 8, 14]. The elevation rises, climatic zones change, and the formation of a zircon as a change of the vertical zonal regularity is equally important. Upper, middle and low mountains of medium height in the Zarafshan river basin; high plains (plateaus and hills), plains of moderate height (proluvial and deluvial plains and alluvial plains), low plains (terraces I and II) and others. Mountain geosystems by their suitability are divided into 3 groups: high, middle and low mountains of medium height. There are no high mountains (above 2000-3000m) at the study site. Geosystems from the average altitude mountains: Chakilkalon, Karatepa, North Nurata, Koratov and Aktov were classified.The classification of geosystems is based on similar features (table 1).

Table 1. Classification of Zarafshan river basin geosystems

classification units	definition of classification units	examples of classification units	
class	morphotectonic properties and diversity of natural zones	mountains	
		landscape	
small class	formation of geosystems (step) formed).	mountains	1.middle high mountains 2. lower mountains
		landscape	1. high plains 2. plains of medium height 3. low plains
group	wet, water and geochemical properties	automorf half hydromorph hydromorph	
type	zonal and bio-climatic signs	mountain, highlands, desert	
middle type	according to the small zonal signs	1. a) upper mountains of medium height, b) middle part, c) lower part	
		2. on the plains: high hill, lower hill	
classification	geological and geomorphological types	denudation, erosion, accumulation	
sorts	relativity and resemblance of relief in geosystems	a) water environment b) private disabilities c) tepalities d) terrases e) the beggars	

The lower reaches of the mountains include Karakchitov, Gubdintov, Gallaorol, and Ziyovuddin-Zirabulak geosystems on the left bank. High Plains Passarghom, Ulus; Mid-Plains Plains Aktash, Malikchul and Karnobchul and Low Plains Alluvial valleys of Zarafshan river basin: Samarkand, Kattakurgan geosystems; erosion-tectonic swamps (Dengizkul lake, Oyagogitma, etc.); hills (Jarkoq, Saritosh, Balandkum, Khamza, Dengizkul, Gazli, Chaponota, etc.), alluvial-delta plains, and sub-classes of geosystems of the region are included.

Geosystems grouped according to their water-geochemical properties, namely moisture. The basis of zonal and bio-climatic features of the separation of geosystems was laid. The subspecies were classified on the basis of small zonal signs, categorizing geosystems according to geological and geomorphological features (denudation, erosion, accumulation). The study considered the types of relief, the lithological composition of the rocks that make up the geosystems.

Types of geosystems are units of individual geosystems that have the same origin and internal structure. They are distinguished by the dominant females in the geosystem, along with their morphological structure (slope of relief). An example is the Miyangal geosystem, covered by typical clay soils in the alluvial valley (Fig. 2 and Table1).

The classification of geosystems of Samarkand, Kattakurgan, Navoi-Konimeh, Bukhara and Karakul oasis was as follows. Zarafshan River Basin is a unique region in terms of meliorative conditions of the oasis geosystems. The geosystems of the oasis are subdivided into sub-classes of alluvial plains of terraces of the plain, the plains of the Zarafshan River and slopes, and alluvial-delta plains. The most important feature of Oasis geosystems is that they extend from east to west. The basin has a moderate continental climate. The border between them crosses the Hazora corridor. Key Climate Differences explained by Kogay and Babushkin [5].

Changes in a certain boundary of the components that make up the geosystem are based on geographic patterns. When considering the principles of stratification of geosystems, it is appropriate to mention the relief plastic card of the research object (Fig 2). In the course of the research, the map of relief plasticity of the Zarafshan river basin geosystems was compiled using the topographic mapping equations of scale 1: 50000 and 1: 100000.

Geosystems grouped on the basis of their moisture content and dynamic movement of groundwater. According to it, hydromorphic geosystems in the basins of the region are located 1-3 meters above ground surface, and automorphic geosystems in the areas below 3m.

The category of geosystems - is grouped by species based on the geological and geomorphological conditions, the genetic types and the geological structure of the relief. They were classified as alluvial plains, deluvial and proluvial plains, a number of sand dunes and more.

The categories of geosystems are divided into species. The landscape species are classified on the basis of the similarity of the most common dominant species in geosystems. The mountainous part of the basin is named after the geological rocks of the earth. For example: limestone-covered karst caves; in the desert, psammofit bushes with sandy soils and so on.

The following principles (principles) are used in conducting the land-reclamation and geographical assessment of soil salinization in the geosystems of the Zarafshan river basin.

1. Genetic (formation and development of geosystems).
2. Combining both zonal and azonal factors.
3. Dynamics (dynamic transformation of salts in geosystems soils and movement of matter and energy within a certain boundary).
4. Symmetry. Each geographic unit on the surface is symmetric or asymmetrical in its shape. These features also influence the stratification of geosystems.

The genetic principle of stratification of geosystems of the oasis is related to their geological history. Most of the oasis of Samarkand, Kattakurgan, Navoi-Konimeh, Bukhara and Karakul have been affected by different geological periods, as well as by differentiation of geosystems in the oasis. The mountains surrounding Samarkand and Kattakurgan oasis are also rich in various chemical elements. Under the influence of atmospheric precipitation and water, these minerals also melt and accumulate in the low-density geosystems of the geosystems by surface water.

The principle of combining the zonal and azonicity factors was followed by a systematic analysis of functional integrity in order to determine the reclamation state of the geosystems of the oasis.

There is a transformation of salts corresponding to the geographical location of the oasis geosystems. These dynamics indicates the distribution of salts in the soil on the basis of certain regularities. Hydrocarbon salts are first found along the river, followed by sulphate and chloride salts.

Most rivers on the right and left sides of the river are symmetric. This enables data to be transferred from one to another and used in a different geosystem. This is based on the relative homogeneity of geometric structures. Rectangular shapes are considered to be the most symmetric state in nature. The triangular shapes in the delta represent the accumulation of matter. The limestone forms are abundant along the river, and the river water does not exert too much effort on their formation (basin relief plastic card).

Cartographic techniques play an important role in describing the formation, development and stratification of geosystems of the region. The spatial features of the spatial development of the functional changes of geosystems are illustrated using the cartographic method. Geosystem mapping techniques are implemented in a variety of ways.

As you know, the cards are divided into general geographical and special thematic cards. and others have mentioned the diversity of geographic maps and their classification for different purposes [26, 33]. When dividing geographic maps into groups, they are based on the space they are describing, the purpose, content, scale and other features of the card. Topographic maps are used as a basis for drawing up geographical maps. General issues such as the mathematical basis of the card developed in the cartography, mapping, analysis, and printing are also first of all topographical [6]. The features, contents and methods of using modern topographic maps are quite complex. Modern computer technologies are used to create and use cards.

On the left bank of the Zarafshan River, mountainous geosystems are represented by mountains of average altitude (Chakilkalon and Karatepa), low mountains (Ziyovuddin and Zirabulak) and lowland plains with low plains (Pasdargom, Kokdala, Ulus). The geosystem class, subclass, group, type, type, sub-type, category, type was isolated in the plains of the valley and separately evaluated (table 1). Classes of geosystems of plains, steppes, hills, foothills, alluvial delta and ancient delta plains geosystems class, subclass, group, type, subgroup, category, type were analyzed separately.

The present irrigated part of the Zarafshan river basin consists of alpine, first and second shelf terraces, mountain slopes, right and left coastal deluvial and proluvial plains, and terraces of the third.

Surface lithology has been studied as a key factor in determining the boundaries of geosystems of the Zarafshan River Basin. The lithogenic basis is the geological, lithological and surface relief forms. The lithogenic basis has been studied as a key factor in the work.

To separate the boundaries of the geosystems of the studied area, the relief was taken as a landscape-forming factor. The boundaries of the geosystems are passed through the aquifers of surface and groundwater basins, which differ in their geochemical changes and geophysical features [16].

The third shelf terrace is separated by steep slopes from the alluvial valley. There are deluvial and proluvial plains between the third terrace and the mountain. Its width extends up to 15 km on the right bank and 8 km on the left. Most of the second terrace is on the island of Miyangol. It consists of alluvial pebbles and loamy sands, and up to 5 m in some places. The first reverse plane occupies the entire area of the Zarafshan river on the right bank, 0.2-1.8 km east of Samarkand. On the right bank of the Black Sea, it is difficult to distinguish the boundary of the Terrace II.

The Zarafshan river basin has a clear natural boundary and can be divided into 3 cascade systems (upper, middle and lower) on a 1: 100,000 scale relief plastic card. They are important in assessing the ameliorative status of oasis geosystems. Studies of the research object as cascading systems are available in published work. The middle part of the Zarafshan river basin has a "long onion" shape on the relief plastic card. This form plays an important role in land reclamation. The direction of flow has arisen and evolved depending on the forms of relief. Hydrodynamic processes with different geochemical and geophysical features complicate the water-salt properties.

The change of water-salt parameters of geosystems soils over time and space was analyzed. The map distinguishes mountain and plane geosystems, and their different geosystems are classified. On the basis of field researches and published literature reclamation geographical assessment was carried out. The criteria used were the geosystems 'salts' quality characteristics, the amount of gypsum, the salts of light salts and the surface water and mineral content.

Saline areas of Bulungur (I) hydrocarbonate (II), sulfate (II) in Kattakurgan and chlorine (III) type in Karakul were identified as salt deposits. Recommendations were developed within the framework of scientific research.

Our recommendations are aimed at establishing a scheme for assimilating various parts of geosystems for farming purposes (Fig. 2).

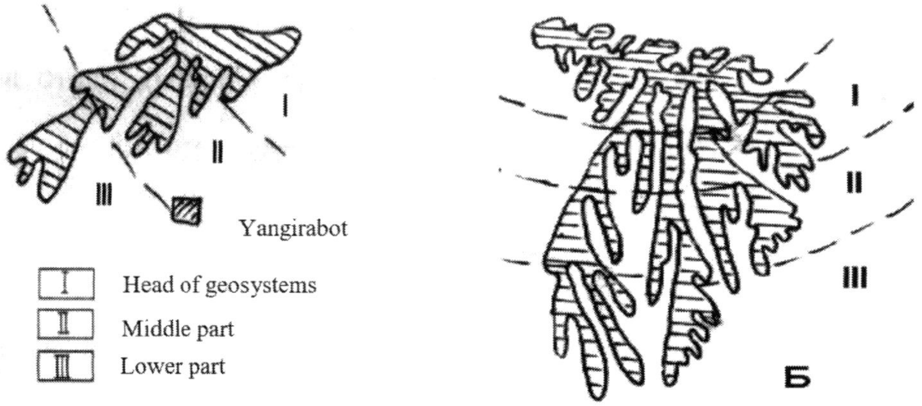

Fig. 2. Scheme of land development (A - alluvial plains, B - as an example of deluvial and proluvial plains on mountain slopes).

Stabilization and improvement of reclamation conditions in the middle and lower parts of the Zarafshan River Basin will ensure the efficient use of existing land resources. As a result of the research, the following scientific and practical recommendations for the reduction and classification of saline lands have been developed.

1. In the first phase of the study of the reclamation situation in the geosystems of the region, topographic maps and aerial photographs should be analysed in a natural geographical manner. The natural geographical conditions of the region's geosystems and the economic activities of people should be studied using a systematic approach within large and small ponds.

2. When applying a systematic approach, the horizontal horizontal and large-scale topographic maps are analysed using the Release plastic method. Topographic maps create images of streams formed by ground and surface water. It is recommended to limit the geosystems using the relief plastic card and to separate the upper, middle and lower parts of the geosystems.

3. From the geosystems isolated using the relief plastic card, the base points (areas) for field research will be identified. It is desirable to determine the reclamation state by analysing the chemical composition of groundwater and surface water salts in the soil layers of these regions.

4. The use of geosystems based on relief plastics can be cost-effective in designing irrigation systems and implementing reclamation measures. In the alluvial plains, the main part of the geosystem should be developed first, the second part II and the third part III (Fig.2 A). In the geosystems of the deluvial and proluvial plains on the mountain slopes, the initial part I (Fig.2, B) is first developed. After that it is necessary to master the middle part II and finally the third part. This is because lithodynamic flow caused by the natural movement of groundwater and surface water moves from top to bottom. As a result, soil salts accumulate in the lower parts of geosystems. Our recommendations are aimed at establishing a scheme for developing different parts of geosystems for farming.

5. In designing of collector and drainage networks, it is necessary to thoroughly analyse the surface relief forms. It is desirable to design a land reclamation network based on a

28. V. P. Philosophers, *Fundamentals of the morphometric method of searching for tectonic structures.* Publishing House - Sarat. University (1975)
29. A. N. Khozhimatov, Oasis landscapes of the desert zone of Uzbekistan: state, assessment and forecast of possible changes (for example, agricultural systems of the lower reaches of Zarafshan and Amu Darya): *Author. DiP Cand. geographer. sciences.* (1996)
30. A. Egamberdiev. *Formation, current status, problems and prospects of cartography in Uzbekistan.* University (2001)
31. Li Xia, W. McCarty, Gregory, Application of Topographic Analyses for Mapping Spatial Patterns of Soil Properties, Earth Observation and Geospatial Analyses https://doi.org/10.5772/intechopen.86109 (2019)
32. C. C. Paz, T. R. Teresa, *Factors of Soil Formation.* Encyclopedia of Earth Sciences Series https://doi.org/10.1007/978-1-4020-3995-9_207 (2008)

Present Software of photogrammetric processing of digital images

Máté Lehoczky[1,2,*] and *Zoir* Abdurakhmonov[3]

[1] Pannon Geodézia Ltd, Veszprémi str 7., 8100 Várpalota, Hungary
[2] iPalota Technology and Science Park, Szent István str. 16, 8100 Várpalota, Hungary
[3] Tashkent Institute of Irrigation and Agricultural Mechanization Engineers, Kori Niyaziv str., 39, Tashkent, Uzbekistan

Abstract. Four open source and three commercial software been used to create point clouds from identical data: images taken at five different locations. The investigation compares elapsed time and the size of the generated point clouds for the different software. Based on the tests, suggestions for users are provided.

1 Introduction

Nowadays photogrammetry and laser scanning (LiDAR) competes in generating point clouds. The present study point clouds generated with photogrammetry is discussed. The aim of the study is to provide an overview of the available commercial and open source software, and also deliver a comparative analysis on their efficiency based on certain aspects. Among the huge number of available software, seven has been selected for the test, which are considered to be most widely used ones.

2 Data and software

As a first step of a photogrammetric processing sequence, for generating a sparse point cloud, most software applies the Structure from Motion (abbreviated as SfM) method [1]. This algorithm presupposes that the camera is in motion while the object is still. The SfM method has primarily been developed for machine vision, though can efficiently be used for automatization of photogrammetric processing as well. The steps of the SfM method (which is not strictly followed by a software):

(1) Automatic search of key points (automatically detectable, uniquely marked points) on each image, with methods such as SIFT, SURF, or ORB, delivering the position of the points (different on each images) and attribute information (independent of size and direction, thus independent on the appearance on the different images) [2].

(2) Relative orientation (optionally absolute orientation as well in case control points are available) providing the relative (or absolute) position of the images to each other and the

* Corresponding author: lehomate@gmail.com

coordinates of the reference points (a subset of key points, which can successfully be determined by the orientation).

The resulted point set is the so-called sparse point cloud, the outcome of the SfM. A sparse point cloud may visualize appropriately the texture, geometrical content is rough (e.g. edges). Even though the SfM derives a 3D model, it does not generate new corners or edges, but applies the edges of the texture [3].

After the SfM, the dense point cloud is generated. Most of the software use telemetry data or the absolute oriented point cloud to generate derived products, such as orthophoto, Digital Terrain Model (DTM) or Digital Surface Model (DSM) [4]. The source of the images can be hand-held camera or cell phone, UAV-borne camera, but also pictures available on social media (e.g. Flicker, Mapillary) can be used (c.f. Fig 1). Mapillary is a project dedicated to data acquisition from social media, which resembles to Google Street View, though Mapillary contains images of front-facing cameras only. The data is collected volunteers mostly with cell phones mounted to the windscreen of cars. The processing software may also make use of the EXIF (Exchangeable Image file Format) content of JPEG files, such as telemetry data, focal length, etc. Generally, 80% or more overlap of the images is expected.

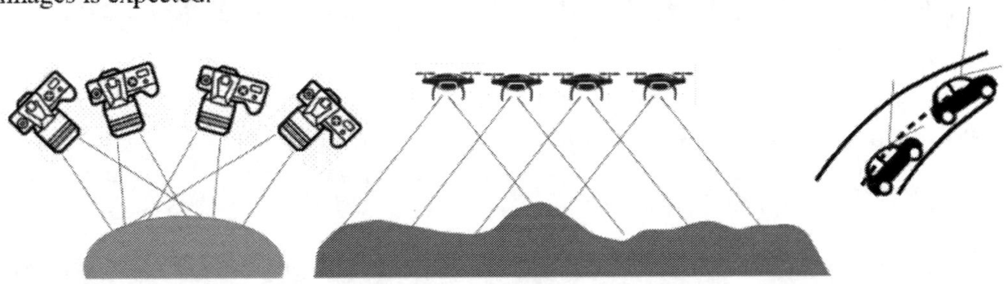

Fig. 1. Ways of data acquisition: terrestrial, UAV-borne, Mapillary.[5]

Several software is based on open source base directories (commercial also), such as OpenCV (Computer Vision) and OpenMVG (Multiple View Geometry). The OpenCV is a C++ program library, which includes methods for search of key points, referencing images, and can solve processing up to the generation of the sparse point cloud. The functions of the OpenCV library can be accessed also by using Python. OpenMVG can be used for generating the dense point cloud from the oriented images.

For most of the software, the steps of the processing is quite similar. (Steps in parentheses are sometimes skipped, usually these are required only for absolute orienting / georeferencing).

1. Start a project
2. Add images to the project
3. (Input the coordinates of control points)
4. Automatized search for key points
5. Triangulation: generation of the sparse point cloud
6. (Mark control points on images)
7. Generation of the dense point cloud
8. (Generation of derived products, e.g. orthophoto, mesh model, DTM, DSM)

In case of an automatized processing sequence, all data input (images, control points) should be entered as a first step, then the processing is done completely automatically.

Fig 2 shows detected key points by one of the tested software. It can be seen clearly that no adequate points were identified on the homogeneous surfaces (see the bottom-left and top-right corners of the figure). The size of the coloured circles displays the accuracy estimated while the marked radius displays the estimated direction of the key point on another image. Each image may contain ten thousands of key points.

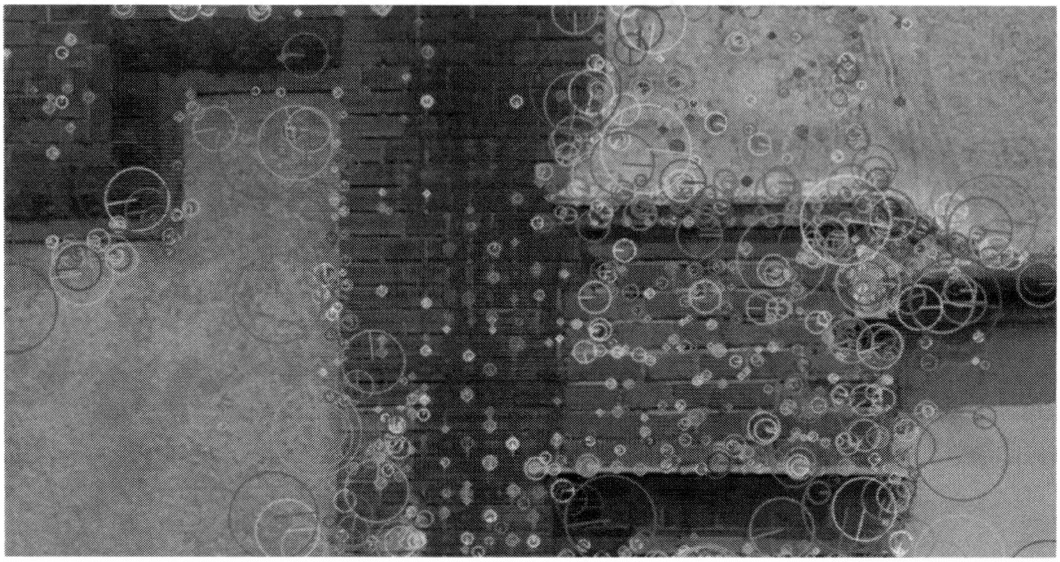

Fig. 2. Identified key points, with the accuracy (size of the circle) and the direction of the point on another image (marked radius of the circle) is displayed. [5]

There are expectations on the Hardware capacities for most of the software, such as number of processors (CPU) and nodes; often 6-8 nodes may be demanded. The different steps of the photogrammetric processing sequence mean different workload for the processors as well. The computation time of some steps depends only on the clock speed, while others can make use of the multiple nodes. For example, generating a dense cloud is efficient with multiple nodes, while building a texture or a mesh model needs generally high clock speed but less nodes. Also, the processing power of the graphics card (GPU) may also be involved in the processing, for example the computation time of densification of the point cloud depends notably on the number (and the type) of GPUs. Since this step is the most time consuming, it is important select all the available GPUs to be used for processing. Approximately, a graphics card with an additional GPU can reduce the computational time by 20-25%, which with one more GPU can be further reduced by ~10%. Additional GPUs, however, won't affect the computation time relevantly (1-3% only), so it does not verify the additional costs.

In case of hundreds or thousands images are processed, the memory (RAM) capacity should be maximized (at least 64GB is needed). The memory demand depends on the number of images, and the demanded resolution. As the use of RAM is approximately linear, by a test run for a subset of data, the computation time can properly be estimated. Normally, generating the dense point cloud and the mesh model is the most RAM-consuming. For processing aerial images, in case of some software there are special settings for the nadir images, which enables more efficient processing. In case of oblique images, however, the demand on high memory capacity is unavoidable.

Furthermore, data storage is also a relevant issue. As primary drive, SSD is suggested to be used, which contains both the operational system and the processing software. High speed SSD enables fast loading and handling of the data.

All in all, for professional photogrammetric processing the optimal computer can be defined somewhere between the Gamer PCs and laptops. For smaller tasks and study, middle range PCs can also be sufficient. Finally, the option of online processing should also be mentioned, as cloud computing is more often supported by photogrammetric software.

Within the frame of the present analysis, 4 open source (VisualSfM, Regard3D, COLMAP, ODM/WebODM) and 3 commercial (3DSurvey, Agisoft Metashape, Recap Photo) software are compared.

3 Used software

For the tests, altogether 7 software has been used [6]. In the following section these are introduced, and their main parameters are summarized in Table 1.

3.1 Open source software

Due to the OpenCV and OpenMVG libraries, new photogrammetric processing software can easily and efficiently be developed. As a result, several such projects have been started from 2010, although in many cases further updating and developments are ceased. For the present study, open source software with stable background (with frequent updates, and relevant user and developer community) are considered. Note, however, that open source software may have a limited applicability, thus it is always to check carefully the software's licence before use.

3.1.1 VisualSfM

One of the earliest open source photogrammetric processing software, which applies SfM and MVG algorithms. The development has been started in 2011, and no new version has been published in the last two years. It has a special licence, indicating that it can only be used for private and research use [7].

For determining the key points, it requires ATI/nVidia/Intel GPU with at least 1 GB memory. For the 3D reconstruction nVidia CUDA or the CPU can also be used. For generating the dense point cloud PMVS/CMVS patch is used. No derived products (e.g. orthophoto) can be generated. Both graphical and command-line interface can be used.

Though the program can efficiently process the images, also the user interface can easliy be acquired, drawbacks are that the updates have been ceased, and also that commercial use is not supported.

3.1.2 Regards3D

It has been started to be developed in 2015. Very user friendly, it guides the user with clear instructions through the processing steps in an interactive manner, enabling setting for different algorithms and parameterization in dialog boxes. Within one processing sequence, plenty of variants of processing can be run in parallel. No special requirements for the hardware is defined, however, it cannot process more than some hundred images at once. It uses MIT-licence, which enables wide range of use, including its use within a commercial software as well.

No reference points can be involved into the processing; accordingly, only relative point cloud can be generated. In order to transform to an absolute coordinate system, some other software (e.g. CloudCompare) should be used. The software is frequently used in cultural heritage management for modelling buildings and sculptures [8]. It is a good starting point for beginners to get acquainted with this technology.

3.1.3 COLMAP

The COLMAP software has been started as a PhD research, it's first licence has been published in 2016 as a BSD-license. The main purpose of the software was pure research, development, and test of algorithms. As a result, the program has a very wide range of options for parameterization, which may be a difficulty for unexperienced users. It is rather for research purposes for experimenter users.

Large (more than 10000 images) can also be processed, although the computational time is not optimal. It can used with graphical and command-line user interfaces as well. The present version cannot be used for generating orthophoto. Even though the software is mainly developed by one person, dozens of developers contribute to his work.

3.1.4 OpenDroneMap (ODM) and WebODM

Among the open source software, ODM is prominent with its dynamics in development (started in 2016) and various services. The software is published under the GPL3-license. The ODM can only be used from command-line, which might be a difficulty for a regular user of present days. It is supported by the WebODM, which allows the use of ODM from a web browser with a graphical interface. With the use of the webODM, also cloud computing can be used.

Originally ODM has been developed for processing UAV-borne images. Although a plenty of parameters can be set, for beginner users also a default setting is suggested for each parameter. WebODM is not only a processing tool, it can also display nicely the results of the computations (c.f. Fig 3). Beyond the dense point cloud, the mesh model and the orthophoto, DTM and DSM can also be generated. It is recommended to be run under the Docker container of the Linux operation system. Note that the instalment is rather complex.

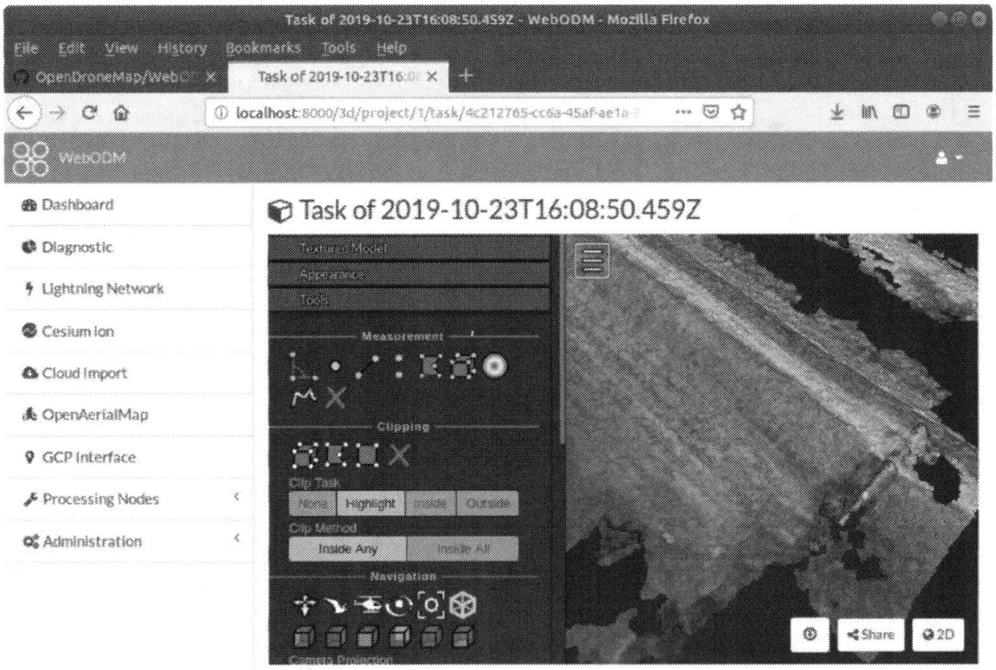

Fig. 3. Coloured point cloud in WebODM.

3.2 Commercial software

3.2.1 3D Survey

The 3D Survey has been started to be developed back in 2011 by a Sloven group of (mainly) surveyors. The software has been first published in 2014, since then its popularity is raising year by year. The company consists of 11 persons, among them 3-5 are purely working on the development. Its success is based on the user-centric and dynamic development policy and adaptive support service. Quite a few setting and parameterization option can be set by the user, which is both an advantage and a disadvantage. Nevertheless, it is an optimal choice for a beginner user.

There are a number of constantly evolving modules, such as more and more refined CAD-functions, point cloud classification, volume calculation, automatic profile- and cross section generation. A unique service of the software is the automatic reference point recognition, for which special, round-shaped markings are provided by the company. The recognition, which is based on machine vision, can notably reduce the manual workload.

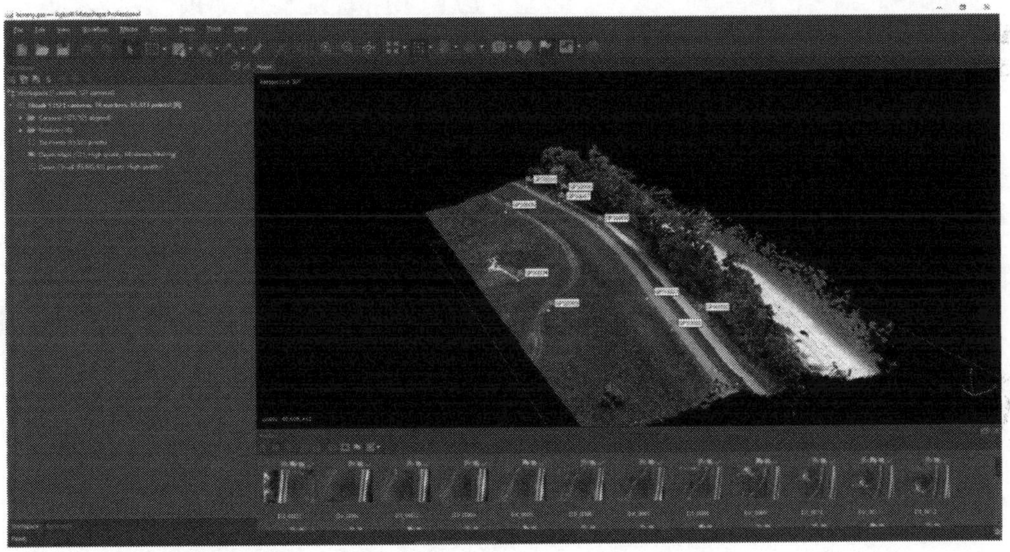

Fig. 4. Dense point cloud in Agisoft Metashape.

3.2.2 Agisoft Metashape

One of the most commonly used photogrammetric processing software, originated from Russia started to be developed in 2006, first published in 2010 under the name Agisoft Photoscan. A huge range of options for parameterization and settings, which also results in an increased number of source of errors. In case a limited capacity hardware is available, it can be used by distributing the task into smaller units, so-called chunks: the processing can be done separately for each chunk, and their outcome can be merged into one. A special feature is that task lists can be defined for a certain processing sequence, yielding an automatization of the processing. Different, sometimes quite special camera calibration data can also be involved in the processing, e.g. shutter compensation. Multispectral images can also be processed by assigning for each channels a colour. This software has one of the best point cloud classification method (Fig 4).

3.2.3 ReCap Photo

A product of Autodesk, which has been appeared as ReCap Photo in 2016, which is a cloud computing service making use of earlier photogrammetric products such as ReMake and 123D. There are nearly no options for parameterization, with limitations on the size of a project: minimum 20, maximum 1000 images can be uploaded. Before uploading, one can choose between to options. In the first option, no map projection system and reference points can be included; this option can be used for object (e.g. building) modelling, appropriate for cultural heritage management. The other option can be used for processing nadir images, where a map projection system can be selected, output data can be defined (point cloud, texture mesh and orthophoto), and also reference points can be defined. The results of the calculations can be stored on an Autodesk cloud storage.

Table 1. Software data sheet. Abbr.: T - terrestrial, A - aereal, CLI - command-line interface, GUI - graphical interface, Pc - point cloud, Mm - mesh model, Tx - texture, Op - orthophoto, DSM - Digital Surface Model

Software	Op. system	Value	A/T	Interface	Licence	Output
VisualSfM 0.5.26	Lin, Win, OSX	+	T, (A)	CLI, GUI	open**	Pc
Regard3D 1.0	Lin, Win	+	T, (A)	GUI	Open MTI	Pc, Mm
COLMAP	Lin, Win, OSX	++	T, A	CLI, GUI	Open BSD	Pc, Tx, Mm
ODM, WebODM	Lin, Docker	++	A	CLI, web	Open GPL	Pc, Tx, Op, Mm, DSM
3D Survey	Win	++	A, T	GUI	Commercial	Pc, Tx, Op, DSM
Agisoft Metashape	Lin, Win, OSX	+++	A, T	GUI	Commercial	Pc, Tx, Mm, Op, DSM
ReCap Photo	Win+cloud	++	A, T	GUI	Commercial	Pc, Tx, Mm, Op, DSM

4 Tests and results

Altogether 5 worksites have been used for testing. For the tests, relatively small worksites were used as the performance of the available hardware was limited. The observations were acquired using 3 different UAVs: DJI Phantom 4, DJI Phantom 4P, UAV Birdie Geo+. The test sites and the information on the measurements is summarized in Table 2.

Table 2. Data of the test acquisitions

settlement	setting	No. of images	UAV	Notes
Horány	rural	12	DJI Phantom 4	
Öskü	urban	42	DJI Phantom 4P	oblique, taken from a tower
Siófok	urban	116	DJI Phantom 4P	taken in a rectangular grid
Horány	rural	121	DJI Phantom 4	
Várpalota	urban	160	UAV Birdie Geo+	

The following hardware have been used:

- N1: i7 core 6, 4,5 GHz, 64 GB RAM NVIDIA GeForce GTX 1080, Windows 10
– N2: i7 core 6, 3,5 GHz, 32 GB RAM, NVIDIA GeForce GTX 1050 Ti, Windows 10
– N3: i5 core 4, 2,3 GHz, 8 GB RAM, NVIDIA GeForce GT 525M, Ubuntu 18.04

The processing with the ODM and WebODM software has been performed on a LinuX operation system (N3), while the Agisoft Metashape licence was available only for the N1 computer.

For each test runs, the computational time in minutes (ct) and the size of the point cloud in million-points (mp) have been recorded. The results are summarized in Table 3 and 4.

Table 3. Results of the test run for the open source software. ct: computational time in minutes, mp: size of the point cloud in million-points

settl.	**No. im.**	**fac**	**Regard3D**		**VisualSfm**		**COLMAP**		**3Dsurvey**	
			ct	**mp**	**ct**	**mp**	**ct**	**mp**	**ct**	**mp**
Horány	12	N2	12	0.9	4	0.5	16	2.8	6	1.2
		N3								
Öskü	42	N2	157	2.0	10	1.0	94	4.7		
		N1							8	7.0
Siófok	116	N3								
		N2	101	2.8	72	1.7	170	2.3	108	9.5
		N1	89	2.8	67				99	8.7
Horány	121	N3								
		N2	94	3.5	51	1.9	74	4.7	52	3.7
		N1								
Várpalota	160	N2							91	11.5
		N1			120	17.9			120	35.7

Table 4. Results of the test run for the commercial software. ct: computational time in minutes, mp: size of the point cloud in million-points

settl.	**No. im.**	**fac**	**Agisoft**		**RC**	**WebODM**	
			ct	**mp**	**ct**	**mp**	**ct**
Horány	12	N2					
		N3				9	0.8
Öskü	42	N2					
		N1	5	21.7	2.0		
Siófok	116	N3				119	3.6
		N2					
		N1	93	59.4	5.0		
Horány	121	N3				124	10.6
		N2					
		N1	38	45.7			
Várpalota	160	N2					
		N1	60	140.2	3.0		

5 Discussion and summary

According to the results (c.f. Table 3 and 4), open source software cannot compete with commercial ones in performance, capacity, and complexity. Note, however that the development of open source software is much dynamic than that of commercial ones as these developments are usually focusing on certain steps of the workflow.

An advantage of open source software is that nearly for any platform and operating system apt software can be found. Also, an advantage is that they allow a huge freedom in parameterization, supporting research activity. Due to the accessible code, the processing manner can be adjusted to a certain task, resulting in a more efficient workflow. Nevertheless, this requires a high-level background in informatics, in general. Among the tested open source software, ODM (with its alternate, WebODM) turned to be the most perspective.

Commercial software are in most cases optimized for Windows (possibly for IOS). Their use is definitely user friendly, and their operation is quite stable. Meanwhile less room for parameterization is enabled. Furthermore, as they work as a "black box", i.e. information on the computational method is hidden, there is very limited possibility for check / verification of the results. By time, commercial software tends to create more and more complex user interface in order to fulfil all demands and deliver a tool for wide range of applications.

According to the tests, Agisoft Metashape was found to be the most efficient providing the largest point density with the shortest computational time. Still, other software also have their speciality, in which they prove to be the best. The support, the customization of the orthophoto, and the volume calculation is outstanding for 3D Survey. ReCap Photo can create the most plastic / statuesque models; its high-level texture is a consequence of the advancement of Autodesk software, in general.

All in all, open source software have relevant potential in development, they provide a proper playground for testing unique ideas and demands. In contrary, they require a more creative and advanced user. Open source software are recommended for learning and research activity. Commercial software are better (at the time of the present investigation), due to their structural and financial background. Users can use these software with very few background in surveying or informatics, although they have very few options to change the settings of parameterization. For commercial use, in order to ensure productivity, the use of these software is still recommended over open source software.

References

1. J. L. Schönberg, J. Frahm, 2016 IEEE Conference on Computer Vision and Pattern Recognition (CVPR), Las Vegas, NV, 4104-4113. DOI: 10.1109/CVPR.2016.445 (2016)
2. D. G. Lowe, International Journal of Computer Vision, **60**, 91–110. DOI: 10.1023/B:VISI.0000029664.99615.94 (2004)
3. S. Bianco, G. Ciocca, D. Marelli, J. Imaging, **4**, 98. DOI: 10.3390/jimaging4080098 (2018)
4. F. Remondino, M. G. Spera, E. Nocerino, F. Menna, F. Nex, Photogrammetric record, **29**, 146, 144-166. DOI: 10.1111/phor.12063 (2014)
5. M. Lehoczky, Z. Siki , Geodezia es Kartografia **2020/2**, 23-27. DOI : 10.30921/GK.72.2020.2.4 (2020)
6. G. Vacca, Int. Arch. Photogramm. Remote Sens. Spatial Inf. Sci., **XLII**-4/W14, 239–245. DOI: 10.5194/isprs-archives-XLII-4-W14-239-2019 (2019)
7. C. Wu, link: http://ccwu.me/vsfm/ ; last retrived: 20 September, 2020 (2020)
8. H. Rahaman, E. Champion, Heritage, **2**, 1835-1851. DOI: 10.3390/heritage2030112 (2019)

Deformation analysis based on GNSS measurements in Tashkent region

Dilbarkhon Sh. Fazilova[1*], and *Lola V.* Sichugova[1]

[1]Astronomical Institute of Uzbek Academy of Sciences, 100052, Astronomicheskaya 33, Tashkent, Uzbekistan

Abstract. This paper presents the results of the GNSS geodetic network deformation analysis in the Tashkent region, as an example of an urban area, where obtaining reliable information for assessing hazard risk is of great importance. A software package in Delphi language has been developed for the assessment of the datum differences between 2009 and 2011 by implementing the 3D Helmert transformation method. The result revealed that there is significant translation and rotation in the network, while the scale of the network remains almost constant during two years period. The area strain was estimated by the finite element method. Most of the Tashkent region can be considered to be in a high compression (negative dilatation) strain state with maximum value -230×10^{-8}. On the contrary, remarkable positive dilatation strain is concentrated on the coastline of the Charvak water reservoir, where large strain is about 351×10^{-8}.

Keywords. GNSS network, strain accumulation, finite element method.

1 Introduction

Tashkent region is located in the Middle Tien-Shan zone at the boundary two large components. It is a part of the post-platform mobile region and is located in the transition zone between the Tien-Shan orogenic territory and the Turan plate. The mountain ranges (Karzhantau, Chatkal and Kurami), covered by young structures in some areas, surround this field and decrease in southwest direction. Recent up thrusts and deflections characterize plain part of the territory [1, 2]. According the results are presented in map of Peak Ground Acceleration, the region belongs one of the two fields in the country of very high seismic activity (up to 4.8 m/s^2) [3]. The magnitude of the maximum ground velocity must exceed values for 50 years 50 cm / s with probability P = 0.99 according the analysis the velocity graphs of M=3.8–6.2 earthquakes [4]. Significant velocity rates between 30 mm/year and 60 mm/year were found to the tectonic fault zones and mountainous part of the area, the flat, foothills zones show velocity rate up to 15 mm / year [5]. Charvak water reservoir with area around 40 km^2 and with the volume of 2 billion m was constructed for hydropower generating, irrigation purpose and for controlling river flow. In 1970, construction of the

* Corresponding author: dil_faz@yahoo.com

Charvak dam with the height of 167 m has ended and started flowage of Brichmulla depression and filling of the Charvak reservoir [6].

The GNSS geodetic network includes 31 stations (fig. 1), collected in campaign mode between 2009 and 2011 by researchers from the Institute of Seismology and the National Center of Geodesy and Cartography to study the modern deformations of the Earth's surface using space geodetic methods.

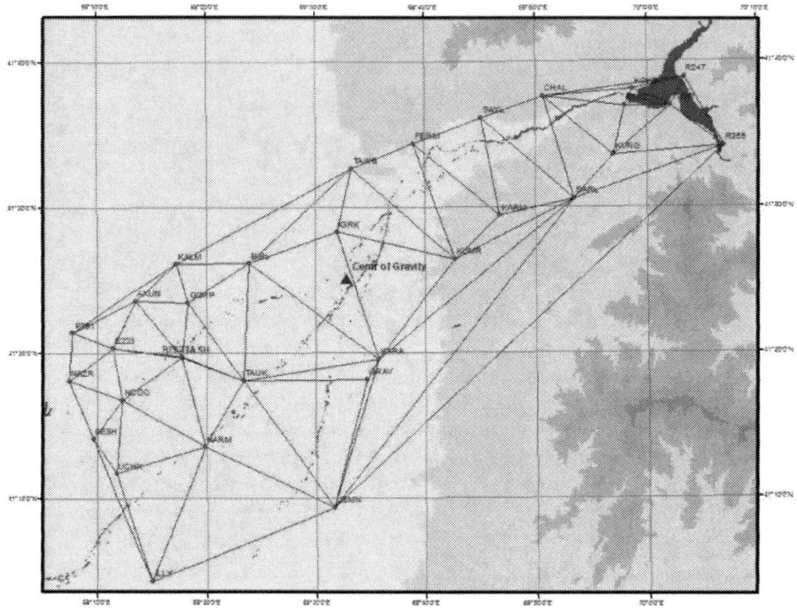

Fig.1. GNSS network of study area

The region is well studied in terms of seismic activity and therefore a geodetic network was established here. The main task was to conduct research on the station's velocities, deformations based on long-term GNSS measurements. The data were to become the basis for constructing a geodynamic model of the region's microplates. However, after 2011 measurements were not taken due to both financial and organizational limitations. Therefore, in this work, we only consider the stability of the network, horizontal and vertical displacements caused by the influence of various datums. Stations installed with different spacing and varies from 5 km to 20 km. The GNSS data were observed over 6 time periods (04.09.2009 – 27.09.2009; 05.11.2009 – 02.12.2009; 25.04.2010 – 06.06.2010; 17.10.2010 – 6.11.2010; 19.04.2011 – 10.05.2011; 20.09.2011 – 6.11.2011) with the observation rate 10 s. Session duration for each station was 3 days. Ashtech Z-Surveyor dual-frequency receivers and ASH 701975-01 antennas were applied in all the surveys. Observations were carried out simultaneously by 12 receivers, and 6 were in stationary mode. Site positions and baselines for each epoch were estimated by aforementioned organizations with Trimble Total Control (TTC) software version 2.73.

The free network solution was introduced for each campaign. For mathematical processing of satellite measurements, at least one point must be connected to a point on the international geodynamic network. Therefore, the absolute coordinates of the initial point NCGC were preliminary calculated relative to the international station TASH. Additionally, to tie the regional measurements to the global reference frame ITRF2008, data from continuously operating IGS permanent network stations (KIT3, MDVJ, POL2,

NSSP) were used. Coordinates of the test network estimated by six solutions have both different epochs and different datum. For accurate calculation of station displacement vectors, it is necessary to take into account the measurement errors that arise as a result of the choice of the reference system. As a result, "rotation" and "deformation" of the network arise. To eliminate the effect of visible "rotation" and "deformations" it is recommended to use the transformation methods. This study was aimed to compare the coordinate difference between the campaign's epochs and referencing them into the first epoch of measurements by using 3D Helmert transformation method for the possibility of an accurate assessment of network deformations and strain accumulation computation.

2 Materials and methods

2.1 Datum investigation with 3-D Helmert transformation

Usually, the processing of geodetic measurements to identify modern movements of the earth's crust is performed on relatively stable geological structures. However, for the study area, there is not yet a sufficiently accurate tectonic model and, accordingly, there is no sure confirmation that the points will remain stable. The territory of the Tashkent region is subject not only to modern tectonic movements, but also a change in the water level in the Charvak reservoir can be the cause of the induced disturbances of the Earth's surface. One of the solutions to the problem is the implementation in network processing of continuously working stations, for example IGS data. Taking into account that our network was processed relative to several high-precision stations of the IGS, in this study, all 31 points were used to perform the transformation procedure. Another reason was to look at how and where there will be more horizontal movement for further redefinition of the regional tectonic model.

The network adjustment error is usually influenced by two factors: orientation and scale of the network. Therefore, a 7-parameter transformation method was chosen in the work to analyze deformation between two epochs and strain parameters in the geodetic network, three-dimensional conformal (it preserves shapes of objects) coordinate transformation is also known as seven-parameter similarity transformation was applied. It converts separate surveys into a common reference coordinate system and it can be thought of as a three-step process: scaling, rotation and translation. The seven parameters of the transformation between convertible *(x,y)* and reference *(X,Y)* systems (epochs) - one scaling (m), three rotations (R_X, R_Y, R_Z) and three rotations translations (T_X, T_Y, T_Z) can be determined uniquely with the minimum of three common points for both systems [7, 8]:

$$\begin{bmatrix} X \\ Y \\ Z \end{bmatrix}_i = \begin{bmatrix} T_X \\ T_Y \\ T_Z \end{bmatrix} + (1+m)\begin{bmatrix} 1 & -R_Z & R_Y \\ R_Z & 1 & -R_X \\ -R_Y & R_X & 1 \end{bmatrix}\begin{bmatrix} x \\ y \\ z \end{bmatrix}_i \quad (1)$$

When more than three stations are present, the solution is obtained by the conventional least square adjustment procedure and the equations system can be written in the matrix form:

$$A_i dp = L + V \quad (2)$$

Where L-coordinate difference, V-vector of residuals. The design matrix A_i for station i of the are defined as

$$A_i = \begin{bmatrix} 1 & 0 & 0 & x_i & 0 & 0 & y & z & 0 & 0 & z & -y \\ 0 & 1 & 0 & 0 & y & 0 & x & 0 & z & -z & 0 & x \\ 0 & 0 & 1 & 0 & 0 & z & 0 & x & y & y & -x & 0 \end{bmatrix} \tag{3}$$

The adjustment of the slightly overdetermined system (2) via normal equations yields the parameter vector *dp* of unknown transformation parameters (scaling, three rotations and three rotations translations) $\boldsymbol{dp = [dT_x\ dT_y\ dT_z dm\ d\omega_x\ d\omega_y\ d\omega_z]^T}$

$$dp = (A^T A)^{-1} A^T L \tag{4}$$

Once parameters of the transformation are determined, the coordinates *(x, y, z)* are transformed to new $(X, Y, Z)'$ coordinate system using eq. (1). The coordinate values for all sessions were transformed to the topocentric coordinate system (E, N, U):

$$\begin{bmatrix} E_n \\ N_n \\ U_n \end{bmatrix} = \begin{bmatrix} -\sin L_0 & \cos L_0 & 0 \\ -\sin B_0 \cos L_0 & -\sin B_0 \sin L_0 & \cos B_0 \\ \cos B_0 \cos L_0 & \cos B_0 \sin L_0 & \sin B_0 \end{bmatrix} \begin{bmatrix} X_n - X_0 \\ Y_n - Y_0 \\ Z_n - Z_0 \end{bmatrix} \tag{5}$$

where (X_0,Y_0,Z_0) and (B_0, L_0) are geocentric and geodetic coordinates (latitude and longitude) respectively of the center of gravity of the network (marked as red circle on fig.1). First epoch (Sept. 2009) of the measurement was choosing as reference for further calculations and the horizontal displacement vectors of pairs first – second (Nov. 2009), first – third (June 2010), first – fourth (Nov. 2010), first – fifth (May 2011) and first -six (Nov. 2011) epochs of each GNSS monitoring station was calculated as

$$R_n = \sqrt{(N_n^{'} - n_n)^2 + (E_n^{'} - e_n)^2} \tag{6}$$

$dU_n = U_n^{'} - u_n$ height component variations will be used for estimation of the network vertical deformation. A software package in Delphi has been developed for the assessment of the datum differences between 2009 and 2011 by implementing the 3D Helmert transformation method.

Data analysis was performed in the ArcGIS 10.1.

2.2 Estimation of strain accumulation by finite element model

The calculation of the network strain by the finite element method was used to assess the deformation effect. The principles and basics of finite element method are generally known and are described in numerous publications [9, 10]. In the first step, the network has been constructed of finite elements (triangles) based on the Delaunay triangulation (Green and Sibson, 1978) (fig.1). Local strain field was computed for each triangle in the network. The strain values (E) on the earth's surface can be expressed in terms of displacements between adjacent stations at initial and final epochs (l_0, l_1) [11]:

$$E = \frac{|l - l_0|}{l_0} \tag{7}$$

This equation is created for each baseline of a triangle. The coordinates of GNSS points were represented in the Mercator projection for calculating distances (radius-vectors) between the nearest stations.

Finally, all individual results are integrated for whole investigated region by creating homogeneous surface with the nearest neighbor interpolation method. It is based on comparison of the distribution of distance d between a point x and the nearest neighboring points y of a set of randomly distributed data [12]:

$$d(x,y) = \|x-y\| = \sqrt{(x-y)(x-y)} = ((\sum_i (x_i - y_i))^2)^{1/2} \quad (8)$$

The method has been investigated and used in recent work for assessment of normal heights surface for Fergana valley territory in Uzbekistan [13]. It showed more reliable results than commonly used polynomial interpolation method for creating surface over the high mountains region.

3 Results

The transformation parameters translations (T_X, T_Y, T_Z), scale (m) and rotations (ω_X, ω_Y, ω_Z) between the first-reference and subsequent (from second to sixth) epochs are given in the Table 1. The result revealed that there is significant translation and rotation in the network, while the scale of the network remains almost constant during two years period (from September 2009 to November 2011).

Table 1. The transformation parameters between different epochs of measurements

Epochs	*Period*	T_x, *m*	T_y, *m*	T_z, *m*	*m*	ω_X, *sec*	ω_Y, sec	ω_Z *sec*
1–2	09.2009-12.2009	-9.96	0.02	4.12	0.99	0.076	-0.294	0.185
1–3	09.2009-06.2010	-5.16	-2.04	-1.90	1.00	0.025	-0.028	0.160
1–4	09.2009-11.2010	2.48	-5.53	6.20	0.99	0.258	-0.006	-0.110
1-5	09.2009-05.2011	-6.99	1.35	9.24	0.99	0.157	-0.295	0.106
1-6	09.2009-11.2011	4.29	8.72	21.16	0.99	0.289	-0.112	-0.044

The horizontal displacement vectors were calculated using the eq. 5 and, for a more visual representation of the deformation of the whole region, the values were interpolated over the study area (eq. 7). The presented in fig. 2a result demonstrate horizontal displacements of the network with rates between 2 cm and 21 cm during whole observation period (first and sixth epochs) from September 2009 to November 2011. Analysis of calculated displacement vectors has been carried out separately for different parts of the region. The investigated field was conditionally divided into three main parts. The first one belongs to the Charvak coastal zone (KOKB, NM12, YUSP, CHAL, KUNG, R247, R255, SAYL, SARL, KARM, FERM) and divided from the others by the Kumbel fault. The second and the third are separated from each other by the largest Karjantau depression and hereinafter called conditionally the Northern coastal zone of the Chirchik river (AXUN, BIRL, E223, GORP, IGRK, KALM, NAZR, R351, TAWB) and the Southern coastal zone of the Chirchik river (KUMR, GRAV, KARA, TAUK, R732, NARM, NCGC, BESH, UCHX, USMN, ILLY). It should be noted that deformations in the Chirchik river valley are less (about 1 cm) than in the coastal part.

The stations located in Northern coastal part of the river have displacement rate from 4 cm to 10 cm and the stations located in the Southern coastal zone have a displacement rates from 2 cm to 17 cm. Moreover, the stations located closely to a coastal part of the Charvak reservoir (R247, NM12, R255, YUSP) have a maximum displacement rates in this zone between 13 cm and 20 cm. There is a tendency to rotational movement along the Kumbel and Karjantau faults (fig2a).

The presented in fig. 2b result demonstrate vertical displacements of the network with rates between - 5 cm and 20 cm during the two years observation period. Station (KALM) located in this region shows relative to other stations a high rate of vertical displacement 20 mm, the cause of which may mainly be instrumental (antenna information, et al.) errors than tectonic effects. Therefore, this station was excluded from further analysis. It can be concluded that for this region, horizontal displacements are a consequence of both tectonic and hydrological processes. While vertical deformations are the result of most tectonic phenomena

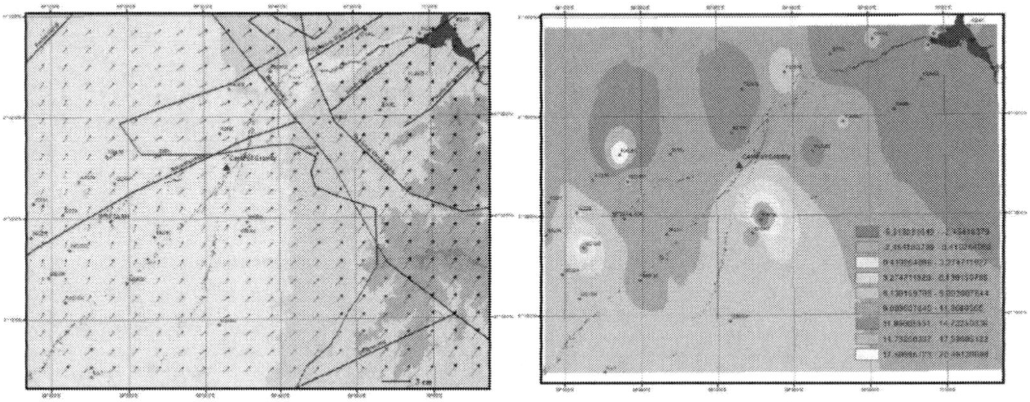

a) b)

Fig.2 Horizontal (a) and vertical (b) deformations of the network during 2009-2011

The result revealed that there are significant horizontal displacements, therefore, it is necessary to estimate the strain accumulation regardless of the datum and the displacement vectors may be used to derive this parameter.

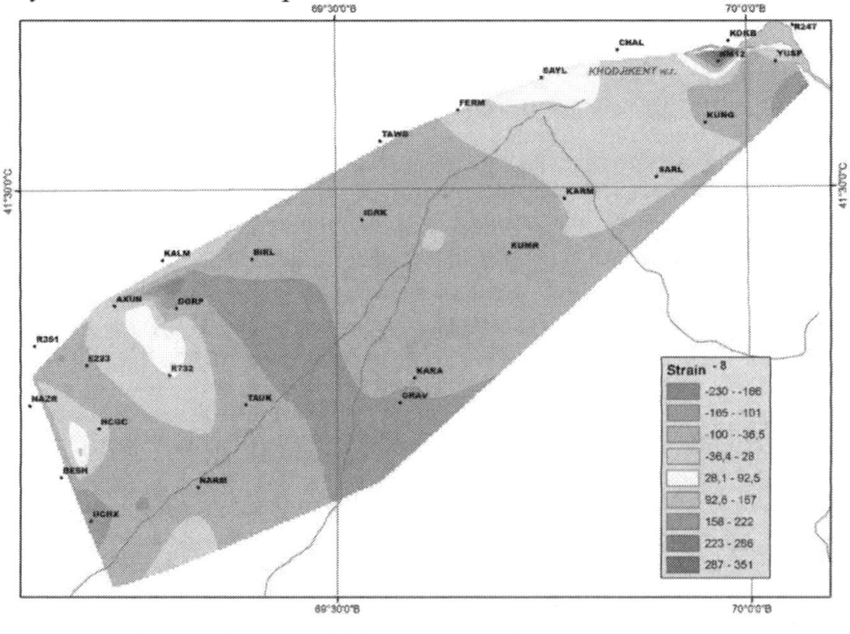

Fig. 3. The map of strain values based on GNSS data over region

Fig. 3 shows the principal strains rate distribution. Most of the area can be considered to be in a high compression (negative dilatation) strain state. Strain rate for the southwestern part near stations GRAV, KARA, AXUN, GORP as maximum value with -230×10^{-8}. Minimum values of extension are found in with a value of 10×10^{-8} in flat areas. The results are very similar to the [5]: the strain rate at intersection of active faults Karzhantau, Kumbel and Ugam confirmed 80% of the value major shear stress caused by the tectonic movements of the whole region. On the contrary, remarkable positive dilatation strain is concentrated on the coastline of the water reservoir, where large strain is about 351×10^{-8}. The result clear demonstrates that, along strain caused by the tectonic movements, the water level variations in the reservoir is also a major factor determining the displacement and stability of the area.

4 Conclusions

The effect of the deformation processes in Tashkent region, was studied by strain deformation calculation with GNSS data. In this regard, as the first step, GNSS stations coordinates were transformed to a datum of the first epoch using the 3D Helmert transformation method, and the horizontal displacement vectors were determined for each station. The result revealed that there are significant horizontal displacements with a maximum value up to 20 cm in the network and vertical displacements with rates between -5 cm and 20 cm. The calculation of the network strain was used to assess the network deformation. The network has been constructed of finite elements (triangles) based on the Delaunay triangulation. We found that most of the Tashkent region can be considered to be in a high compression (negative dilatation) strain state confirmed, surely, of the value major shear stress caused by the tectonic movements of the whole region. On the contrary, remarkable positive dilatation strain is concentrated on the coastline of the water reservoir, where large strain is about 351×10^{-8}. We can conclude that more GNSS campaigns in combination with gravity measurements are needed in this area in order to better constrain the relatively small deformation rates.

This work was carried out within the scientific and applied project FA-Atab-2018-57 of the Astronomical Institute of Uzbekistan with the financial support of the Ministry of Innovative Development of the Republic of Uzbekistan. We are very grateful to Institute of Seismology and National Center of Geodesy and Cartography, who organized GNSS experiments and made the data available to this study.

References

1. A. R. Yarmukhamedov. *The morphostructure of the median Tien Shan and its relationship with seismicity*. Tashkent: Fan, 163 (1988) (In Russian). https://www.geo-fund.am/files/library/1/15247392644395.pdf
2. T. Rashidov, L. Plotnikova and S.Khakimova. *Seismic hazard and building vulnerability in Uzbekistan // in Seismic Hazard and Building Vulnerability in Post-Soviet Central Asian Republics.* edited by S.A. King, Vitaly I. Khalturin, B.E. Tucker. NATO ASI series. **Vol.52.** (1999)
3. D. Giardini, G. Grünthal, K. M. Shedlock, and P. Zhang. *The GSHAP Global Seismic Hazard Map.* In: Lee, W., Kanamori, H., Jennings, P. and Kisslinger, C. (eds.): International Handbook of Earthquake & Engineering Seismology, International Geophysics Series 81 B, Academic Press, Amsterdam, 1233-1239 (2003) http://www.gfz-potsdam.de/GSHAP

4. T. U. Artikov, R. S. Ibragimov, T. L. Ibragimova, K. I. Kuchkarov, M. A. Mirzaev. Quantitative assessment of seismic hazard for the territory of Uzbekistan according to the estimated maximum ground oscillation rates and their spectral amplitudes. Geodynamics & Tectonophysics **9 (4)**, 1173–1188 (2018) doi:10.5800/GT-2018-9-4-0389.
5. H. L. Hamidov. Identification of morphokinetic indicators of modern geodynamics of western Tiang Shan // Modern geodynamics of Central Asia and dangerous natural processes: quantitative research results: Materials of the All-Russian meeting and youth school on modern geodynamics (Irkutsk, September 23–29, 2012). - In 2-ht. - Irkutsk: IZSORAN, **T. 2.** 211 (2012) www.crust.irk.ru/images/upload/newsfond143/299.pdf
6. M. Juliev, A. Pulatov, J. Hübl. Natural hazards in mountain regions of Uzbekistan: A review of mass movement processes in Tashkent province. International Journal of Scientific & Engineering Research, **Vol. 8**, Issue 2, February (2017)
7. R. P. Wolf, and D. C. Ghilani. *Adjustment computations: statistics and least squares in surveying and GIS.* Wiley, New York, 347 (1997)
8. I. Deniz, & H. Ozener. Estimation of strain accumulation of densification network in Northern Marmara Region, Turkey. Natural Hazards and Earth System Science, **10(10)**, 2135–2143 (2010) https://doi.org/10.5194/nhess-10-2135-2010
9. A. Dermanis. A method for the determination of crustal deformation parameters and their accuracy from distances. J. Geod. Soc. Japan, **40**, 17-32 (1994)
10. O.C. Zienkiewicz, R.L. Taylor, and J.Z. Zhu. *The Finite Element Method: Its Basis and Fundamentals.* **6th edition**, Butterworth-Heinemann. 752 (2005)
11. A.O. Agibalov, V.A. Zaytsev, A.A. Sentsov, A.S. Devyatkina. Assessment of the influence of modern crustal movements and the recently activated Precambrian structural plan on the relief of the Lake Ladoga region (the southeastern Baltic Shield). Geodynamics & Tectonophysics **8 (4),** 791–807 (2017) doi:10.5800/GT-2017-8-4-0317
12. P. D. Dumitru, M. Plopeanu, and D. Badea. *Comparative Study Regarding the Methods of Interpolation.* In: Proceedings of the 1st European Conference of Geodesy & Geomatics Engineering (GENG '13), 32-37 (2013) http://www.wseas.us/elibrary/conferences/2013/Antalya/GENG/GENG-00.pdf.
13. D. Fazilova, H. Magdiev. Comparative study of interpolation methods in development of local geoid. International Journal of Geoinformatics, - **V.14,** №1. 29–33 (2018) http://journals.sfu.ca/ijg/index.php/journal/article/view/1114/599

Application of GIS on Research of Horizontal Refraction in Polygonometry on Network

Abdusali Suyunov[1,*], *Shukhrat* Suyunov[1] and *Olim* Urokov[1]

[1]SamSACEI, Department of Geodesy and Cartography, 140147, Samarkand, Uzbekistan

Abstract. The article examines the influence of refraction on the results of measuring the horizontal angle in special places. In the conditions of the hot and arid climate of Uzbekistan in the polygonometric network of river banks, according to the results of field experiments, the new version of the developed methodology of the GIS makes it possible to increase the accuracy of measuring horizontal angles and lengths of lines. Significant temperature changes along the water and the coast, on rocks, in the upper layers of sand-concrete pavements are also explained by the properties of the heat and radiation balance.
Keywords. Electronic technologies, high-precision geodetic measurements, the influence of refraction, atmospheric measurement, uneven distribution of the light beam, modern devices, positive opportunities.

1 Introduction

One of the most important reasons limiting the accuracy of measuring horizontal angles using modern technology of high-precision angular measurements is not instrumental errors, but an error called horizontal refraction of the atmospheric atmosphere caused by external environmental influences.

Studying this phenomenon and eliminating its harmful effects, triangulation and polygonometry is the first task in science and industry to improve the accuracy of networks.

This problem is taken into account when conducting geodetic measurements in developed countries, including Russia, Ukraine, Germany, the United States and other countries, to improve the accuracy of measurements, taking into account the horizontal refractive properties, which significantly affect the measurement results.

In large-scale cartographic work, geodesy makes all the requirements related to the accuracy of the geodetic base, especially angular measurements, forcing to revise the newly created and developed systems for constructing the geodetic base, and requires the development of a new method.

In this regard, one of the important tasks is to ensure that the external environment is not always in a stable state, but to change it from time to time in established ways.

The new basic geodetic survey system shortens the sides of triangles and polygon measurements as it requires more compaction of points. As a result, we observe a slight

* Corresponding author: suyunov303@inbox.ru

deterioration in the state of light transmission along the uneven line. In such cases, it is necessary to improve the accuracy of the angular measurement results. To overcome such problems, it is necessary to study new cases of the influence of horizontal refraction on the accuracy of the results of measuring horizontal angles in urban polygonometry, i.e. laws of subordination of the atmosphere to height..

To accomplish these tasks, it is important to determine and improve a modern solution for measuring the impact of the external environment on horizontal angle measurements.

In the Strategy for the Further Development of the Republic of Uzbekistan for 2017-2021, the Republic of Uzbekistan "On Geodesy and Cartography", Decree of the President of the Republic of Uzbekistan dated May 31, 2017, No. UP-5065 "On Further Improving the Activities of the State Committee for Land Resources, Geodesy, Cartography and cadastre of the Republic of Uzbekistan" other regulatory documents [1].

With the development of this area, the scientific problems of urban polygonometry using horizontal angle and linear distance measurements are expanding, which requires taking into account the effect on horizontal angle measurements in hot, very hot and dry climates.

It is important and urgent to find a solution to these problems, especially in the context of the climate of the Karshi city of the Republic of Uzbekistan, in dry and hot climates, the variability of the atmosphere is very high, which leads to significant errors in the measurement of the horizontal angle.

Methods for taking into account the influence of the atmosphere on measurements of the horizontal angle have been tested to one degree or another in the middle latitudes of the normative climate and in other relatively lower regions of the republic's climate, but in an arid climate, they have not been studied at all.

The influence of horizontal refraction on the results of geodetic measurements was studied in scientific research of scientists of the republic A.S.Suyunov and T.M.Abdullaev, and they have achieved certain positive results [3,4,6,9].

However, the work of the aforementioned scientists and researchers did not study the implementation and improvement of the accuracy required to measure these objects in modern new geodetic instruments, as well as its regional organization.

Much attention in the country is paid to the construction of new buildings and structures on the basis of comprehensive geodetic control with high precision, accuracy and standards. However, no method has been developed to improve accuracy by determining the degree of influence of the near-earth layer of the atmosphere on the results of geodetic measurements. Thus, this article to some extent meets the above goals and objectives.

Sokoba's research on reforestation and riverbank research is based on a very short (six-hour) experience with a number of drawbacks [5,7,8,11].

Therefore, the effect of refraction on geodetic measurements has not yet been studied in urban polygonometry.

In fact, the solution to this question is not enough for Marchenko [5] investigated in his polygonometric system of wall mounting, specifying the refraction effect in this case, without saying a word about the effect of light refraction. At the same time, polygon paths along city streets and river banks are approaching additional established signs, which leads to the fact that horizontal angles are distorted and significantly increase.

In particular, ehe lateral refraction effect is especially noticeable in the lower classes of polygonometry when it comes to the accuracy of measuring horizontal angles. [6]. The reason is as follows:

(a) Polygonometry usually develops in a plane, in a closed space, that is, according to the temperature (density) of the air, as well as the refraction of the light beam under conditions of a strong influence of the air temperature below and the height of the air atmosphere up to 1.5 m and in places with obstacles. Especially in polygonometry of fields

3 and 4 and urban polygonometry, the influence of a light beam on the state "not between obstacles, but between obstacles" should be noted separately. Triangulation is typically used primarily in open, intersecting, or semi-enclosed areas, that is, in conditions where it is visible at all distances or partially so that light passes through the lower surface of the earth.

b) In polygonometry, the azimuth transfer error is approximately 1.3 times greater than the triangulation error [2,10,12-16].

c) In the lower classes of polygonometry, when the line length is 0.5–1 km, small, well-defined refractive fields have a strong refractive effect on the measured geodetic results of all polygonometric trajectories. Thus, the values of the effect of the refractive field in the lower classes of the polygonometric trajectory will be very close to the indices of the refractive field in polygonometry of the 1st and 2nd classes.

The microclimatic regime in the surface layer of the atmosphere is characterized, first of all, by superadiabetic vertical temperature gradients during the day and nighttime inversion of the temperature distribution and their significant differences in the horizontal plane and even greater gradients in the near-surface layers.

As is known from the courses of meteorology, the physical basis of these properties stems from the heat and radiation balance in the near-surface layer of the atmosphere.

The influence of the local climate of water bodies and rivers and the microclimate of coastal areas is known.

Consequently, it reduces the daily and annual amplitudes of air temperature near water bodies; in summer, the temperature decreases along the coast and above the water surface during the day, and at night it is higher than in places far from water bodies; water bodies heat up during the winter months, and the adjacent drylands are relatively cooled; The range of influence of shores stretching for tens, and sometimes hundreds of kilometers, depends mainly on the size of the reservoirs.

Throughout the water and along the coast, on the rocks, in the upper layers of sand-concrete pavements, there are obvious temperature drops;

Again, this is due to the properties of heat and radiation balance.

Based on the results of Suyunov's and Abdullaev's installation of meteorological stations in the upper water layer in the region of 5 geodetic points (5-10 m) around the Koratepa reservoir, the effect of lateral refraction was determined. did not take into account the influence of the external environment in field experimental studies.

2 Methods

2.1. Study area

In the experimental region, the refractive index n of the air atmosphere has a predominant effect on the horizontal temperature gradient. In coastal soil conditions, horizontal moisture and temperature gradients can have a significant impact on coastal areas. From a geodetic point of view, we found answers to the following questions to study and account for the effect of lateral refraction to improve the accuracy of horizontal angle measurements:

a) values of horizontal gradients of temperature and humidity of air at a distance of the first ten meters from the surface of the earth and water of the atmosphere near the earth;

b) in each direction, in different weather conditions, horizontal gradients of temperature and humidity during the day, the path of change in hours;

c) density of the range of spatial distribution of the light beam above the surface layer of the atmosphere, etc.

The answers to these questions were found during a survey conducted by the authors in June 2018 and July-August 2019 at the geodetic polygon in the Khudoizod massive which is located in Karshi city of the Republic of Uzbekistan.

2.2. Methods

Points for meteorological observations were selected from open areas on the left bank of the Beshkent River. The glasses were set 12 meters from running water. This river flows out of the Beshkent Canal, not far from the Khudoizod massive (0.8-1.0 km). This place is distinguished by its unusual width, landscape, and natural features. Observation platforms of low (2-4 m) relief, complex, are located at a depth of several tens of meters wide and in hills. The width of the massif directly with the observation deck is 3.5 meters towards the lower reaches of the river, the width of the coastline is 30-35 meters, the height of the coastline is 3-3.5 meters. The azimuth of the river flow is 125°. It is asymmetrical on the left bank of the Beshkent River, stretches 1-1.5 km to the north and north-west, and is located at the foot of the hills, then turns into a gentle hill, therefore it is located at a distance of 700-900 m from Beshkent. the absolute height of the points is close to the right bank of the river valley and depends on the river bed. In 2018, at the main observation site, the temperature was measured from the earth's surface, at a height of 0.75 m, and temperature and humidity - at a height of 1.5 m. Atmospheric pressure, wind speed, and direction were also determined. The state of heaven was recorded. At the same time, air temperature and humidity were observed on the water surface at the above-mentioned heights and its readings were recorded. For this, a special boat was built and a mast was attached to it. Modern electronic psychrometers were installed on the masts. All measurements were made by the members of the expedition.

Thus, at the aforementioned heights, meteorological data were measured at the same height above land and water at a distance of 3.5 m from the coast, from the water. Aneroid, hand-held anemometer, psychometers were used to measure meteorological elements. Before all measurements, the measuring instruments were inspected.

At the same time, it was taken into account that the height of the aiming beam when viewing the place itself is legally greater than at ground level.

For the measurement of meteorological elements: psychrometers-thermometers with fabric protection (coefficient of continuity 200 seconds) and scale values of 0.1 ° and 0.2 ° C. Particular attention was paid to the measurement of air temperature. All thermometers were checked and compared every hour. Their readings were brought to the readings of a psychrometer installed at a height of 1.5 meters in the main zone. This method significantly improves the detection of gradients with psychrometers and emphasizes their accuracy. The reason is that, without correction by using thermometers, artificial ventilation becomes absolutely necessary. Finally, due to the variability of solar radiation, there will also be significant (up to 1 ° S) daily variability, which depends on the strength and direction of the wind speed.

In each measurement, thermometers were counted three times, and the second and third counts were made at intervals of 2–4 min, respectively. The result is the arithmetic mean of these numbers. We did not measure temperature, although the science of meteorology recommended observing air temperature readings at microclimate altitudes (20 cm and 50 cm).

It should also be noted that each psychrometer has its own radiation errors, and in meteorology, it is generally preferable to measure gradient values with a single psychrometer, regardless of the fact that these measurements are not simultaneous.

We, on the other hand, used two psychrometers and counted them simultaneously, which reduced radiation errors by comparing thermometers to each other every hour.

Compared to the meteorological measurements of our 2019 field experiments, the observations made in 2018 were closely related to simultaneous angle measurements. In these measurements, the main observation point is located at a distance of 7 m from the edge of the water boundary.

In addition, the horizontal gradient of air humidity in 2018 has not been determined. In 2019, air temperature, humidity, and wind speed were determined.

Accordingly, all measurements were made on two additional psychrometers. Psychrometers are located on the left bank of the embankment at a distance of 15 meters from the water and 50 meters from the station, as well as at the station on the right bank. In turn, in 2018, additional air temperatures were measured in the direction transverse to the river at a distance of 400 m from the station. During the observation, the openness of the sky was recorded on a five-point scale.

All observations were usually carried out from morning to evening, and sometimes during the day (24 hours). To facilitate further processing and understanding, the main point of meteorological observations has been moved to the origin of the spatial coordinate system.

The system plane coincides with the horizon plane. The main arrow is directed perpendicular to the stream of the river. As a result, the main axis runs in the direction of the riverbank, and in fact, it is a parallel line equal to the temperature and humidity of the air [17].

Thus,

$$\frac{dT}{dY}, \frac{dT}{dZ}, \frac{de}{dY} and \frac{de}{dZ} \quad (1)$$

(1) Differences in temperature and humidity are proportional to horizontal and vertical gradients.

3 Results

We can arbitrarily call gradients positive or negative if the temperature and humidity of this coordinate axis fall or fall in their direction. The average values of the horizontal temperature (t) and air humidity of interest to us are calculated using the results of measurements in the middle of the river and river coordinates ($x = 0$, $u = 50$ m, $z = 0$).

Fig. 1 The results of field observations in 2018 are presented in the middle of the river, at a distance of 5 m above the water level on land and at a height of 1.5 m along the river bank at the water level, demonstrating a picture of daily changes in air temperature drops.

All of the above observations were carried out with a cloudy, opaque (variable), and open sky (sunny weather) wind at a speed of 0.5 m / s in the direction of the river flow. During the day, the wind force in this direction increased significantly and reached 4-5 m/s [18,20].

The air temperature difference shown in the graphs is formed by the following formula:

$$\Delta T = T_{\text{к}} - T_{c} \quad (2)$$

Where, T_K is the air temperature on the river bank; T_C - air temperature near the groundwater level. These values significantly decrease in the south of the Republic of Uzbekistan on sunny days, to very large 2^o5 C), when the sky is variable cloudy cool days (up to 0°,3-0°,5 C). Differences in ΔT are positive for negative values in the morning, evening, and night, between 7:30 am and 8:30 pm after sunrise and between 7:00 pm and 20:00 pm before sunset. (During the field experiment, sunrise was 6:30 am and sunset was 8:40 pm. (Fig. 1.)

The uneven distribution of real moisture in the atmosphere is very interesting. During the day, at a height of 1.5 m on the river bank, at a distance of 15 m from the river bank, the humidity is relatively high, the wind blows weakly in all directions up to 5 m above the river surface. In the morning, evening and night, on the contrary, the water on the surface confirms that the humidity is quite stable and very high.

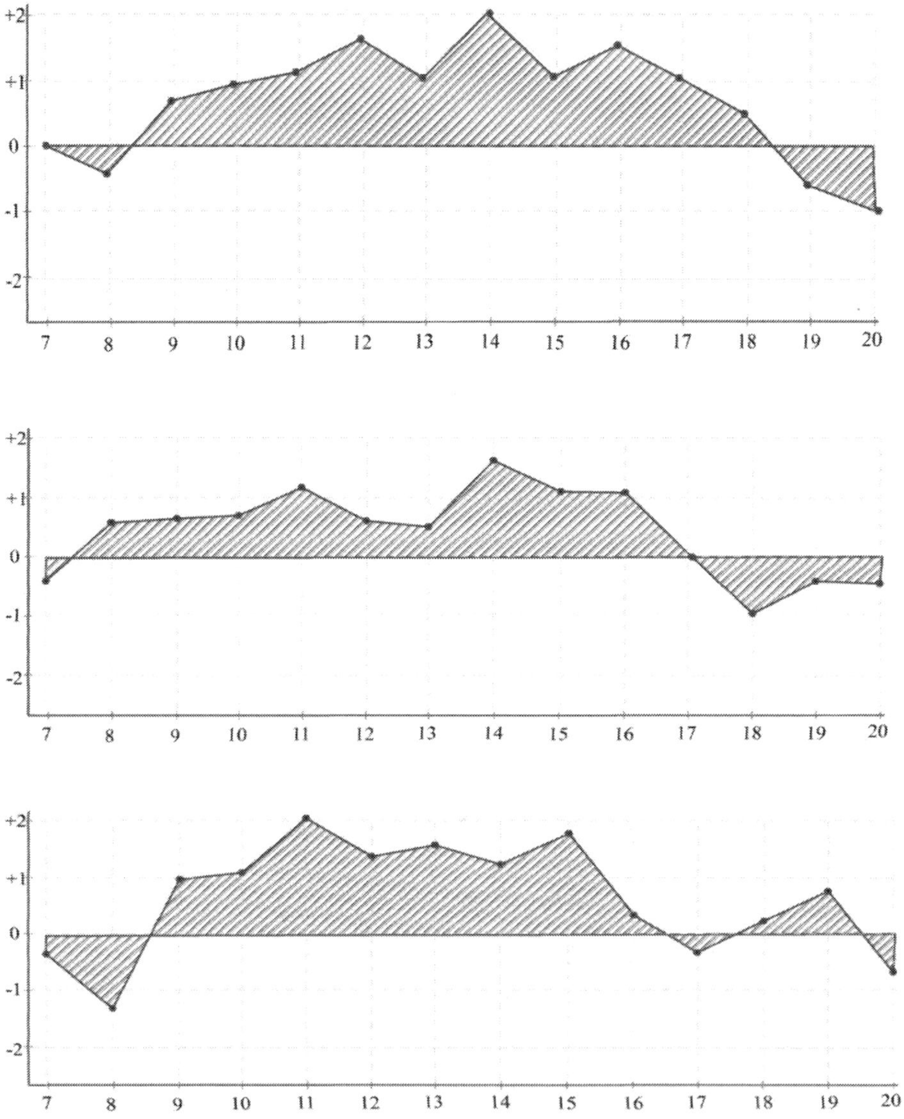

Fig. 1. The daytime observation results

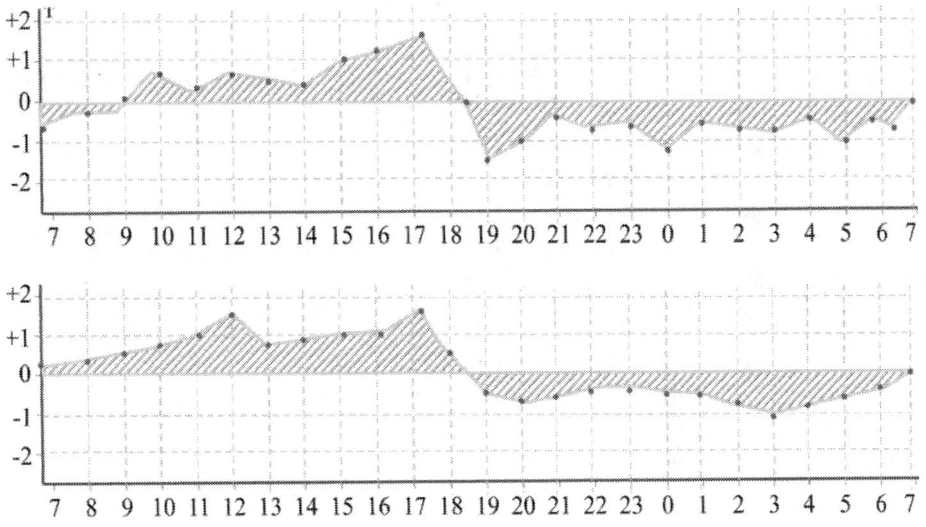

Fig.2. Results of daily monitoring

4 Discussion and conclusion

GIS is an existing multi-faceted environmental dataset based on electronic computing technologies [10,19]. GIS is designed to provide easy viewing of data available at different levels of human economic activity. The main stage in the development of GIS is the last 7-8 years. This is due to the fact that the computer-aided design of all work processes based on geographic data is organized technologically and ideologically.

GIS integrates information of different content and distribution into one system, creating a coherent system for analyzing geographic data. Regardless of how important it is to understand and manage the various activities and resources, if GIS are not linked, this issue remains unresolved.

We came to the following conclusions in the experiment:

- based on the measured meteorological data of the atmospheric atmosphere under the conditions under study, the possibility of studying the regularities of the distribution of the microclimate that creates a refractive field has been established:
- during the day, we showed that the influence of lateral refraction on the results of measuring the horizontal angle and line in hot and dry climates can be established in river conditions in the south of the country, every two hours of the day and in various climatic conditions:
- using the meteorological elements identified in the field, the geodetic measurements made in the river bank polygon as corrections were applied to the lateral refraction results and the results were improved:

These are new solutions used by GIS to combat external influences and improve the accuracy of polygon networks.

The authors are grateful to the staff of the Department of Geodesy and Mining of the Karshi Engineering-Economics Institute for their contribution to the collection of scientific results in this article.

References

1. Decree of the president of the Republic of Uzbekistan, "About the strategy for further development of the Republic of Uzbekistan", PF-4947 dated February 7, 2017
2. A. S. Suyunov, Sh. A. Suyunov, On the accuracy of geodetic measurements in the conditions of the Republic of Uzbekistan (2017)
3. T. M. Abdullaev, The study of patterns of action and the search for methods of accounting for lateral refraction when measuring horizontal angles (Tash STU, 2005).
4. Sh. A. Suyunov, U.E. Karimova, Sh.Sh. Fayziev, Special volume, J. Bull. Geograph. Society Uzbekiston (2018)
5. Sh. A. Suyunov, Doctoral research disser., (2018)
6. Sh. A. Suyunov, S. A. Tashpulatov, G. Mullajanova, Modern scientific challenges and trends. Warsaw, **12** (2) (2019)
7. A. S. Suyunov, Sh. I. Fayziyev, Intr. J. Adv. Res. Sci., Engin. Tech. **6** (11) (2019)
8. A. S. Suyunov, Sh. A. Suyunov, G.M. Mullajanova, Intr. J. Adv. Res. Sci., Engin. Tech.. **6** (11) (2019)
9. A. S. Suyunov, Sh. I. Fayziyev, Studying the experience of full breakage on the streets of karshi in kashkadarya region. Intr. J. Adv. Res. Sci., Engin. Tech. **6** (11) 11873-11875 (2019)
10. A. S. Suyunov, Sh. A. Suyunov, G.M. Mullajanova, Analysis of the qualification developmentof natural hazards at geoecological monitiring of landslide hazardeus territiries. Intr. J. Adv. Res. Sci., Engin. Tech. **6** (11) 11534-11537 (2019)
11. A. S. Suyunov, A. A Salahiddinov, Sh. A. Suyunov, Analysis of the influence of the atmosphere surface layer on the measurement made by electronic total stations. Intr. J. Adv. Res. Sci., Engin. Tech. **4** (1) 3098-3101 (2017)
12. I. S. Trivoy, The nature of the temperature field in the city and the phenomenon of anamal lateral refraction. "Geoecological problems of urbanized territories" *(Характер температурного поля городе и явление анамальной боковой рефракции. "Геоэкологические проблемы урбанизированных территорий")* 15-16 (1999)
13. A. S. Suyunov, Sh.A. Suyunov, Refraction accounting methods for geodetic measurements *(Методы учета рефракции при геодезических измерениях)* 158 (2020)
14. O. B. Vshivkova, Profile aggregation as a means of accounting for the influence of the surface layer of the atmosphere with minimal costs *(Профильное комплексирование как средство учета влияния приземного слоя атмосферы с минимальными затратами)* **4** 3-11 (2017)
15. A. S. Suyunov, Sh. A. Suyunov, M. K. Isakov, Geodetic refraction in Central Asian. *(Ўрта осиё шароитида геодезик рефракция)* conditions Materials of the XIV Republican scientific-practical conference. **2** 36-39 (2017)
16. A. S. Suyunov, Sh. A. Suyunov, U. Berdikulov, A new way to take into account refraction in trigonometric leveling in Uzbekistan *(Ўзбекистон шароитида тригонометрик нивелирлашда рефракцияни ҳисобга олишнинг янгича йўли)*. J. Irr. And Mel. **2** 55-57 (2018)
17. A. E. Fedorishchev, L. S. Xijak, Investigation of the temperature field in order to take into account refractive distortions in the boundary layer of the atmosphere. *(Исследование температурного поля с целью учета рефракционных искажений в пограничном слое атмосферы)*. J. Geo. Car. **21** 66-69 (1995)

18. M. Hennes, Zum Refraktionseinfluss auf terrestrische geodдtische Messungen im Kontext der Messtechnik und der Instrumentenentwicklung. Fldchenmanag. Und Bodenordn. **2** 73-86 (2016)
19. W. Gurtner, G. Beutler, The use of meteorological data in large scale GPS networks. Spring Meet, Baltimore. **3** 23-28 (1994)
20. P. Harzer, Berehningder Ablenrungender Lichtetrahlenin der Erdefufrein meterologiech – physikalische Grundlage. Publ. Der Sternwartein Kiel, 24 (1982)

Application of modern geodetic tools in the operation of railway reconstructions

Yuriy Ten[1], *Rustam* Oymatov[2*], *Kholmurod* Khayitov[2], *Guzal* Saydalieva[2], *Ulugbek* Nulloev[2] and *Izzatillo* Nematov[3]

[1]SUE "UZGASHKLITI", 100096, Katartal str., 38, Tashkent, Uzbekistan
[2]TIIAME, 100000, Koriy Niyozy str., 39, Tashkent, Uzbekistan
[3]Karshi branch of TIIAME, 19, Khanabad str., Karshi city, Uzbekistan.

Abstract. This research considers the survey of the circular curve of the operating railway using an electronic total station, which determines the coordinates of points on the curve and calculates the coordinates of the main points of the circular curve. The goal is obtaining information with sufficient accuracy about the spatial position of the parameters of the railway plan, i.e. the device of the railway plan: turning angles, radiuses, lengths of straight lines and curves. Recommendations for determining the elements necessary to locate the main points of curves and their coordinates are considered. The advantage of this method is the simplicity of calculation, which allows the compiled program and high efficiency of determining the main points of curves using a modern geodetic device-the electronic total station Trimble VX (USA).
Keywords: Curve, parameter, program, calculation, angle, total station.

1 Introduction

Currently, in the tasks of inventory of Railways and railway stations, it is possible to create automated databases, which will be based on digital and electronic plans of stations and crossings within the railway right-of-way. For these purposes, modern measurement tools are widely used in the form of electronic total stations, GNSS, laser scanners, unmanned aerial vehicles (UAVs) and aerospace survey methods. They allow you to get the necessary information about the area and objects most objectively, accurately and quickly [9-11].

On Railways, to ensure the safety of train traffic, the technical operation rules provide for constant monitoring of the condition of the track, track devices and structures. The control can be performed using track measuring devices, as well as by conducting special geodetic measurements [4-8].

One of the most important stages of shooting with electronic total station of operated Railways is the shooting of its curved sections. It is performed for the purpose of drawing a curve on the plan, determining the elements of the curve, as well as the values of straightening (straightening) the path to give it a calculated shape [1-3, 13-14].

* Corresponding author: zohid3095@gmail.com

2 Methodology

The correct position of the track in the plan ensures smooth running of trains and the least impact of rolling stock on the track. Therefore, periodically check the curves of the path, and according to the results of the check, they are straightened (straightening). Survey of the planned position of the curve geodesic method is usually performed in three Methods:

1. measuring deflection arrows with a chord length of 10-20 m;
2. using theodolite, method of I. V. Gonikberg;
3. sighting the curve using an electronic total station [10-12].

We will consider the third method - shooting a circular curve of railway tracks using an electronic total station Trimble VX (USA) polar method from the point of the total station with the direct determination of the coordinates of its points on twenty - or ten-meter segments. At each point of the railway track to be removed, a rack with a retroreflector is installed and the distance to this point, as well as the horizontal and vertical angles to it are measured and recorded. These measurements are used to determine the planned and altitude position of these points relative to the total station (Fig.1).

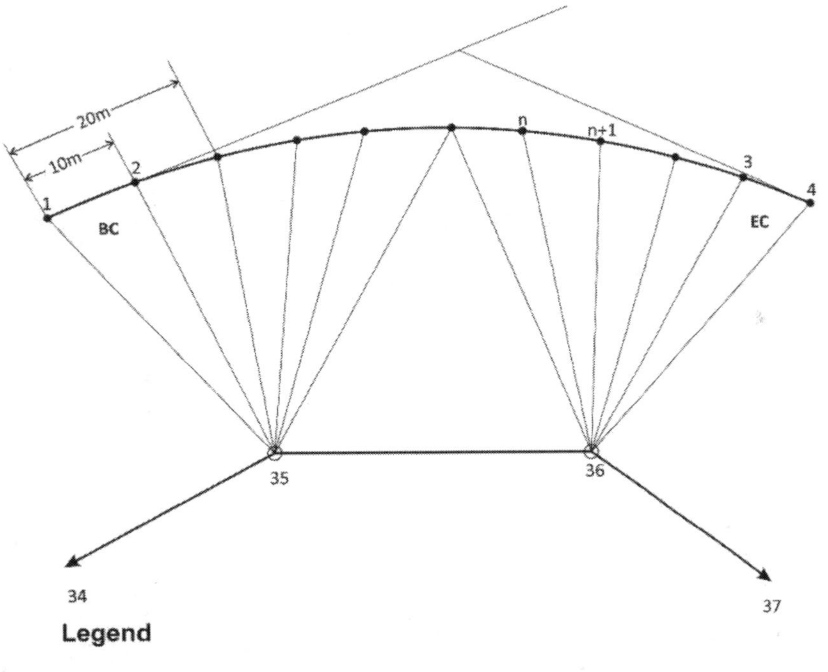

Fig.1. Sighting the railway using the polar method

Staying points of the total station must be determined from the theodolite course of accuracy 1: 3000 to obtain survey results in a single coordinate system and heights. The accuracy of this method of sighting any point of the curve is no more than 5 mm, the average square error of measuring the angle and length of the line is respectively: the angle - 1", line - ± (2 mm+2 mm/km) electronic total station Trimble VX (USA), which makes it easy to bring the shifted curves to design data [10].

Performing survey work with the installation of an electronic total station outside the railway track not only increases labour productivity due to the need to remove the device and clear the way for passing trains with its subsequent installation and performance of work on orientation to the initial direction, but also, most significantly, significantly increases the safety of geodetic activities [11-13].

The coordinates of points are used to determine the elements of the radius, rotation angle, and main points of the circular curve, as well as the coordinates of the circular curve (fig.2).

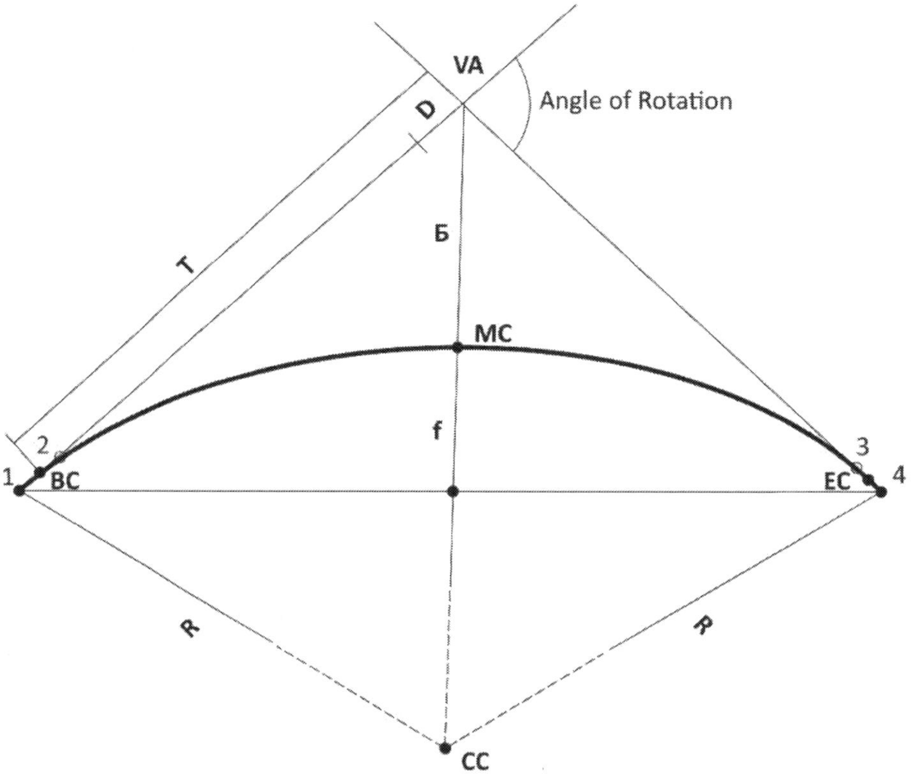

Fig. 2. Scheme of the positions of the main points of curves

Descriptions of the scheme:

- VA – vertex angle of the rotation;
- 1, 2, 3, 4 – measuring points;
- BC – beginning of the curve;
- EC – end of the curve;
- MK – middle of the curve;
- CC – center of the curve;
- f – deflection;
- R – radius of the circular curve;
- B – bisector;
- T – tangent;
- D – domer;
- Line 1 – 4 – string.

The calculation is performed for the case of an asymmetric location of the midpoint of the curve described [4].

The coordinates of the vertex of the rotation angle are calculated using the following formulas:

$$x_{VA} = \frac{(x_2 - x_1) * [x_3 * (y_4 - y_3) - y_3 * (x_4 - x_3)] - (x_4 - x_3) * [(x_1 * (y_2 - y_1) - y_1 * (x_2 - x_1)]}{(x_2 - x_1) * (y_4 - y_3) - (x_4 - x_3) * (y_2 - y_1)}$$

(1)

$$y_{VA} = \frac{(y_2 - y_1) * [x_3 * (y_4 - y_3) - y_3 * (x_4 - x_3)] - (y_4 - y_3) * [(x_1 * (y_2 - y_1) - y_1 * (x_2 - x_1)]}{(x_2 - x_1) * (y_4 - y_3) - (x_4 - x_3) * (y_2 - y_1)}$$

where: x_{VA}, y_{VA} – coordinates of the vertex of the rotation angle;
$x_1, y_1 - x_4, y_4$ – coordinates of the measured curve points.

The angle of rotation of the curve α is calculated using the formulas:

$$A_{2-1} = arctg\frac{y_1 - y_2}{x_1 - x_2} \tag{2}$$

$$A_{3-4} = arctg\frac{y_4 - y_3}{x_4 - x_3} \tag{3}$$

$$\alpha = A_{3-4} - A_{2-1}$$

The distances from the angle of rotation to points on straight sections of the path located near the beginning and end of curves are calculated using the following formulas:

$$d_{(2-VA)} = \sqrt{(x_{VA} - x_2)^2 + (y_{VA} - y_2)^2},$$

(4)

$$d_{(VA-3)} = \sqrt{(x_3 - x_{VA})^2 + (y_3 - y_{VA})^2}.$$

To determine the deflection arrow f, the distances between measurement points 1 – 2 and 3 – 4 are analysed (fig.3). The triangle with the largest distance is taken into consideration. The deflection arrow is calculated using the area of the triangle.

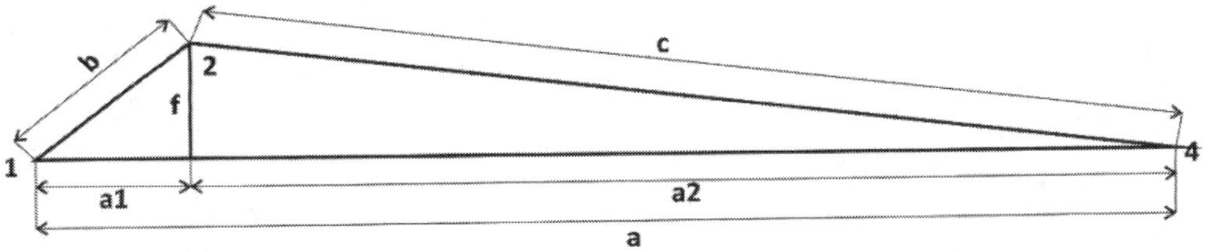

Fig. 3. Scheme for determining f and R

The lengths of the sides of the triangle are determined:

$$d_{(1-2)} = b = \sqrt{(x_2 - x_1)^2 + (y_2 - y_1)^2},$$

$$d_{(2-4)} = c = \sqrt{(x_3 - x_2)^2 + (y_3 - y_2)^2}, \tag{5}$$

$$d_{(1-4)} = a = \sqrt{(x_3 - x_1)^2 + (y_3 - y_1)^2}.$$

The area of the triangle is calculated:

$$S = \sqrt{p(p-a)(p-b)(p-c)}, \tag{6}$$

where: $p = \frac{a+b+c}{2}$.

The value of the deflection boom f is determined by the formula:

$$f = \frac{2S}{a} \quad (7)$$

The value of the radius of the curve is calculated through the deflection arrow f when the midpoint is not symmetrical according to the formula:

$$R = \frac{a_1 a_2}{2f} \quad (8)$$

where: a_1 *and* a_2– distances from the extreme points to the base of the deflection boom.

According to the value of the rotation angle α, the radius of the curve R, circular and transition curves are calculated [6]:

- tangent (T) according to the formula – $T = R\ \mathrm{tg}\frac{\alpha}{2}$;
- length of curve – $K = R\ \alpha$,

where: the angle α in radian indicator;

- domer – $D = 2\,T - K$;
- bisector – B = R (1/ (cos α /2) – 1).

The coordinates of the main points of the curve are determined using the following algorithms:

- Coordinates of the beginning point of the curve BC are calculated from the point VA by the directional angle of the line 2-1 and the length of the tangent T;
- The coordinates of the point at the end of the EC curve are calculated from the point VA by the directional angle of the line 3-4 and the length of the tangent T;
- Coordinates of the centre point of the curve CC is defined as the intersection point of two circles with radius R and centers at points BC and EC;
- The coordinates of the point midpoint of the MC curve are calculated from the point CC by the directional angle of the CC-VA line and the length of the radius of the curve R.

Based on the above formulas, the program "Calculation of railway curve parameters" in the Delphi 7 integrated environment has been compiled [5].

3 Results

To perform calculations using the program "calculation of parameters of the circular curve of the railway", information need to be prepared in the form of a text file. The data separator in the file is a semicolon (;). The sample string line looks like below.

m637;67674,05;16444,84;426,46;

where: m637 – name (number) of the point;

67674,05 – the X coordinate;

16444,84 – the Y coordinate;

426,46 – point mark (may be omitted).

To calculate the curve parameters, select points near the beginning (BC) and end (EC) of the curve. For example: according to the scheme fig.4. there are points 1, 2 and 3, 4. In this sequence they are entered for calculation.

Fig. 4. Scheme of the shooting curve

Applying program in the calculation is performed in the following order:

1. Run the program (see fig.5);

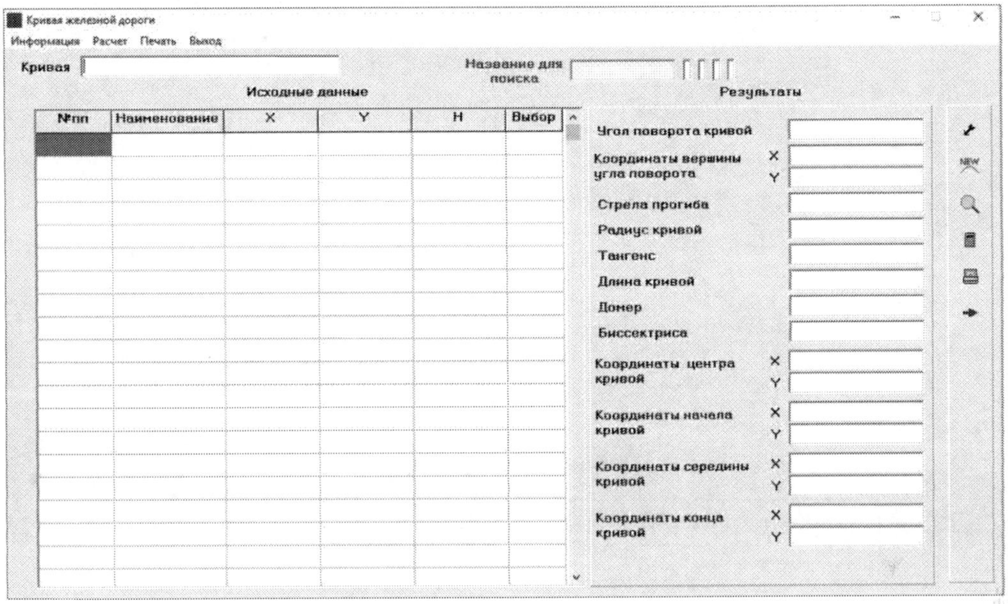

Fig. 5. The interface of the program

2. Enter information (Главное меню – Информация – Чтение данных) or select on the right side of the panel. Information can be entered directly into the table on the screen (for each item: name, coordinates X and Y);
3. Type the name of the item in the "Название для поиска" window and select in the panel. The required item will be selected from the table, and it will be assigned the sequence number required for the calculation. To the right of the "Название для поиска" window, there is an indicator that indicates how many points are selected;
4. Select (Главное меню - Расчет) or on the panel calculate the curve;
5. After analyzing the received data, we print the results (Главное меню - Печать) or on the panel . The curve name is formed from the names of points in the Curve window (fig.6 and tables 1-3). This name can be changed at will;
6. To calculate a new curve, select (Главное меню – Информация – Новая кривая) or NEW on the panel. Then follow the steps starting from point No 2 or No 3.
7. Exit from the program is performed (Главное меню - Выход) or on the panel.

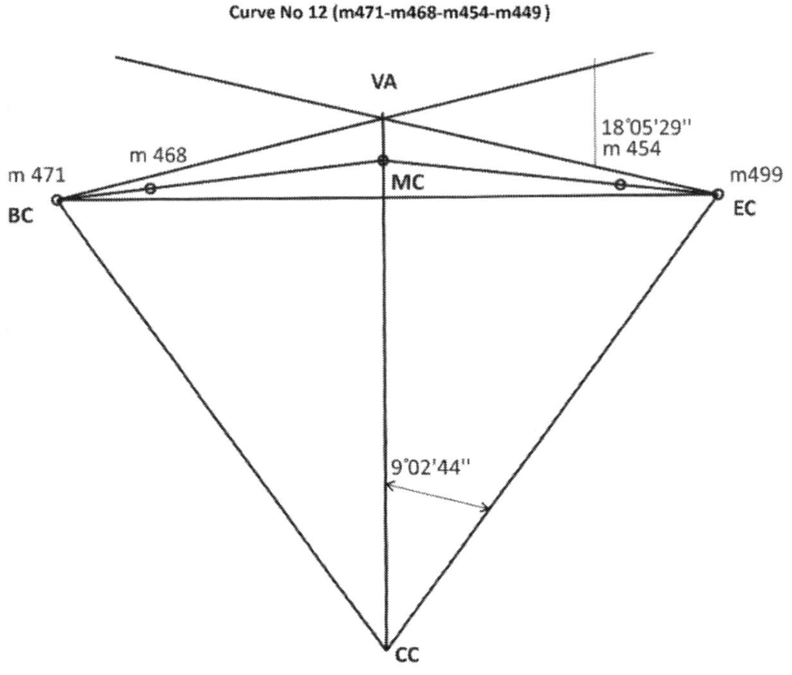

Fig.6. Calculation of circular curves (example)

Table 1. Entered data

Name or number of points	X	Y
m468	67 703.92	16 439.79
m471	67 696.77	16 439.16
m454	67 733.44	16 437.94
m449	67 744.35	16 435.41

Table 2. Result of calculation

Rotation angle of way	$18^0 05'29''$
Radius of the circular curve	122.57
Tangent, T (m)	19.51
The length of the curve K (m)	38.70
Domer D (m)	0.32
Bisector of curve B (m)	1.54
Chord length (m)	40.67
Deflection arrow f (m)	1.34

Table 3. Coordinates of the main points of the circular curve

Name of points	X	Y
Vertex of the rotation angle, VA	67 719.53	16 441.17
Center of the curve, CC	67 710.85	16 317.36
Beginning of the curve, BC	67 700.09	16 439.45
Middle of the curve, MC	67 719.42	16 439.63
End of the curve, EC	67 738.54	16 436.76

4 Discussion

Obtaining information with sufficient accuracy about the spatial positions of the parameters of the railway plan, i.e. the device of the railway plan: turning angles, radiuses, lengths of straight lines and curves. The advantage of above mentioned method is the simplicity of calculation, which allows the compiled program and high efficiency of determining the main points of curves using a modern geodetic device - the electronic total station.

References

1. KMK 3.06.05-98, *Railway, road and hydraulic tunnels. Metros. Production and acceptance of works* (1998);
2. ShNK 1.02.07-15, *Engineering and technical surveys for construction.* Fundamentals (2015)
3. V. I. Poletaev, E. G. Tolstov. *Sighting railway curves,* Saint-Petersburg, p. 37 (2011)
4. A. Khomonenko, V. Hoffman, E. Meshcheryakov, V. Nikiforov, *Delphi 7,* Saint-Petersburg, pp. 1201 (2003).
5. V. N. Ganshin, L. S. Khrenov. *Tables for splitting circular and transition curves. 4th ed.*, Kiev, p. 432 (1974)
6. A. Inamov, J. Lapasov and S. Hikmatullaev. *Engineering surveying,* Tashkent, p. 200 (2017)
7. E. Nurmatov, U.Utanov. *Geodesy,* Tashkent, p.234 (2002)
8. Sh. Avchiev. *Applied Geodesy*, Tashkent, p.166 (2007)
9. T. Abdullaev, U. Islamov, U. Mukhtorov and A. Inamov. *Advanced Geodesy,* Tashkent, p.224 (2017)
10. W .Schofield. *Engineering surveying*, Oxford, p.534 (2001)
11. R. Abdiramanov, A. Ninamov, A. Dadaboeva. Internauka. 12 **(46)** (2018)
12. S. Tashpulatov, B. Nazarov, N. Shavkatova. *Modern geodetic instruments*, Tashkent, p.300 (2018)
13. H. Muborakov, S.Ahmedov. *Geodesy and Cartography*, Tashkent, p.304 (2002)
14. Z. Okhunov. *Proctices from Geodesy*, Tashkent, p.200 (2009)
15. S. Tashpulatov, U. Islamov, A. Inamov. *Space geodesy,* Tashkent, p.121 (2018)

Improvement of the method for comparing subsidence of structures using the Fischer's F-test and the Foster-Stuart test

Abdusali Suyunov[1,*], *Shukhrat* Suyunov[1], *Malika* Aminjanova[1] and *Kamola* Rakhmatullaeva[1]

[1] Department of Geodesy and Cartography, Samarkand State Architectural and Civil-Engineering Institute, Lolazor 70, 140147, Samarkand city, Uzbekistan

Abstract. To improve the quality of construction and increase the durability of engineering structures under construction, complex geodetic works should be performed, including geodetic observations of deformations of structures. These observations are carried out during the construction of buildings and structures and their operation, mainly before the period of deformation stabilization. In this regard, a reliable statistical definition of deformations close to the limit is necessary, based on the data of geodetic observations. The research helps to improve the definition of deformations of structures using the Fischer's F-test and the Foster-Stuart test, based on analysis of the measurements of horizontal and vertical monitoring of industrial structures. According to the results, the magnitude of the subsidence plays a more significant role from than its absolute value, thus the value of the deformation intensity is of primary importance in justifying observation periodicity.
Keywords: deformation analysis, dispersion, Cochrane criterion, Foster-Stewart test, subsidence of structures, Fisher's F-test

1 Introduction

In General, the term "deformation" refers to a change in the shape of the object of observation. In geodesic practice, it is customary to consider deformation as a change in the position of an object relative to the original one.

Under constant pressure from the mass of the structure, the soils at the base of its foundation are gradually compacted (compressed) and a displacement occurs in the vertical plane, called the settlement of the structure. In addition to pressure from its own mass, subsidence (sediment) can be caused by other reasons: karst and landslide phenomena, changes in the level of groundwater, the work of heavy mechanisms, traffic, seismic phenomena, etc.

To improve the quality of construction and increase the durability of the engineering structures being erected, complex geodetic works should be carried out, including geodetic observations of the deformations of structures. These observations are carried out at the stage of construction of buildings and structures and at the stage of their operation, mainly

* Corresponding author: suyunov303@inbox.ru

before the period of deformation stabilization. In this regard, a reliable statistical determination of deformations is necessary, close to the limiting ones, according to the data of geodetic observations [5].

The method of performing geodetic works when observing subsidence and deformations of engineering structures is well studied and reflected in the works of Uzbek and foreign scientists geodesists [1, 2, 4], as well as in the relevant regulatory documents [7, 8]. According to most recent investigations, the precise observation of the subsidence (vertical monitoring) is very critical not only from the precision of the observation technology, but also from the variations of the environment, such as temporal variations of the reference horizontal level, or the geoid [9, 10].

In this article, vertical monitoring analysis, thus repeated measurements of subsidence of a construction are compared to each other is performed, using the Fisher's F and the Foster-Stewart tests, where the choice of time interval between the measurement campaigns depends on the type of structure, the period of its operation, the rate of deformation change among other factors.

On average, during the construction period, systematic observations are performed once or twice in a quarter, and during the operation period once or twice a year. In case of urgent observations, they are performed before and after the appearance of a factor that dramatically changes the usual course of deformation [1].

2 Methodology

The method considered in the article was used to analyse the results of geodetic observations of the subsidence of an industrial complex structure.

To determine a sufficient condition for the stability of a structure using Fisher's F-test, it is necessary to determine the standard deviation of the differences $\Delta(k)_i$ from the average unevenness

$$m^2_\Delta = \sum_{i=1}^{n-1} (\Delta(k)_i - \Delta(k))^2 / (n-2) \tag{1}$$

and compare it with the general variance estimate

$$\acute{m}^2 = \sum_{i=1}^{n} m^2_{Si} / n \tag{2}$$

According to Fisher's criterion [3]. The uneven settlement of the structure in this case is recognized as identified if at the level of significance p the inequality is fulfilled [2]

$$(m^2_\Delta / \acute{m}^2) > F_{1-p}, \tag{3}$$

where, F - distribution is considered with f1 = n-1 and f2 = n degrees of freedom.

Expression (3) - one-way analysis of variance - is valid only when the general variance of reproducibility σ2 is the same at all levels n. To test this assumption, it is necessary to consider the ratio of the maximum variance to the sum of all the others (Cochran's test) [4]

$$g = m^2_{Si} / \sum_{k=1}^{n} m^2_{SK} - m^2_{Si} \tag{4}$$

The distribution of g depends on n and r (the number of degrees of freedom - redundant measurements in the levelling network). In the case of a significant change in the estimates of the variances m^2S_k $(1 \le k \le p)$, e. $g > g_{table}$, it is necessary to select the appropriate transforming function [4].

Another useful criterion for identifying patterns in the differences in elevation or settlement of adjacent structural marks is the Foster-Stewart test [6]. A distinctive feature of this method is that it allows to simultaneously identify significant changes in variance estimates and determine possible trends on average. The test is based on two characteristics

$$[c] = \sum_{i=1}^{n} c_i, \tag{5}$$

$$[d] = \sum_{i=1}^{n} d_i \tag{6}$$

in which the values of c_i and d_i are calculated by the formulas

$$c_i = u_i + l_i, \tag{7}$$

$$d_i = u_i - l_i \tag{8}$$

The u_i and l_i values are determined by sequentially comparing the benchmarks on structures, and

$$u_i = \begin{cases} 1 \text{ for } H_i > H_{i-1},\ H_{i-2}, \ldots H_1, \\ 0- \text{ in other cases,} \end{cases} \tag{9}$$

$$l_i = \begin{cases} 1 \text{ for } H_i > H_{i-1},\ H_{i-2}, \ldots H_1, \\ 0- \text{ in other cases,} \end{cases} \tag{10}$$

When $c_i \in (0,1)$, and $c_i = 0$ if H_i is not an extreme among all previous values, in the opposite case $c_i = 1$ [5], i.e.

$$0 \leq [c] \leq n - 1 \tag{11}$$

Comment 1. If the elevations (subsidence) of all points are equal, ($H_1 = H_2 = \ldots H_{n-1} = H_n$ or $S_1 = S_2 = \ldots = S_{n-1} = S_n$), then $[c] = 0$, if they change monotonically or their oscillations alternate, systematically increasing or decreasing, then

$$[c] = n - 1. \tag{12}$$

In turn, the quantity $d_i \in (0, 1, -1)$, whence

$$-(n - 1) \leq [d] \leq n - 1. \tag{13}$$

Comment 2. The lower limit corresponds to a monotonically decreasing, and the upper one - to a monotonically increasing series of elevation values (subsidence).

Cases are interesting for practice if the value $d = 0$:

1) $H_1 = H_2 = . . . = H_n \ (S_1 = S_2 = . . . = S_{n-1})$,

that $\sum_{i=1}^{n} ui = 0, \quad \sum_{i=1}^{n} li = 0$

(which indicates the complete absence of uneven settlement of structures);

2) $\sum_{i=1}^{n} u_i = \sum_{i=1}^{n} l_i$ - two periods in the behavior of structures of this structure

with opposite deformation tendencies;
3) $d = 0$, when the motion (uplift or subsidence) of the H_i (S_i) levels alternate [8]. It should be noted that the indices *[c]* and *[d]* are asymptotically normal and have distributions independent of each other (it is obvious that their distributions depend only on the order of arrangement of benchmarks on the structures) [7]. The indicator *[c]* is used in this case to statistically detect changes in variance, while *[d]* is used to determine possible trends in a trend. Testing the hypothesis about whether the differences *(d - 0)* and *(c - M)* can be considered random (*M* is the expectation value of *c* for a random distribution of benchmarks on a structure) is carried out using the Student's t-test

$$t^{(2)}{}_{1-p} = \frac{[c] - M}{\sigma_1}, \qquad (15)$$

where, the values σ_i $(i = 1, 2)$ are determined from the expressions

$$\sigma_1 \approx \sqrt{2 \ln n - 3.42532}, \qquad (16)$$

$$\sigma_2 \approx \sqrt{2 \ln n - 0.8456} \qquad (17)$$

3 Test and results

The method introduced above was used to analyse the results of surveying/geodetic observations performed in order to moitor horizontal and/or vertical motions (deformation analysis) of industrial complex structures. In the frame of this study, subsidence of a pile-founded, reinforced concrete building has been performed.

Table 1. Initial values of research at this stage

H	M	σ_1	σ_2		H	M	σ_1	σ_2
10	3,858	1,288	1,964		26	5,672	1,800	2,379
12	4,125	1 ,243	2,027		28	5,831	1,841	2,413
14	4,392	1,361	2,105		30	5,990	1,882	2,447
16	4,659	1,456	2,153		35	6,294	1,956	2,509
18	4,927	1,535	2,216		40	6,557	2,019	2,561
20	5,195	1,602	2,279		45	6,790	2,072	2,606
22	5,354	1,660	2,313		50	6,998	2,121	2.645
24	5,513	1,712	2,346		60	7,360	2,201	2,713

The foundation inside the frame was built in the form of a monolithic reinforced concrete slab, lying on the soils of the poured base. The equipment installation technology has imposed strict requirements on the stability of the structure's base: the base roll tolerance was 0.001.

As a preparatory analysis, the effect of the magnitude of the subsidence has been investigated by the Fischer's F-test. Assuming different amount of the subsidences, the values of M, σ_1 and σ_2 are estimated, and provided in Table 1.

According to the results, the value of the subsidence intensity plays a more significant role from the point of view of judging the ongoing process than its absolute value, which leads to the conclusion that the value of the deformation intensity is of primary importance in justifying the accuracy of the observation periodicity.

Observations of deformations show that each individual tolerance value cannot serve as a consistent assessment of the stability of a structure. Therefore, for a complete judgment about the stability of the structure, the observation technique must be developed based on the complex effect of all characteristics that have limiting values.

References

1. V. I. Bandurka, N. A. Loshkarev, J.Geodesy and aerial photography, **4** (1972)
2. P. I. .Bright, E. N. Medvetsky, *Measurements of sediments and deformations by geodesic methods.* (1959)
3. T. A. Gura, A. O. Biryukova, J. Young scientist. **30** (2016)
4. E. I. Pustylnik, *Statistical methods of analysis and processing of observations*, (1966)
5. A. S. Suyunov, M.B. Aminjanova et al., *ICT youth and their social well-being* (2013)
6. E. M. Chetyrkin, *Statistical methods of forecasting* (1975)
7. GOST 8.207-76. GSI. Direct measurements with multiple observations. Methods of processing the results of observations. Fundamentals.
8. *Guide to observations of deformations of the foundations of buildings and structures.* (Moscow, 1975)
9. Cs. Égető, L. Földváry, T. Huszák, *The effect of tunnelling on repeated precise levelling measurements for vertical deformation control of the Metro4 project*, Journal of Geodetic Science, 3(2): 95-102 (2013)
10. Cs. Égető, N. Rehány, L. Földváry, *Variations of the gravity field due to excavations of the Budapest Metro4 subway line*, Periodica. Polytechnica Civil Engineering, 58(2): 131–136 (2014)

Advantages of formation non-agricultural land allocation projects based on GIS technologies

Allabergan Babajanov[1,*], *Rashid* Abdiramanov[2] , *Ilhom* Abdurahmanov[1], and *Utkir* Islomov[2]

[1]Tashkent Institute of Irrigation and Agricultural Mechanization Engineers (TIIAME), Land Use Department, 39 Kari Niyazov Street, Uzbekistan
[2]Tashkent Institute of Irrigation and Agricultural Mechanization Engineers (TIIAME), Geodesy and Geoinformatics, 39 Kari Niyazov Street, Uzbekistan

Abstract. Based on the collection and study of official data, the main criteria for selecting a land plot using GIS (Geographic Information System) technologies for the construction of a water treatment plant are presented. The main stages of data analysis are determined when solving the question of the location of the building using ArcGIS, ArcMap, ArcCatalog and it is scientifically proven that the use of such programs when choosing land plots for such types of construction can save in large amounts of material and labour resources. Together with these, in the conducted study, a methodology was developed and presented for the formation of a database necessary for compiling similar GIS projects, as well as a methodology for a feasibility study for the implementation of land allotment projects for the construction of various economic facilities. The advantages of using GIS technology for non-agricultural land use projects, which are the main form of inter-farm land use, which is a modern measure aimed at rational land use, are described. Key factors and key indicators, as well as expected project results.
Keywords: land management, design, land allocation, land selection, land use, geographic information systems (GIS) technologies.

1 Introduction

In the system of organization of land use in the Republic of Uzbekistan, a special role is given to inter-farm land management, since this is the creation of various agricultural or non-agricultural land or land use, the boundaries of existing ownership of agricultural land, for various purposes. Allocation of land plots, allocation of corresponding changes in the boundaries of territorial units, allocation of special land plots for large hydraulic structures and structures. A number of measures have been taken to protect the soil and the environment.

Of course, all this is based on existing and comprehensive projects. Therefore, in the shortest possible time, the use of GIS technologies for the development of integrated (economic, social, technical, environmental) interagency land management projects, for

* Corresponding author: r.abdiramanov@mail.ru

example including the land allotment projects for sewage treatment plants in large cities, is particular importance. Because the general features of such a complex project, including a specific goal, determination, efficiency, the need to take into account a large number of interrelated factors and their coordination, require an integrated approach to time and financial constraints.

Recognizing that improving the supply of clean drinking water to the country's cities and villages is one of the main directions of the republic's domestic policy, the need for comprehensive land allocation projects for the construction and operation of such stations is once again demonstrated.

The purpose of the study is to identify the advantages of using GIS technology in the development of inter-farm land management projects related to land allocation and comprehensive justification for the construction of a wastewater treatment plant, which provides for wastewater treatment and increasing the efficiency of drinking water use.

2 Problem statement

Geo-information system technology is rapidly entering for all aspects of human society. Today, it remains difficult to find an area where these technologies have not penetrated. Its rapid penetration allows us to find the simplest and most effective ways to positively address the existing problems in this area. Therefore, the development and implementation of various projects based on or using such technologies is a great practical importance for sectors of the economy. Thus, the theoretical study and research of control and monitoring of GIS-based projects and the necessity for these measures form the basis of the following issue.

3 Research methods

Today, the design and research projects based on GIS technologies, which are implemented in Uzbekistan by various design or research organizations, are becoming increasingly important, since GIS contributes to financial and labour costs. GIS is considered as a modern computer technology that can be used for mapping and analysis of real objects, as well as for investigating events in the life and activities of the population living in this area. [1,2,3]. Today, GIS is successfully used in various areas of our lives, especially in industry, transportation services, agriculture, and location of complex engineering communications in the regions, field development and management of such activities [4]. Therefore, the proper planning of GIS projects related to inter-farm land use, and especially the allocation of relatively large land plots for non-agricultural purposes, will ensure the effective development of inter-farm land management projects and their effective implementation tomorrow. Thus, when developing and justifying inter-farm land management projects using of GIS technologies, we apply the methods of variability, analytical analysis and monographic research.

4 Research results and discussions

When planning any GIS project, the project goal is determined, the project database is formed, and the GIS functions are created using the analytical model to achieve the project objectives and analysis results, that is, the project consists of several stages. Therefore, it would be advisable to take several steps in GIS projects for the allocation of land for the construction of large urban wastewater treatment plants.

Step 1. Definition of the main aim. This is the initial step in the design process, the purpose of which is to determine the main goals and objectives. To define tasks, the following questions must be answered: What is the task? How is this solved today? What are the alternative ways to solve GIS? What should be the end result of the project? Is there enough data and maps for a quality project? Who are the users of the design results? Do others use this design information? What are the design requirements?

This step is important for GIS-based land acquisition projects, as the answers to these questions will determine the scope of the project and what methods should be used for the project.

Step 2. Creating a project database. Information is the basis of the system, without which the system does not work and does not exist at all [5]. This means that to develop a project, it is necessary to collect information and create a database. It should consist of three stages: creating a database, entering and loading the collected data, and managing the database.

Creating a database includes analysing the spatial data that we need for analytical requirements, determining the necessary attributes of objects, determining the boundaries that need to be investigated, and determining a coordinate system. Data entry includes the digitizing previously collected and processed information and changing them to other formats that we can use, as well as data verification and error correction. Defining a coordinate system and combining adjacent layers represent database management. As a rule, creating a database is the most complex and time-consuming part of any project, especially a land acquisition project, as it provides the information needed to select a land plot, including area, topography, soil, and the previous owner, environmental conditions, the availability of engineering infrastructure in the area and a number of other data. The accuracy of the analysis results is determined by the accuracy of the data used for this period.

Step 3. Data analysis and evaluation of results.

It is well known that data analysis using GIS technology ranges from simple mapping to complex spatial modelling. According to scientific sources, the model is a reflection of the real world used to model processes, predict results, or analyse problems. The use of one or three categories of GIS functions in a number of spatial data leads to spatial modelling. Such GIS features include:

- geometric modelling functions;
- measuring distances, creating buffers and calculating areas and perimeters;
- cost modelling functions;
- integration of data sets to search for graphs of values that overlap;
- approximation modelling functions
- separation of location, search for ways to change it, positioning.

The results of the implementation of GIS projects for inter-farm land use, especially for non-commercial land use, show that quick and accurate results of previously impracticable or manual long-term analysis (land selection, project allocation, and multiple allocation of land) will be possible. Several alternative land acquisition projects can be created by changing methods and GIS usage parameters and repeating the analysis.

Evaluation of the results obtained in the analysis process is also one of the important steps of such projects. In this case, the use of economic-statistical methods, in particular, the method of correlation, which reflects the correlation of the results with the factors, has a positive effect.

Step 4. Presentation of the results.

The results of the analytical work performed the using GIS technology in the process of inter-farm land use, in particular, non-agricultural land distribution, are often reflected on the map of the land management project. However, when using a GIS, it is important to know the charts and report types for the selected data reflection data. You can create graphs, charts and reports separately, insert them into documents created in other applications, or place them separately on maps.

Thus, we have already learned about the successive stages of the development of inter-farm land use projects, especially non-agricultural land use projects using GIS technologies. Based on this introduction, we can now begin planning GIS projects at specific sites. It is desirable that project planning be carried out with reference to an object clearly identified at the above stages. With this in mind, we will consider the use of GIS technology in the design of a site for a new wastewater treatment plant in Tashkent. This work usually begins with planning.

Planning is an important step in any project [6,7,8]. This helps to spend as a little time and effort as possible on the development, analysis and display of the required databases. The stages of the project will determine the main objectives of the project, the areas of analysis and the data necessary for the analysis. At the same time, it is advisable to choose approaches to analysis and the type of final product.

Once this is done, it is advisable to begin the process of developing a project database that will be developed, and then work on small GIS analytical projects. You can also learn how to plan large GIS projects and use ArcGIS, ArcMap, and ArcCatalog programs during such projects. The choice of location for new construction, the steps that will be taken to conduct certain types of analysis, and the tools used will be typical of most GIS projects. In the above example, the main content of the project is the selection of a land plot suitable for the construction of a wastewater treatment plant.

As you know, the city of Tashkent is growing and expanding every year [9]. The construction of a new wastewater treatment plant and the improvement of secondary water supply are some of the issues that need to be addressed in order to ensure its further development at such a high rate. The municipality plans to protect water resources and increase the level of secondary water use to meet expected water consumption.

The main purpose of this GIS analysis is to identify and select a plot of land suitable for a wastewater treatment plant. The city administration has never used GIS technology to solve this problem.

The location of the existing wastewater treatment plant was established many years ago using a paper map with transparent diagrams on it and consultations with the chief architect of the city and other leading experts. This process was timely, but required considerable effort and time and did not allow the general public to make decisions. Currently, the problem is becoming more complicated due to the growth of construction in the city and increasing requirements for the environment and human health. The city administration decided to use GIS technology to speed up the process and ensure compliance with all requirements.

The municipality understands that the location of wastewater treatment plants is the most discussed issue, and it wants to identify all suitable sites and then the most suitable plots based on more stringent criteria. There is wider public participation in the discussion of all possible land plots. Therefore, you need a special map to clearly indicate which plots are suitable for this purpose. The government recommended a list of eligibility criteria. A review of the official data obtained showed that such criteria include:

- it should not be above 400 m above sea level to prevent the flow of water;
- it must be located outside the drinking water zone to ensure the normal discharge of fresh water;
- it should be at least 1000 m from the river to minimize the cost of flow through the pipeline;
- treatment facilities should be located at least 150 m from residential areas and parks in order to reduce discomfort for urban residents;

- Where possible, there should be a free, non-agricultural plot of land for the station.

In addition, the role of the land selected and allocated to reduce future costs should be:

- at least 500 m from the sewer networks;
- at least 50 m from the road network

Fig. 1. Land allocation for a wastewater station based on GIS technology.

Calculations showed that at least 20 hectares of land would be required to locate the station. To determine its location, you must first study the maps and other materials to determine the location of the site. This area will be the project work area. Here, the analysis of GIS technology allows you to combine all the criteria for finding the right plot of land for the construction of the station (figure 1).

The next step in a GIS land acquisition project is to create a project database.

Creating a database for a land acquisition project would be desirable, in our opinion, in two parts. The first part is designed to collect and analyze existing data, and the next section is to prepare data for analysis. Only accurate data can be used, while others require additional processing. It is important to note here that the process of searching for the necessary information and its analysis based on GIS is very important, since it is the key to a positive solution to the problem. Typically, there are several steps in preparing data for analysis. They can be divided into:

- verification of data quality (accuracy and relevance of data);
- change the data format;
- data entry, numbering, scanning, modification, geographical reference;
- determination of the coordinate system;
- providing a new coordinate system;
- Joining adjacent layers.

The use of basic GIS technologies, in particular ArcGIS and ArcCatalog, has a positive impact on all these stages. In addition, data processing using these programs will allow for accurate analysis from both a scientific and a practical point of view, as well as make scientifically sound decisions. Processing data on land acquisition for the construction of a new wastewater treatment plant using the ArcGIS program showed that it was 12 km from the city. It was found that an area that is not suitable for agriculture, 2 km from the existing station, is suitable land for the construction of such a station. The location of the new station in this section is fully consistent with the above requirements and does not have a negative impact on the environment.

It should be noted that the allocation and full grounding of such lands will take almost 30 days in traditional conditions. The use of GIS technology has reduced the time required to solve this problem by 15 times. This means that, of course, with a reduction in time, material costs will also decrease by several times.

Of course, the stages of processing and analysis of GIS data for the design and allocation of land for the construction of a sewage treatment plant and recycling in Tashkent are almost identical to all similar land management projects. Therefore, we can assume that the importance of such technologies in the future is extremely important.

Of course, it is important not only to develop projects, but also to justify them. A GIS-based feasibility study is the project's feasibility study. This document is primarily necessary for the investor, as the investor makes the final decisions on capital investments for the creation and restoration of fixed assets, based on this document. An investor is always thinking of investing in his or her unused cash flow in one sphere. It is important for him that he invested his money in the project, how important this project is for him, and how long the funds invested in the project will begin to benefit. These issues will be considered using the feasibility study of the project, in particular, the inter-farm project for land management.

If capital investments in fixed assets are fully or partially financed from the state budget or its extra-budgetary funds and own financial resources of state enterprises, then the feasibility study is a mandatory document. The decision on the feasibility study of projects financed from other sources of financing is independently made by the investor (customer). Based on the results of our study, we tried to determine the primary data necessary for the preparation of a feasibility study for the implementation of a GIS project related to the allocation of land for the construction of an industrial enterprise. This list can be found in the feasibility study for other GIS-based projects:

1. Forecast of data on the long-term socio-economic development of the republic (for example, by 2030).
2. A copy of the decision of state and local authorities on the construction of the facility.
3. Information about the natural conditions, infrastructure, recreational and protective zones of the territory on which the land plot is allocated for construction.
4. Information about the possibility of using imported equipment at the facility.
5. Information on various environmental impacts resulting from the operation of the facility under construction.
6. Information on determining the optimal capabilities of a facility under construction.
7. The results of previous scientific and experimental design work on technological processes, devices and their improvement at the construction object.
8. Design and estimate documentation approved in the established manner, indicating the construction area of the facility and the location of the facility that will be connected to networks and communications.
9. Information on buildings and structures that will be demolished as a result of construction of the facility on the allotted land, information on the number of resettled citizens and families.
10. Other information describing the characteristics of the facility under construction.

As a rule, the feasibility study of the project is intended to provide creditors or organizations with credit guarantees, financial stability and solvency, as well as tax and other incentives, subsidies, etc [10]. The following is a summary of the feasibility study of the land allocation project:

• the basis for the development of a feasibility study;
• purpose of investment;
• description of the investment object;
• information on the status of resources involved in the economic activities of the future enterprise;
• initial assessment and research results;
• information about the client, investor, potential contractors, suppliers of construction materials, as well as the contractor of the feasibility study;
• location of the enterprise;
• basic technical solutions (options);
• environmental impact assessment;
• human resources and social development;
• schedule for the implementation of the land allotment project;
• economic evaluation and analysis of the project;
• conclusions and recommendations;
• applications (engagement documents, graphic materials)

In the conditions of modern market competition, it is important to evaluate and justify the implementation of the project. In our opinion, this rule is especially important for GIS projects, since geographic information systems are modern, complex and highly accurate in tasks, and therefore a special business plan will be developed to support them.

It is well known that the economic and social reforms carried out in the country in recent years are associated with the development of entrepreneurship and the creation of various production and service capacities based on large-scale investments in the economy. This, in turn, is directly related to the selection, distribution and legalization of suitable land for such facilities. At the same time, it is very important that these facilities are not damaged in the agricultural sector and in engineering infrastructure. Thus, the use of GIS technologies, such as those described above, to more effectively solve these problems associated with land acquisition and separation, is very effective because it allows you to quickly analyze a large amount of GIS data and identify the best solutions, as well as the feasibility of the project. It is possible to develop a basic version.

5 Conclusion

Thus, it can be concluded that the implementation of inter-farm land management measures, especially the selection, distribution and registration of land for large non-agricultural purposes, with GIS-based projects can take some time and effort. This allows you to justify the allocation of land. At the same time, these technologies make it possible to draw sound conclusions, develop several options for choosing land and distributing land, as well as put into operation such projects and ultimately increase the efficiency of land use in the area.

The authors would like to express their deep gratitude to the organizers of the scientific conference in advance for acquainting and accepting the results of this modest scientific research on the development and justification of inter-farm land management projects based on GIS technologies.

References

1. E. Safarov, *Technology of geographic information systems.* Tashkent University, (2009).
2. L. Guryanova, *Downloaded to GIS.* Mn., BSU, (2009).
3. I. Lurie, *Geo-information Cartography.* M., Moscow State University, Book House, (2008).
4. S. Saidkasimov, *Technology of geographic information systems.* Tashkent, Economics Finance, (2011).
5. S. Lipsky, The problem is that all your problems are in the background. M., IP Kim L.A., (2019).
6. A. Babajanov, *Project organization and management.* Tashkent, Economics and Finance, (2016).
7. V. Shapirov, *Modernization of projects* / Ed. I.I. Mazur, – 6th edition - M.: Omega-L, (2010)
8. *A Guide to the Project Management Body of Knowledge.* Project Management Institute, (1996).
9. *National Encyclopaedia of Uzbekistan.* Volume 7. Tashkent, "National Encyclopaedia", (2006).
10. S. Avezboev, S. Volkov, *Land management design (Textbook).* Tashkent, New Generation, (2004).

E3S Web of Conferences **227**, 05002 (2021) https://doi.org/10.1051/e3sconf/202122705002
GI 2021

Improvement of cadastral information provisioning system in an administrative region

Kosimdjon Rakhmonov[1], *Mokhigul* Abdurakhimova[1,*]

[1]Tashkent Institute of Irrigation and Agricultural Mechanization Engineers (TIIAME), State Cadastre Department, 39 Kari Niyaziy Street, Uzbekistan

Abstract. Improvement of the land cadastre information system in the administrative district will make more efficient use of the land fund. More than 42,000 land users in Kibray, which is the object of the research, can quickly manage land resources through the creation of land information systems. As a result, the use of specialized software at the district level increases the quality of the specialists' work compared to traditional methods, and time consuming 3 and 5 times the amount of data transmission to consumers. The collected and regulated data will serve as the main source of the unified system of land registration in Kibray district and the creation of a special land fund and its rational distribution among land users. The method developed to compile a digital cadastre map based on an example of an administrative district is the basis for the registration of land parcel and, in turn, the land fund distribution. Information on the state of land resources is mainly collected in the State Land Cadastre. Registration of the rights to the land parcel, the main component of which is the re-registration of new land users and the existing ones, requires the most up-to-date information on the land cadastre.
Keywords. Land cadastre, land fund, land database, land records, land balance.

1 Introduction

In the world practice a particular attention is paid to the problems of realization of many scientific and theoretical matters on the creation of a database. Since the period of automated systems development, theoretical problems have remained as the main driving force. Despite the technological challenges in the country, including in administrative districts, there are many problems in the implementation of these systems and the implementation of land information management using land cadastre data. The process of constantly changing the land fund, the area of land categories, the number, boundaries, area, and purpose of land users and their land parcels in all administrative districts of the Republic necessitates the creation and operation of the land cadastre system. The purpose of developing land and information databases associated with maintaining the land cadastre in the district is to create a software package that allows the processing, storage, analysis, and transfer of information on a particular land user, as well as individual facilities located

*Corresponding author: mohigul_9429@mail.ru

on land parcels. It is well known that at the district level, the process of land use management is the main source of the land cadastre database. This entails the systematic formation of information about the district land fund. While preparing the data of land cadastre, regional land resources and department of state cadastre are the main sources in the formation of land- information systems.

2 Materials and methods

As professors A. A. Varlamov and S. A. Galchenko described, "The Land Cadastre is one of the information resources of the state, which is a system of information reflecting the legal, natural and economic status of the land fund" [1,2]. For this reason, it was determined that detailed information on the use of district land resources would be needed if the survey did not provide sufficient data. This will give a clear impetus to the establishment and improvement of the land information system in maintaining the land cadastre of the administrative district.

Professor A.S. Chertovitsky and A.K.Bazarov have described it as "The cadastral information systems have several features that are inherent to it: the complexity of cadastral indicators, the provision of information in two types - text and graphics, capabilities and necessity of an automated system." In order to create this concept of the land cadastre information system in the administrative districts at the level of modern requirements, it is necessary to address the following issues:

- Provision of public authorities with information on land use;
- Formation of the district land balance and statistics on the status and use of land resources at the level of modern requirements
- Organization of work on registration of rights to the land parcel by all land categories;
- Achieve proper registration of documents certifying land use and ownership, long-term use of land, lease of land and private ownership of land;
- Preparation of documents for preliminary agreement on the allocation of economic facilities, withdrawal and allocation of land, entry of land into urban areas and changing the boundaries of administrative-territorial units;
- Achievement of development goals of urban and rural areas, comprehensive development of agriculture, regulation, and coordination of land and water relations;
- Improved land records and land balance and the normative value of agricultural land;
- For proper implementation of goals of compensation of losses and losses on land acquisition;
- To encourage the rational use of land resources and to determine the methods of payment for land use;
- for the development of the real estate market and mortgage and other useful purposes.

In the administrative district land-information system, a complete database of each land parcel is collected, stored, processed and communicated to the public and private consumers. Z. S. Abdullaev acknowledges that "Cadastral assessment of land resources cannot be improved without reliable and qualitative information" [3]. However, this definition does not fully explain the nature of land cadastre information.

3 Discussions

In our opinion, the automated land information system in the district can be described as a set of software and hardware-based databases that enable the entry, processing, storage and retrieval of land cadastre data. The main automated database is a system that combines data with other, aggregated data within the system.

Currently, it is necessary to create a unified database for systematic collection, updating, processing, and dissemination of land cadastre data. Increasing the use of new technologies, particularly computerization and specialized electronic software, not only enhances the development of such systems but also creates many conditions and limits. In the market economy, land-related information plays an important role in the intelligent use and development of district land resources. There is a need for a transition to traditional land registration, land surveying and land surveying using software programs.□We consider it necessary to work in the district on each mass, village and community gathering in communities, on farms, and all land plots. It is required that the specialist of the land cadastre responsible for each area should submit the data on the territory to the Department of State Land Resources and the State Cadastre. One or two experts in the department should be able to enter data into a computer and complete the data production process [4, 5].

We consider that the main tasks in developing the land cadastre information system in the district are:

- Creation of accurate and general information about the district land fund;
- Provision of information on targeted state activities to coordinate land reform activities [6].

The district land and information base is the lower level of the geo-information system, which is the basis for systematic improvement of the land cadastre designed for the management, analysis, and reflection of information related to land real estate and property rights.

The process of creating a district land and information system requires full registration of all land parcels owned by all land users. The task of registration of land parcels is to create a district land information system to obtain a reliable and clear basis for the right to own and dispose of the land. The data of the district land cadastre information system includes the location, boundaries, cadastre number, location, results of land ownership and land valuation, privileges and restrictions, and other information related to land servitude.

Regardless of ownership, every parcel of land in the district is subject to state registration. This will guarantee the land owner's ownership of the property and will improve fair and reasonable land and real estate taxation. The district aims to create a system of automated registration, to provide users with high-quality, graphic, written, digital, and cartographic information by simplifying data acquisition, systematization, and classification of land resources. Figure 1 summarizes the automated land cadastre information system created in the district.

The scheme of functioning of the land cadastre information system, shown in the above picture, can significantly improve the efficiency and quality of the work of specialists in the establishment of the Office of the Department of Land Resources and State Cadastre. Modern tools for collecting and processing land cadastral information in the administrative district should be created based on computing and telecommunication techniques. At the same time, the technical structure of the land cadastre information system should be formed as a result of the disclosure at all stages of the system of software and hardware systems.

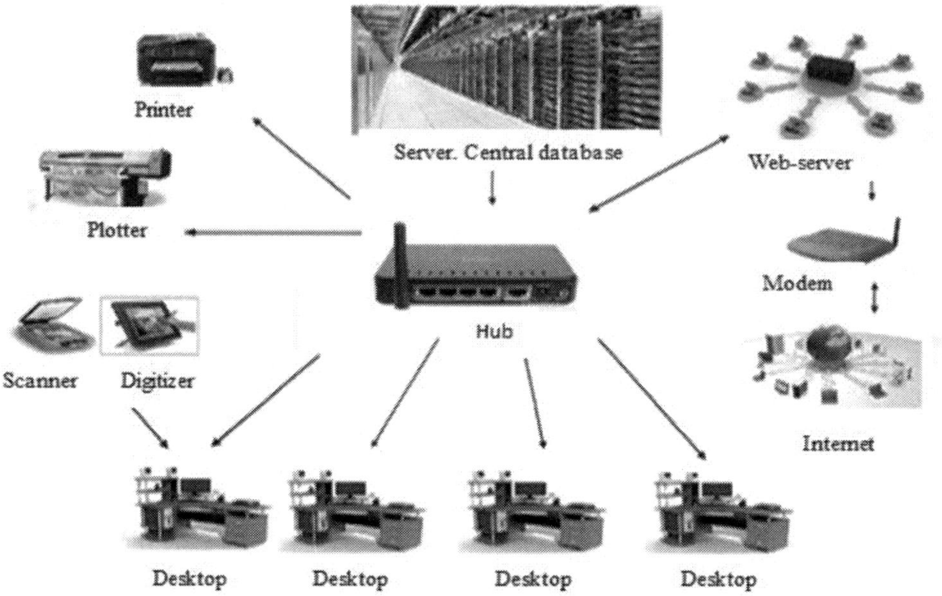

Fig.1. Scheme of operation of automated land cadastre information system of the district

Depending on the degree of opening of software and hardware complexes, the internal computing networks of various structures are one-sided, that is, the information that is distributed on the users' computers, which can be accessed from the internal network of the organization. Today, in the Kibray district of Tashkent region and other districts of the Republic, the research revealed that the above-mentioned work and use of advanced technologies were not organized at the level of modern requirements.

The total area of Kibray district is 55473 hectares, which is divided into 8 categories (as of January 1, 2019), most of which (45%) is allocated for agricultural purposes. 51% for industry, transport, defence, communication, and other purposes, and the remaining land belongs to other categories of land. The information on land users is presented in Table 1 below.

Table1. Distribution of Kibray district land fund by land categories

№	Categories of land fund	Total land area, hectares	Total agricultural land in production, hectares	Other lands, hectares
1	Agricultural lands	25246	17774	7472
2	Lands of reserve	271		271
	2.1 Chirchik River	258		258
	2.2 State Reserve	13		13
3	Lands of the Forest Fund	791	27	764
4	Lands of settlements	176		176
	4.1 Kibray district center	80		80
	4.2 Salar settlement of the hydroelectric power station	64		64

	4.3 Institute of Nuclear Physics	14		14
	4.4 Yangiabad rural citizens' gathering	10		10
	4.5 Housing (Property and Energy Company)	8		8
5	Lands for industry, transportation, communications defense, and lands for other purposes□	28434	242	28192
6, 7	Lands for nature protection, healthcare and recreational purposes and Lands for historical and cultural heritage	94	22	72
8	Lands of the water fund□	540		540
	Total land	55552	18065	37487
	Land used outside the district administrative boundaries	79	76	3
	Total land within the administrative district	55473	17989	37484

Land use data in the district show that the main agricultural lands are now available for use by farmers and other land users of various specialties. Currently, there are 18 village gatherings in the district with 25246 hectares of land. Apart from that, some lands are also allocated to other businesses, institutions and organizations in the district. First and foremost, newly created or newly created land and land users, in land management and improvement of land use efficiency in the district, The following illustration shows the existing layout of arrays located in the Kibray district of the Tashkent region. molten. The current status and analysis of the registration of rights to land parcels in Kibray district are explained as follows.

It is well known that at present there are a large number of land users or other legal entities to another type of land parcel. Also, several new landowners and land users are emerging each year. It also allows these entities to register their rights to land and issue special legal documents, and to use their land to enable them to carry out their activities in a particular order. To date, over 5 million 900 thousand land plots have been distributed among land users in the Republic of Uzbekistan. In the Tashkent region and Kibray district, this figure is 410 thousand and 38 thousand respectively. To date, there are 466 major land uses in the district. It does not include household plots, trade points, and other real estate objects. Therefore, setting up a database on land is a pressing issue [7].

The number of agricultural land uses in the district is 3079 (as of January 1, 2019). Their average area is 6.9 hectares and is located on a total of 25,246 hectares. The number of land users is increasing year by year. The main reason for this is the increase in the allocation of land to farmers and dehkan farms and other types of land use. 50% of the land fund of the district fund is owned by the Holding Company "Uzmevasabzavotuzumsanoat", which has a total area of 10334 hectares. By land weight the enterprises of the Ministry of Agriculture and Water Resources (7.7%), subsidiary agricultural enterprises (20.6%), agricultural research institutions and educational institutions (14.2%), and land users of regional subordination (14.2%), organizations of State joint-stock company of uzbek and dairy industry (0.7%) and other agricultural enterprises (2.5%) were allotted land plots (as of January 1, 2019).

When analyzing the legal status of land parcels in Kibray district, the extent of their use and the distribution of land resources, the following was identified:

1. There is no uniform system of land registration in the offices of the district land resources and the state cadastre;

2. The distribution of land fund by land users by land categories is not scientifically-methodical;

3. Basis of cadastral map mapping for land parcels registration has not been studied methodically;

4. The share of agricultural land in the district is 45% of the total area and there is no proportionality in the distribution of land by agricultural specialties;

5. There are differences in the total area of land in residential areas. For example, 5110 land users are registered in the Baitkurgan village. Of these, 4,530 are land users and 580 are other land users. Of the 47 documents on land allotment is not fully completed, 16 are not drafted, and 15 are not included in the register;

6. There is no complete consistency in the documentation of land parcels in district land resources and the state cadastre services.

The process of selling the rights to a land parcel through auction also leads to an increase in registration and land allocation. As a result of real estate transactions of citizens and legal entities on the parcels of the Republic, the volume of registration and documentation is growing. During the period 2010-2018, the district carried out an average of 221 sales and re-registration of the right of inheritance [8].

4 Conclusions

Based on the results of the study, it is advisable to include the following in the "Land Resources and State Cadastre" section of Kibray district:

1. Centralized storage and processing of data and the creation of continuous access to the system's users;

2. Creating the necessary conditions for the processing of graphic and written information in the automated workplace by consumers;

3. Provision of primary internal data computing systems and data entry tools;

4. Availability of printing and duplication of transmitted data;

5. Communication and communications equipment should be provided to support the adopted information processing and archiving technologies.

The land information system software consists of operating systems for technical and information support, ie database management systems, as well as special application software for automating the process of processing, analyzing, storing and displaying land cadastral information. It is well-known that at present the modern development of automated information-information systems requires implementation in scientific, methodological, technological and implementation directions. Electronic program "Yeravtobank" (Landautobank) (2015), developed by the author, will provide software for creating land and information base in the district:

- processing of cadastral information on land parcels and individual objects prior to the formation of the balance sheet and ensuring the prompt transfer of information to the district land and state cadastre service;
- search and retrieval of information on cadastral numbers, contact addresses or names of users and parcels of land;
- search and selection of information about groups of land parcels by type and functional importance;
- development of electronic reporting forms for land surveying, land registration and land surveying using electronic software and transfer to district land resources and the State Land Cadastre;
- development of user manuals for software packages.

Implementation of the proposed software program in Kibray district will allow the creation of a complete database of all land parcels required by land users operating in the area. This system includes filling out the required tables, forms, report forms, and classifiers. This allows you to divide this information into multiple archives, transmit and extract the required data using electronic forms, and analyze or print the data in the form provided by the reports. Along with the improvement of land cadastre information, provides reliable, comprehensive information on various levels of administrative and economic services dealing with land plots and real estate in the administrative district and addressing various issues of management, planning, and control of these areas. It also serves to provide. The proposed electronic land cadastre electronic program and its scientific and methodological results will improve the efficiency of the land cadastre and create quick methods for obtaining the necessary information about land parcels. We think it is advisable to use the land cadastre for improving the information supply.

References

1. A. A. Varlamov, S. A. Galchenko, *Geography and Information Systems.* - M.: Kolos S., 400 (2005)
2. K. Chertovitsky, A. K. Bazarov. *Land cadastre.* TIIM, T., 302 (2012)
3. Z. S. Abdullaev. *Fundamentals of information support for evaluation of land resources. Monograph.* Tashkent. Fan Publishing, 123 (2008)
4. Q. Rakhmonov, Sh. Narbaev, Z. Mukimov, *Land management.* Tashkent. 154 (2018)
5. A. A. Varlamov, A. V. Sevostyanov, *Land Cadastre.* **Volume 4**. Kolos, 265 (2006)
6. S. Avezbaev, S. Volkov, *Economics of land management.* Yangi asr avlodi, 640 (2002)
7. U. P. Umurzakov, K. R. Rakhmonov, Effectiveness of creating an Earth-database. Journal of Irrigation and melioration. **No 3 (9).** 52 (2017)
8. A. I. Martynenko, Yu. L. Bugaevsky, S. N. Shibalov, *Fundamentals of GIS, theory and practice.* Moscow, 232 (1995)

The essence, features and prospects of homestead land use in Uzbekistan

Abdullo Ashurov[1,*] and *Guzal* Saydaliyeva[1]

[1]Tashkent Institute of Irrigation and Agricultural Mechanization Engineers (TIIAME), State Cadastres Department, 39 Kari Niyazov Street, Uzbekistan

Abstract. The socio-economic role of homestead land plots and private farms is studied, definitions of their terms are given, the role of the homestead land Fund in ensuring food security of the country and increasing the export potential of fruit and vegetable products is considered. The reasons for the higher efficiency of growing agricultural products in private farms in comparison with other forms of farming, as well as the features of the use of homestead land plots, are studied.
Keywords. Land Fund, types of land, resource, Land plots, household, subsidiary farm, dehkan farm, property, inheritance, service life, efficiency, use, rent, massive

1 Introduction

The article emphasizes the importance of rural settlements, in particular, the role of the land Fund of households in the national economy. Indicators of agricultural production, animal husbandry and their low profitability, food shortage and ways to solve these problems are given. Managers of dehkan farms in different regions analysed the results of surveys, the costs of agricultural production and organization of production, as well as the average volume of production on land plots. Prospects for continuous development of land use in farms. In addition, the prospects of using GIS in the management of land use and farms are highlighted. One of the key factors for sustainable land use is the efficient management of natural resources. Dehkan farms should be developed on a systematic basis, including the creation of normal living conditions for the population, provision of agricultural products to rural families through the introduction of resource-saving technologies and optimization of the use of private land, reasonable year-round crop rotation, export of surplus industrial products, and restoration of soil fertility.

2 Situation Analysis

The area of homestead lands of the Republic is 697.3 thousand. Ha, including 453.7 thousand acreages. Ha, of which 422.9 thousand. Ha of irrigated land, which is a significant part of the irrigated acreage from their total area in the Republic - 12.96. % [1]. homestead land is intended for housing and farm buildings of citizens living in rural areas, as well as

* Corresponding author: a.ashurov2810@gmail.com

for gardening in order to meet the needs of the family in crop production and animal husbandry based on the use of personal labour of family members. Surplus products can be sold on the market at the discretion of the family.

At present, there are 4,596,415 million rural families in the Republic with homesteads and homesteads (LPH). Homestead plots of land are granted to the head of the family on the basis of an inherited lifetime right, their size varies in the irrigated area of agriculture from 0.06 to 0.25 hectares, in the rain-fed zone from 0.25 to 0.25 hectares, in the desert. zone - from 0.50 to 1.0 ha [2]. Despite the relatively small share of irrigated land in the household Fund, their contribution to ensuring the country's food security is quite large: as of January 1, 2001 the share of dehkan (personal subsidiary) farms in all categories of cattle farms was 86.2 percent, as of January 1, 2017, 94.1 percent. In the analysed period, there was also an increase in this proportion in the following species: cows - from 89.9 to 94.5 percent, sheep and goats - from 67.4 to 83.8 percent, pigs - from 44.3 percent to 78.7 percent, horses - from 70.2 percent to 85.0 percent, camels - from 33.1 percent to 62.4 percent, bee colonies - from 50.0 percent to 74.1 percent. The share of dehkan (private subsidiary) farms in all categories of farms in meat production in 2000 and 2016 was 91.1 and 94.4 percent, respectively, milk-93.7 and 95.6 percent, eggs-59.9 and 57.3 percent, wool-74.3 and 85.4 percent. Karakul skins - 32.1 and 83.7 percent, honey-55.3 and 79.2 percent. [3].

Every year, the share of these farms in the export of agricultural products increases. However, practice shows that the production potential of the land of the economic Fund in a market economy is not fully used. Not all homesteads are re-sown (re-sown); only a small portion of homesteads receive three crops a year (mainly in the Surkhandarya region). There is no qualitative assessment of these lands; when calculating the amount of land tax from household plots, the average score of the soil bonus of the administrative district is used, which to a certain extent distorts the amount of payments [4]. Private land that plays a significant role in the socio-economic development of the Republic is not classified as an independent category, the nature and features of its use are not studied, and there are no programs for the development of this land Fund. they are not developed or implemented at the state or local level, and the effectiveness of their use is not monitored. For various reasons, a significant part of the land of the household Fund is used inefficiently; with rational use, much larger volumes of agricultural products can be obtained from these lands. In this regard, the President of the Republic of Uzbekistan and the Cabinet of Ministers of the Republic of Uzbekistan have taken appropriate measures to improve the efficiency of land use in the country [5,6,7,8,9, etc.].

The use of the land plots of the Foundation were studied by many Russian scientists (e.g., [10, 11, 12, 13.14, etc.] and foreign (15.16, 17, etc.). However, currently there is no single clear concepts and definitions of the terms "personal land", "private farming" rural families, and there is no clear difference in terms of "private farms" and "private farm. The land code of the Republic of Uzbekistan also does not define " private land "and" private land for own farming. In view of the above, we consider the nature and features of the use of land of the economic Fund in General and private land, especially.

3 Methodology

Homestead plots are plots of residential buildings (estates) of rural families. They are issued for life the head of the rural family to build a house, outbuildings, and farming based on family labour to meet the needs of the family necessary products. The terms "household plot" and "personal subsidiary farm", which are typical mainly for the CIS countries, including Uzbekistan, were introduced in the former Union in the standard Charter of an agricultural cooperative, adopted in 1930 under conditions of complete collectivization.

The new Model Regulation on agricultural cooperatives, discussed and adopted by the second all-Union Congress of collective farmers, confirms the transfer of personal plots to the families of collective farmers. At the same time," Personal subsidiary farming of collective farm yards on household plots seems to be an objective necessity" [18].

The reason for the appearance of the terms "homestead land" and "personal subsidiary farms" is that in the former Soviet Union, all land was nationalized, in agriculture it was transferred to collective farms - collective farms and state farms for the production of commercial agricultural products. In this regard, in the conditions of continuing collectivization of agriculture and insufficient state economic support for the rural population to meet the personal socio-economic needs of the rural family, the need to maintain a small personal subsidiary farm to meet the needs of the family. was recognized. Without a household, the rural population would not be able to live normally due to the lack of natural and monetary income from social production, especially during the economic crisis. On the lands of a specially created personal Fund, a new organizational and economic form of personal family economy (LFH) was created, from which each family was provided with homestead plots.

In the unified land Fund of the former Soviet Union and in the Land funds of the former Union republics, including the Republic of Uzbekistan for state registration of land plots of the land Fund included a graph "private land" introduced the reflection area of private farms, household plots of land. Currently, in the Republic of Uzbekistan, this column has been preserved in the Land Fund, which reflects the quantitative indicator of the homestead land Fund of rural areas. The official language it is called "Tomarka yer uchastkasi".

Thus, a household land plot (ROM) is a land plot attached to a house (estate) intended for personal subsidiary farming based on the use of personal labour of family members. Private farming (LPH) rural family economic use of plots of land based on family labour to ensure the provision of housing, essential food crops and livestock. Moreover, if the surplus of agricultural products increases, the family has the right to sell them on the market at its discretion. However, it cannot be interpreted as a commodity economy. Consequently, the personal subsidiary economy of a rural family is based on the use of household land and the labour of the family itself; this is not a commodity economy. In scientific sources, the term "personal subsidiary farms" is used to refer to the activities of households in the field of crop production and animal husbandry. "Household" is a form of economic activity that covers economic objects and processes that occur in the places of permanent residence of a person and family [19].

The concept of "private farm" is formulated in the Law of the Republic of Uzbekistan "On dehkan farms" [14], which, unlike private farming is treated as a small-scale farm, because along with providing a family of agricultural dehkan farm produces marketable products. Activities in the dehkan economy are considered to be entrepreneurial activities and can be carried out at the request of members of the dehkan economy with or without a legal entity. "A dehkan farm is a small family farm engaged in the production and sale of agricultural products based on the personal labour of family members on an infield plot of land and on additional plots outside the community, granted to the head of the family for life by inheritance" [14]. Activities in the dehkan economy are considered to be entrepreneurial activities and can be carried out at the request of members of the dehkan economy with or without a legal entity. A dehkan farm cannot use wage labour on a permanent basis in its activities. Members of a dekhkan household are the head of the family, his spouse, children, including foster children, foster children, parents, and other relatives who have reached working age, live together and have a dekhkan household. At the same time, it should be taken into account that the dehkan economy is based on the use of a private land plot.

The lands of the Republican Economic Fund have a deep socio-economic essence, and it should be noted that the value of these lands is constantly increasing. The social significance of land increases with increasing demands for increased investment in human potential. Economic importance is growing in the context of the development of a market and innovative economy, which requires a significant increase in the efficiency of agricultural land use, especially irrigated land. The area of household plots in rural settlements has increased by 341.7 thousand over the past 28 years (from 1990 to 2018). as of January 1, 2019-697.3 ha. the number of farms reached 4.596 million. The size of homesteads ranges from 0.08 to 0.35 ha in irrigated agricultural land, up to 0.5 ha on land, and up to 1.0 ha in the desert. The average size of LPH is 0.15 ha, and in the irrigation zone 0.12 ha. They are granted on the right of hereditary possession to the head of the family [2]. It should be borne in mind that the dehkan economy, which is a small-scale economy and high-commodity farms, simultaneously also use the homestead land plots provided for their family needs.

At the same time, land plots provided for individual housing construction (IHS) cannot be considered homestead plots, since they are not aimed at providing the family with the necessary agricultural products. In connection with the above, it should be recognized that the concept of a personal land plot (LPH) and a personal subsidiary farm of a rural family (LPH) applies exclusively to the CIS countries, including the Republic of Uzbekistan.

Currently, the land of the economic Fund is located in several categories of land of the Unified State Fund (EGF), mainly in the category of agricultural land. This indicates that homestead lands do not have a status in the DHF structure and are not systematized within a single category of land. We believe that the land of personal homesteads in the strict sense of the word is the land of rural settlements and should be considered in the category of land of localities, they do not have a target agricultural commodity function. homestead plots and homestead plots based on them for rural families have played and continue to play an important role in social support and material support for rural families since they were granted official status. Their importance is especially growing in the context of the development of market relations in land use and innovation economy. A significant role in this regard is assigned to the lands of the economic Fund in the Concept of land and water resources use in agriculture [17] and in the Draft Strategy for the development of agriculture of the Republic of Uzbekistan for 2020-2030. [18]. the Functioning of personal lands has the following features in their use:

- the concept of a household plot (ROM) and a personal subsidiary farm of a rural family (LPH), their existence and functioning is typical only for the CIS countries, including the Republic of Uzbekistan, since their concept and terms originated in the former Soviet Union;
- homestead plots are granted on the right of inherited ownership only to families permanently residing in rural areas, and they are granted not only to families engaged in agricultural production, but also to citizens of other professions permanently working in this rural area. (teachers, doctors, locksmiths, electricians, etc.);
- in the Republic of Uzbekistan in the absence of a land market and HOMESTEAD land cannot be sold on the land market, the sale of a residential house on the plot right to lifetime inherited possession of a land plot shall be transferred to the new owner of the house;
- personal owner of land may transfer land to a family member or a close relative, right of disposition of land - sale, barter, donation - he has no;
- homestead plots are allocated within the boundaries of rural settlements in accordance with their master plan for development or reconstruction, since there is no system of farm settlements in the Republic;
- for many rural families, income from household plots is either the only source or a priority, which dictates the need for efficient use of land;

- land plots of the Republic's forest Fund make up a significant percentage of the total area of irrigated land, which requires a comprehensive increase in land use efficiency in market conditions and water scarcity;
- in the conditions of Uzbekistan the utilization of agricultural land on private land is higher than the state cultivated land, and approximately equal K = 2; at the same time in the southern regions of the country it may be increased to K = 3;
- in the areas of irrigated agriculture of the Republic, private land plots must be provided with irrigation water for their full use, which is stipulated by the Land code of the Republic of Uzbekistan;
- since homestead plots are provided to all families permanently residing in rural areas, the heads of farms and dehkan farms, in addition to commodity production sites, have land plots on which they also maintain a personal subsidiary farm;
- the quality of land (productivity) of HOMESTEAD plots in rural settlements is usually higher than the quality of public land. Owners of small homesteads pay much more attention to the reproduction of soil fertility: the introduction of organic fertilizers in a much larger amount per unit area, timely prevention of salinization and waterlogging of the land, prevent soil erosion, as a rule, qualitatively and timely carry out pre-sowing tillage, significantly more invest in high-quality manual labor to care for crops.

The development of a market economy requires the rational use of all types of resources and, above all, land. The land of rural settlements and especially the land stock of households play an important role in the country's land use. Its effective use contributes to increasing the production of agricultural and livestock products in the country, the development of small businesses and private entrepreneurship, improving the supply of food to the population, increasing employment and increasing incomes of the rural population. "The rural worker is a strong pillar of life, a pillar of our well-being," President of the Republic of Uzbekistan Shavkat Mirziyoyev said at a solemn meeting dedicated to the day of the agricultural worker on December 9, 2017. in his speech: "in the field of agriculture, there are still untapped opportunities that are waiting for their solution to problems and shortcomings." In his speech, "445 thousand hectares of the most fertile land were transferred to the population in private plots". But the use of these sites is small." [6]. The use of one of these opportunities, indicated in the speech of the President of the Republic, is largely determined by a comprehensive increase in the efficiency of land use by dehkan farms. Despite significant progress in the development of dehkan farms, in the land use of rural settlements and, above all, dehkan farms, there are still problems: improvement of land reclamation, issues of providing feed, pasture areas, etc. the provision of land in the sublease [12]. Real vesting of the right of lifelong hereditary ownership of a private land plot with a pledge obligation, mortgage registration, improvement of roads and transport communications, processing of grown products, implementation of scientifically based crop rotation on a private land plot, etc. All this requires a solution and further improvement of the legal framework for land use in rural settlements.

An analysis of the state and use of existing land in rural settlements indicates insufficient efficiency of its use, including the land of dehkan farms. Free land use, which for many years determined the inappropriate use of land in our country, contributed to the inefficient, irrational use of this valuable resource, which led to the hypertrophied growth of the territories of settlements and the unjustified withdrawal of agricultural land for construction. into extremes. The length of engineering and transport communications has increased, the environment around localities has degraded, and the cost of all types of development of their territories has increased [8]. All this had a negative impact on the efficiency of the use of personal land and the quality of the population. The main reasons for this situation were the lack of reliable accounting of rural settlements ' land, the necessary urban planning documentation for their integrated development, and the deterioration of land reclamation in rural areas [11].

4 Materials and Methods

During the agrarian and land reform, a new organizational and legal management structure was formed in the village - a small family dehkan farm that produces and sells surplus agricultural products (based on the personal labor of family members on the provided private land plot). As a lifelong legacy of mine ownership) [14]. The dynamics of production shows that the share of this type of farms in the total gross agricultural output is constantly increasing and amounts to 81% of potatoes, 56% of melons, 66% of vegetables, 50% of grapes; almost all the country's livestock is concentrated here - 6.5 million heads (92.8%), including 2.8 million cows (94.5%); they account for 95% - meat, 97% - milk, 57.9% - eggs, 71.7% - wool produced in the country [13].

At the same time, an analysis of the current state and use of private land in the Republic shows that their potential is far from exhausted. On the homestead plots, they receive mostly no more than 2 crops, due to which the dehkans meet their needs for potatoes and vegetables, and sell the surplus on the market. Crops such as early (winter) cabbage, beets, and some others are still rarely used. Livestock indicators are quite low; the lack of feed is due to a small area of land. Food crops are mainly grown in the garden. High-quality land plots are low. Mineral fertilizers are often not applied in sufficient quantities, irrigation is not always carried out in the best agrotechnical terms and with the required rate, the land of villages is often located on massive saline lands, and the collector and drainage network within the village is not removed in a timely manner. There is no optimization of the structure of crops and their economically feasible alternation in the garden.

Surveys of dehkan farms in different regions of the country showed that the cost of doing business is 1000-2000 thousand soums. (excluding animal husbandry costs) the cost of production is 4000-6000 thousand soums, income is 2000-4000 thousand soums. or an average of about 3000 thousand soums per family. More than 85% of dehkan farms contain livestock and poultry. About 70% of households keep cash cows, including 35% of fattened calves. The productivity of cows on farms is quite low - about 1700 kg of milk. On average, about 1,100 kg of milk and 150 kg of meat are consumed per family annually, which is 3.3 and 5.25 million rubles. soums, respectively. The cost of agricultural products sold from animal husbandry is an average of 3.7 million rubles. Sum, and in General, taking into account crop production, 4.9-5.0 million. Sums. The average consumption of farmed products by a family is 60-65% (approximately 3.2 million people). Soums), the rest of the products are sold on the market. The experience of advanced dehkan farms and research shows that their efficiency can be increased three to four times.

Decree of the First President of the country I. A. Karimov No. PP-308 dated 23.03.2006 in order to further develop dehkan farms, strengthen their role in agricultural production and provide state assistance to the rural population, persons engaged in cattle breeding in private subsidiary farms have the right to create employment records and register work. experience. In addition, it provides for the allocation of cash cows to low-income families through sponsorship. The number of livestock is increasing, but providing them with feed remains a problem. The provision of animals with their own feed is insufficient (25-30% due to corn, barley meat and vegetable stalks) [1]. Pastures and pasture feed are particularly lacking. Currently, almost all agricultural land in the Republic is distributed among farms, and not every farmer allows representatives of dehkan farms to graze cattle even on the outskirts of their plots or collect hay on the farm. Summer and winter pastures are located very far from populated areas.

5 Results and Discussion

One of the key factors for sustainable land use is the efficient management of natural resources. Dehkan farms should develop on a systematic basis, including creating normal conditions for the population's life, providing rural families with agricultural products through the introduction of resource-saving technologies and optimizing the use of the territory of the household land plot, reasonable year-round crop rotation, export of surplus industrial products, and restoration of soil fertility. The following aspects of long-term development of land use of dehkan farms can be distinguished:

1. Rational organization of the territory of dehkan farms; This aspect provides for rational placement of residential and economic premises, perennial plantings and crops (vegetable garden) and land on the territory of a personal land plot. The most suitable form is rectangular, which provides the most effective layout of the territory for construction, placement of a garden and vegetable garden. The layout of the plot of land should provide maximum comfort for family living, household management and recreation, necessary sanitary and hygienic conditions and fire safety, rational use of land and water in harmony with the improvement of the village.

2. Allocation of additional territories from the number of irrationally used and unused lands (including pastures); Not all dehkan farms have the corresponding areas specified in the Land code of the Republic. If we analyze dehkan farms that are successfully functioning and have good indicators, their areas can be expanded to normal. It is necessary to develop a mechanism for allocating pasture territory. The expansion of the land use area of a dehkan farm should take into account the resources available to the family (financial and labor), which guarantee the effective use of land.

3. Rational distribution of crops and land; it provides for the selection and placement on personal plots of such types and varieties of crops that provide the necessary optimal alternation all year round. At the same time, the constant alternation of crops should not Deplete the land, but rather restore and increase soil fertility. Provide for the restoration of soil fertility on the basis of a rational farming system, the introduction of soil protection technologies and the preservation of humus. The land, when used correctly, restores its productive properties. The development of society, the continuous growth of its population and needs objectively require an expanded nature of agricultural production, which is the reason for the need for expanded reproduction of soil fertility [10].

4. The use of geographic information systems (GIS);

The problem of managing the land resources of rural settlements and, above all, the personal land Fund has always been relevant. The increase in the number of economic entities on earth as a result of increase in population plots of the land Fund has begun expanding the circle of participants in land relations. Modern works on territorial planning, land management, cadastre and environmental assessment necessarily involve the use of geographical information systems (GIS), which will make fundamental changes in the field of information support for management decisions [7].

As the object of research, we selected the lands of subsidiary farms and dehkan farms in Surkhandarya region, which are an ideal testing ground for developing an information and management system to support the system using GIS based on land inventory for making operational management decisions. To create a system of land management of households and dekhkan farms using GIS technologies, we used the method of system analysis of land use rationality and allocation of land plots of irrational land use, illegal use, methods of land observation: yard, inventory geodetic surveys of areas, methods of data organization by creating a geodata database, methods of spatial and logical modeling. The automated cartographic system (EX), which is the core of GIS for land management of households and dehkan farms, has been developed and tested for a fragment of land in homesteads and dehkan farms and is a set of software tools that ensure the creation and direct use of maps.

It is advisable to create a complete GIS for managing the territories of subsidiary and dehkan farms in 2 stages: preparatory and production. The preparatory stage includes: collection, analysis and systematization of archival land cadastre data, planning and cartographic materials, land management documentation for land plots (allotment materials, land inventory materials, land books), as well as data on the composition and population. To Refine and update the planning and cartographic material, as well as to speed up and reduce the cost of work, you can use remote sensing data that is refined at the production stage during geodetic surveys. In addition to surveys and bypass the yard, at the production stage are carried out surveying and cadastral works, in which the application and correction (adjustment) of boundaries, boundaries, parcels, areas of farms and dehkan farms, boundaries of individual land mass and land.

In addition to surveys and bypass the yard, at the production stage are carried out surveying and cadastral works, in which the application and correction (adjustment) of boundaries, boundaries, parcels, areas of farms and dehkan farms, boundaries of individual land mass and land. As a result, digital maps of the area are created and a GEODATA database (EGDB) is created.

Land plots are formed according to topological rules, which allows you to exclude incorrect data, such as overlapping land plots or, conversely, a gap between them.

The geodata database is a spatial-logical model that can be used for:

- inventory of land of subsidiary and dehkan farms;
- monitoring the state of engineering communication networks;
- support, updating and development of the General plan of a rural settlement, construction of detailed planning projects;
- information support for selecting a location for the construction of warehouses, warehouses of commercial objects (for example, purchasing organizations) that require a detailed spatial analysis of the territory;
- information services and consultations for potential users of land and real estate: spatial information about the reference and market value of land plots, cartographic services of real estate databases.

6 Conclusion

A large amount of accumulated information and integration of geographic information systems (GIS) allows you to quickly and fully meet a wide variety of information needs both in content and in form - in the form of reports, thematic maps, analytical results in electronic and paper form for information support of management decision-making. Testing the results of a study conducted on the example of homestead and dehkan farms in Surkhandarya region showed that GIS is an effective tool for managing land resources in rural settlements, for example:

- in the field of Economics-determining the current state of land use, identifying opportunities to optimize and improve the use of land in private and dehkan farms in the region by changing the territorial and sectoral structure of land, analyzing its state and development trends.
- in the field of ecology-identifying the features of the ecological state of land and preventing the development of negative processes.

Reliable and complete information about land resources helps to increase budget revenues from land payments, organize its rational use and protection, promptly regulate land relations and introduce a regulated land market.

We thank the staff of the Department of land resources and state cadastre of Sherabad, Denav, Sariosiyo and Uzun districts of Surkhandarya region (Uzbekistan) and the scientific supervisor of this study, doctor of Economics, Professor A. S. Chertovitsky.

References

1. A. F. Ashurov, Economic potential of land use in rural settlements "AGRO ILM" Scientific Appendix to the journal "Agriculture of Uzbekistan", **Volume-1**, P. 89-90 (2009)
2. Yu. S. Balandin, Peasant economy. Moscow, Agropromizdat (1992)
3. A.V. Chayanov, Peasant economy. Selected works. Moscow "Economy" (1989)
4. Decree of the President of the Republic of Uzbekistan "On measures for the effective use of land and water resources in agriculture" dated June 17, No. up-**5742** (2019)
5. G. Kudratov, E. Akramov, Reform of agriculture in the Republic of Uzbekistan Personal subsidiary farms. Agricultural journal of Uzbekistan, **No. 3**, P. 2-3 (2001)
6. Sh. M. Mirziyoyev, "Speech at the solemn meeting on the occasion of the day of agricultural workers on December 9 (2017)
7. A. L. Paul, F. G. Michael, J. M. David, V. R. David, Geographical information systems and science. Great Britain, **2nd edition** "Zhohn wili and sons, Ltd.", p.517 (2005)
8. V. P. Proshlyakov, Inter-Farm land management in conditions of irrigated agriculture P. 94-95 (1964)
9. Resolution of the Cabinet of Ministers of the Republic of Uzbekistan "On additional measures to support homeowners and develop scientific and practical potential in the field of agriculture of the Ferghana region " dated October 10, No. 846 (2018)
10. G. A. Talipov, Monograph "Land resources of Uzbekistan and problems of their rational use", Tashkent (1992)
11. P. K. Tatur, Monograph "Settlement System and principles of settlement planning for agricultural enterprises", **Volume-3**, p.29 (1975)
12. Concept of land and water resources use in agriculture. Decree of the President of the Republic of Uzbekistan dated June 17 (2019)
13. Land code of the Republic of Uzbekistan, approved by the Law of the Republic of Uzbekistan of April 30, 1998 N 598-I; Entered into force on July 1, 1998. The resolution of the Oliy Majlis of 30 April 1998 No. 599-IR, as amended by Law of RU of August 30, 2003 N 535-II, Law of the Republic of Uzbekistan dated 30.04. 2004 No. 621-II of the law of the Republic of Uzbekistan dated 03.12.2004 No. 714-II, Law of the Republic of Uz from 28.12.2004 G. 2007 N LRU-138 (1998)
14. The law of the Republic of Uzbekistan "On dehkan farms", Tashkent, with changes and additions (1998)
15. A. Uralov, D. Nozilov, A. Farmonov, S. Matyazov, Agricultural Planning, Tashkent "Uzbekistan" P.65-75 (1994)
16. E. E. Rumyantseva, New economic encyclopedia. M.: Infra-M. p.541 (2011)
17. The Concept of using land and water resources in agriculture. Appendix No. 1 to the Decree of the President of the Republic of Uzbekistan dated June 17, "up-5742" (2019)
18. Strategy for the development of agriculture of the Republic of Uzbekistan for 2020-2030. Project, Tashkent, Ministry of agriculture of the Republic of Uzbekistan, (2019)

19. I. V. Bochkov, P. N. Pershin, M. A. Snegirev, V. F. Sharapov-History of land relations and land management, State publishing house of agricultural literature, Moscow, P.162-163. Literature, Moscow.

The Value of the Cards in Water Basins with the Installation of Solar Power Plants in Yangiyul District of Tashkent Province of Uzbekistan

Ilhomjon Musaev[1,*], *Abdujalol* Bokiev[1], and *Mukhtaram* Botirova[2]

[1]Tashkent Institute of Irrigation and Agricultural Mechanization Engineers, 39, Kory Niyoziy, 1000000 Tashkent, Uzbekistan
[2]Sector of the master plan and transport. Research Institute "Uzogirsanoatloyiha" Tashkent, Uzbekistan

Abstract. The article considers the possibility of using the surface of various reservoirs characteristic of regions when allocating land for low-power solar power plants in Uzbekistan. When implementing low-power solar power plants in Uzbekistan, a certain amount of land resources is required (for installing solar panels). From the point of view of respect for natural capital (saving land resources), the method of installing solar panels directly above the surface of water basins based on the corresponding maps of water basins is technically and economically effective. The methodological basis of the research is the results of researchers work on the problems of land relations and land management. In the research, a systematic approach to the analysis of land use has been chosen, in which a variety of tools were used to determine the essence of the analyzed phenomena, processes and patterns. In recent years, industry has been rapidly developing in rural areas of the Yangiyul district of Tashkent Province of Uzbekistan. In this regard, the problems of effective use of land resources, especially targeted use of irrigated land, has become urgent. This means that in the future there will be problems when allocating land for the installation of low-power solar power plants.
Keywords. Land resources, irrigated land, seasonality, renewable energy sources, efficiency, resource conservation, water basins, map.

1 Introduction

The purpose of the study is to develop proposals and recommendations for the use of the surface of various water basins, when allocating land for low-power solar power plants in Uzbekistan. To achieve this goal, the following tasks are set:

- Study theoretical and methodological issues of improving the rational and efficient use of agricultural land in accordance with current conditions;
- Study of land allocation features for alternative energy sources;
- Analysis of the state of the land Fund and the current state of its use;

* Corresponding author: ilkhommus@mail.ru

At the present stage of economic reforms in Uzbekistan, the problems of land and energy resources are being solved through the introduction of modern resource-saving and energy-efficient technologies. When implementing low-power solar power plants in Uzbekistan, a certain amount of land resources is required (for installing solar panels). From the point of view of respect for natural capital (saving land resources), the method of installing solar panels directly above the surface of water basins based on the corresponding maps of water basins is technically and economically effective.

Many researchers have worked in the theoretical and practical development of solar energy. Yandra Shivrat [1] in his article on "cost analysis of small solar and wind power systems" hypothesizes that in the next few years, when the cost of solar photovoltaic modules falls below $ 1 per Watt, Piyali Ganduli, Akhtar Kalam Aladin Zayeg [2] in their research note that a renewable energy system consisting of solar and wind energy is an environmentally friendly and cost-effective option for feeding rural areas compared to traditional sources. Sheriazov Saken Koishybayevich [3] in his dissertation work, as the object of research, chose a combined system of energy supply to agricultural consumers using solar and wind power plants. The methodological basis of the research is a systematic approach to the analysis of a complex power supply scheme. In the work of Tashimbetova M. A [4] the complex use of renewable energy in the Republic of Kazakhstan is considered. Based on the example of DJambyl region of southern Kazakhstan, a justification for the choice of a combined energy complex based on RES for the purpose of power supply of a rural locality with a required capacity of 100 kW was carried out.

In the work of Ahmed Torki and Ahmed Jailanp [5] the subject of research is to provide electricity to autonomous rural consumers based on the use of renewable energy sources in order to develop autonomous systems for combined electricity supply to farms in Egypt using renewable energy sources and a backup liquid fuel power plant. Yarmukhametov Ural Rashitovich [6] in his dissertation work, as the subject of research, chose the regularities of changes in the performance of ship power plants depending on geographical latitude, climatic conditions, design parameters and the degree of orientation of the receiving surface to the sun. In addition, we used a number of literature on the preparation of special maps and plans for energy facilities, in particular solar power plants by remote sensing methods of domestic and foreign [9,10,11,12,] scientists.

One of the most effective is the experience of great Britain, Japan, China and Lithuania in implementing a system for using alternative energy sources installed on the surface of water basins. For example, on a lake in the Berkshire region (Great Britain), rafts are launched on the water, on which 800 photovoltaic panels are installed. Their total power is 200 kW. Such structures are not tied to large areas of land, and they can be built on almost any reservoirs [13]. The state-owned enterprise Lietuvos Energijos Gamyba (LEG) will create an floating solar energy systems together with scientists from the Kaunas University of technology (KTU) LEG and KTU plan to install an experimental 60-kilowatt floating solar power plant in the upper basin of the Kruonisskoi PSPP. The total capacity of the future project should be 200-250 megawatts. [14]. Kyocera TCL Solar LLC launched a 13.7 MW floating power plant at the Yamakura Dam reservoir in Ichihara in 2018. Interest in floating solar power plants was aroused after the rapid development of alternative energy in Japan, and it became increasingly difficult to allocate space for new projects. The water surface is suitable for this purpose. [15] in early June 2017, China launched a photovoltaic plant capable of producing 40 megawatts of energy, which floats on an artificial lake in Anhui province in China are described in Figure 1 and Figure 2 [16].

Fig. 1. A floating photovoltaic plant at the Yamakura Dam in Ichihara

Fig.2. A floating photovoltaic plant on an artificial lake in Anhui Province in China

International experience of installing solar power plants on the surface of water basins shows that there are many effective methods of installing solar power plants on the surface of water basins. Floating photovoltaic installations can be divided into three categories:

- photovoltaic installations consisting of modules mounted on pontoons;
- photovoltaic modules mounted on plastic and galvanized steel rafts;
- Photovoltaic modules installed on rafts are made entirely of plastic.

At the Asian clean energy summit in Singapore (October 2017), the world Bank Group gave two figures for solar floating installations worldwide: 453 MW in 2017 and a forecast of 750 MW for 2018 [19]. With the right scientific-based approach, there are great opportunities in the Republic of Uzbekistan when allocating land plots for solar power plants. In Yangiyul district, there are many natural and artificial water basins (rivers, soybeans, channels), and their banks are very convenient for installing solar panels.

2 Methods

The theoretical and methodological basis of the research is the results of researchers work on the problems of land relations and land management. In the research, a systematic approach to the analysis of land use has been chosen, in which a variety of tools were used to determine the essence of the analyzed phenomena, processes and patterns.

In the experimental part of the study, calculation and analytical methods, correlation and regression methods were widely used. The use of methods that are adequate to the studied phenomena allowed to ensure high reliability of the results obtained. A special program called ArcGIS was also used to create land maps for low-power solar power plants

3 Results

The main land plots of the Yangiyul district of the Tashkent region are fertile, irrigated land. In turn, the efficiency of using these lands is higher than the average for the Republic.

In recent years, industry has been rapidly developing in rural areas, including in the Yangiyul district. In this regard, the issue of effective use of land resources, especially the targeted use of irrigated land, has become urgent. This means that in the future there will be serious problems when allocating land plots in the district for the installation of energy facilities, in particular solar power plants. Based on this, the studied territories of towns such as Gulbakhor, Bozsuv, Nov, Karsadok, Kovunchi, villages such as Khalkabad, Navbahor, Niyazbashi, Kushyagach, Shuralisoy and Eski- Kovunchi (Table 1).

https://doi.org/10.1051/e3sconf/202122705004

Table 1. Water basins of Yangiyul district

№	Names (river, sai and canals)	Length (km)	№	Names (river, sai and canals)	Length (km)
1	Northern Tashkent Canal	10,1	11	Shuralisoy Canal	0,3
2	Boysen channel	4,8	12	Kesken Canal	9,7
3	Jun Channel	20,5	13	Tuyabugizsoy Canal	16,7
4	Upper Tashkent Canal	19,5	14	Canal B-45	6,8
5	Dam channel	5,8	15	Canal P-16	11,9
6	Niyozbosh Canal	11,7	16	Canal P-14	6,4
7	Kirtqurgon 1 Canal	12,1	17	Salargurunchi Canal	9,4
8	Kirtqurgon 2 Canal	7,6	18	Kichik Jun Canal	7,9
9	Chortoqshox Canal	5,9	19	Novoprokopevskiy Canal	5,8
Total:		**178.2**			

Table 2. Open drainage networks of the district

Names	Amount	Length (km)
Total drainage networks:	**531**	**1486.8**
- trunk drainage networks	38	106,4
- interdistrict drainage networks	72	201,6
- inter-farm drainage networks	146	408,8
- drainage networks on the balance of the water user association	275	770

Special attention was paid to the rivers and channels that flow through the district, and not only their banks, but also the possibilities of water surfaces were studied. For example, using 15-20% of the surface of rivers and channels flowing through the district, it will be possible to produce much-needed clean electricity.

The main attention was paid to the possibility of constant cooling of solar panels and cleaning their surfaces. Figure 2 shows a map of the most suitable areas of the banks and surface of rivers and channels flowing through the territory of Yangiyul district for installing solar power plants.

In carrying out the above-mentioned research works, it is of great importance to use geo-information systems data and their rational placement on agricultural crops, as well as to increase the projected volume of agricultural production and the volume of irrigated land. As a result of these studies, a map of the most suitable territories of the banks and surface of rivers and channels flowing through the territory of the Yangiyul district for installing solar power plants using GIS was compiled (Fig. 3). The results of studying the GIS database can provide an opportunity to conduct the following research:

- environmental and energy analysis based on data on irrigated land;
- development of long-term strategic plans;
- forecasting the size and yield of reserve land;
- implementation of construction in the area for installation of solar power plants with minimal costs;
- to prevent the process of salinization of the lands or save it in the same condition, to predict, etc.

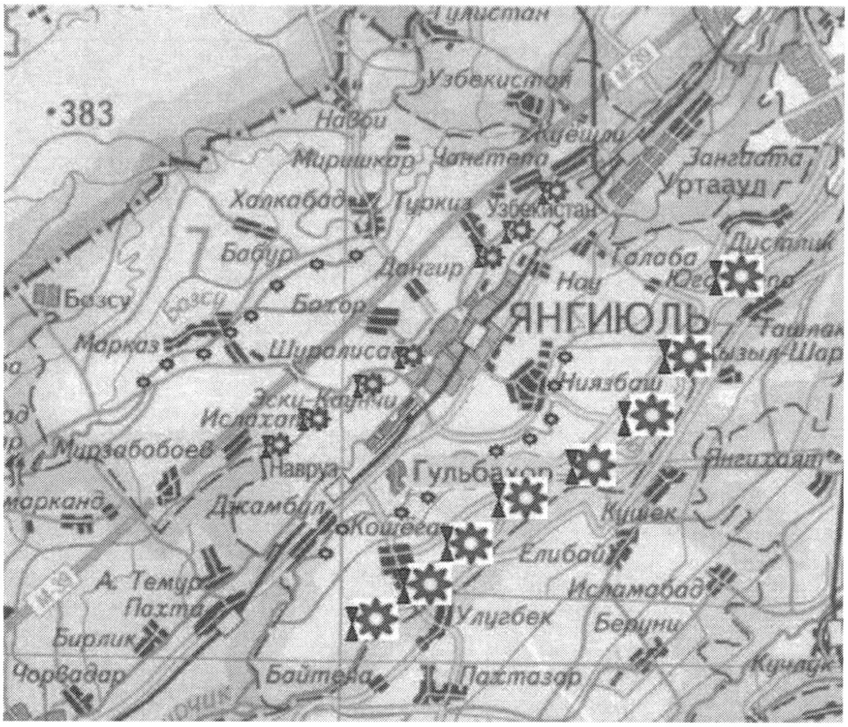

Fig. 3. Map of the most suitable territories of the banks and surface of rivers, channels flowing through the territory of the Yangiyul district for the installation of solar power plants.

From the above it is seen that on the banks of the river Chirchik, each of the protective tributaries can be installed 8 solar power plants with a capacity of 300 kWh, located on the banks local canal "Moscow" from the city of Tashkent in the North-Eastern side of the district 8 solar power plants with capacity of 50 kWh, directly on the banks of the canal Jun flowing through the city of Yangiyul 7 solar power plants with a capacity of 100 kWh.

In accordance with all environmental and other requirements, along the riverbed lines that flow through the settlements of Niyazbash and Gulbakhor, there are 9 solar power plants with a capacity of 30 kWh. The total output is 3770 kWh for the district. As a result, a three-stage program project and a corresponding map for 2020-2025 were developed to identify optimal zones for the construction of small solar power plants in the Yangiyul district.

Taking into account the above, the construction of solar power plants directly along or on the surface of the water basin (Fig. 4, 5 and 6) will eliminate these problems and create opportunities for the development of animal husbandry in these regions, processing of fruits and vegetables, and cereals directly in the field.

https://doi.org/10.1051/e3sconf/202122705004

Fig. 4. Ways to install solar panels along the water basin. 1-control panel; 2-base frame; 3-jumper cables; 4-photovoltaic panels; 5-cooling system

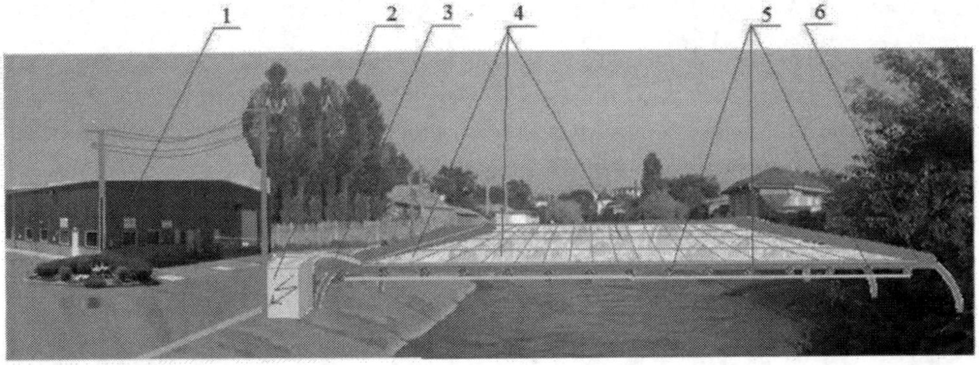

Fig. 5. 1-small enterprise; 2-controller (inverter); 3-jumper cables; 4-solar panels; 5-cooling system; 6- ground rods

Fig. 6. 1-solar panels; 2-jumper cables; 3-ground rods; 4-controller (inverter); cooling system

4 Discussion

One of the promising areas of energy supply for decentralized remote agricultural facilities is the use of solar energy for heat supply by solar collectors and electricity supply by solar photovoltaic installations [6]. The gross potential of solar energy coming to the territory of Uzbekistan annually exceeds the energy potential of all the country's hydrocarbon reserves and amounts to 50973 million t. e [7]. The demand for food, agricultural products of humanity is growing from year to year in connection with growth of world population, the rapid development of industry, the environmental degradation of the planet as a result of global climate change, limited natural wealth, such as land and water resources, rational use of land resources, the development of agriculture.

In recent years, the Republic of Uzbekistan has been conducting research aimed at introducing renewable energy sources, in particular solar and wind power plants. In this sense, the legal framework for allocating land plots for these energy facilities requires improvement in terms of effective use of land plots. [8]

Photovoltaic power plants, sometimes centralized, distributed, tower, roof, and so on, but these are some ground-based photovoltaic power plants. After all, land resources on earth are limited, and the degree of their availability is limited. Therefore, there is a growing interest in floating solar photovoltaic power plants on water. The structural composition of a floating photovoltaic plant mainly consists of photovoltaic panels plus a floating structure. The cost of using a floating installation is about 1/4 of the cost of a steel bracket (the cost of a steel bracket is about 5% -6% of the total cost of a ground-based photovoltaic power plant). [17]

A 2018 World Bank report estimated that the global potential of floating solar panels on artificial water surfaces will exceed 400 gigawatts. [18]

In addition to funding public and private investment, the World Bank Group is committed to supporting the development of floating solar energy. Publications and tools planned for the "Where the sun meets the water" series:

- Report on the floating solar energy market
- Guide to floating solar energy for practitioners
- Global mapping of floating solar energy potential (geospatial tool)
- Proposed technical projects and project structuring. [20]

In Uzbekistan, the introduction of methods for installing solar power stations on the surface of water basins should be associated with the development of maps of water basins. This requires the association of scientists surveyors and power engineers to conduct joint research in this direction

5 Conclusions

Based on the conducted research, the following conclusions can be drawn:

- To improve land use efficiency, it will be important to locate alternative energy sources based on science-based methods, placing them on land that is not being used to the maximum extent possible.
- Installing such devices along or on the surface of water bodies in densely populated agricultural areas saves valuable land resources.
- It is necessary to develop science-based systems of measures for the implementation of water basin maps that provide for the installation of solar power plants on the surface of water basins, training of service personnel, etc. This, in turn, increases the level of reliability and efficiency of power supply, creates favorable conditions for the widespread introduction of modern resource-saving electrical technologies in remote areas.

- It is important to use data geographic information systems in the preparation of maps and plans in rational placing of agricultural crops, as well as an increase in the projected volume of agricultural production and the amount of irrigated land, most suitable areas of the coast and surface of rivers, canals flowing through the area to install solar power plants using GIS.

The authors are grateful to the "Yergeodezkadastr" of the Republic of Uzbekistan, the Khakimiyat of the Yangiyul district of the Tashkent region for providing practical assistance in this research work.

References

1. Sh. Yandra. Cost Analysis of Small Scale Solar and Wind Energy Systems. Intl J. Modern Engineering Research (IJMER) **2**, 6 pp. 4587-4590. (2012)
2. P. Ganguly, A. Kalam, A Zayegh. Solar–wind hybrid renewable energy system: current status of research on configurations, control, and sizing methodologies. Integration, Developments and Control. Woodhead Publishing Series in Energy. pp. 219-248. (2018)
3. K. Sheryazov. Methodology of a rational combination of traditional and renewable energy resources in the energy supply system of agricultural consumers. PhD dissertation. Krasnoyarsk, Russia (2011)
4. M. Tashimbetova. Combined use of power plants based on renewable sources for power supply to local consumers. PhD dissertation. St. Petersburg, Russia (2005)
5. A. Torquay, A. Jailanp. Autonomous power supply systems of Egyptian farms using renewable energy sources PhD dissertation. Moscow, Russia (2010)
6. U. Yarmukhametov. Solar power plants with a tracking system for the sun for the energy supply of agricultural consumers. PhD dissertation (2005). St. Petersburg, Russia (2008)
7. A. Rajabov. Problems and prospects for the development of renewable energy technology in agriculture. Proceedings of the International Conference Prospects for the Development of Renewable Energy Sources in Uzbekistan, Tashkent. March 28-29, (2018)
8. A. Bobozhonov, M. Botirova. Effective land allocation methods for solar power plants in irrigated agriculture, Materials of the International scientific-practical conference Problems of increasing the efficiency of electric energy use in agricultural sectors, Tashkent. November 28, 46-53 pp. (2018)
9. I. Musaev, Sh. Shokirov. Remote Sensing. Tashkent, “Moliya-iqtisod” (2015)
10. Sh. Suyunov, I.M. Musaev. Determination of the theoretical values of the angle of inclination in both states of the Atmosphere near the ground on the basis of geodesic and meteorological measurements. Journal of Irrigation and melioration” **4 (10)** pp.19-21, (2017)
11. S. Musayev, J. Mellor, I. Musaev, M. Nuretdinova. Impact of climate uncertainties on agriculture in Fergana Valley of Uzbekistan. International journal for innovative research in multidisciplinary field. **10**, pp.85-91, (2019)
12. S.Musayev, E.Burgess, & J.Mellor. A global performance assessment of rainwater harvesting under climate change. Resources, Conservation and Recycling, **132**, 62-70 p, (2018)
13. www.russianelectronics.ru/v-velikobritanii-ustanovlena-plavuchaya-solnechnaya-ferma/ (last retrieved: 12.03.2020)

14. www.cleanenergo.ru/2019/v-litve-zapuskayut-proekt-plavuchej-ses-na-250-mvt-ona-utroit-obshhuyu-moshhnost-solnechnoj-energetiki-strany/ (last retrieved: 12.03.2020)
15. www.toneto.net/news/tehnologii/Kyocera-postroit-13-7-mvt-solnechnuyu-elektrostantsiyu-na-vode (last retrieved: 12.03.2020)
16. www.ubratana.com/chitalka/novosti/v-kitae-est-samaya-krupnaya-v-mire-plavuchaya-stanciya-na-solnechnyx-batareyax.html (last retrieved: 11.03.2020)
17. www.energypost.eu/ten-highlights-of-floating-solar-pv-power-station-on-water/ (last retrieved: 12.01.2020)
18. www.scientificamerican.com/article/putting-solar-panels-on-water-is-a-great-idea-mdash-but-will-it-float/ (last retrieved: 7.03.2020)
19. www.en.wikipedia.org/wiki/Floating_solar (last retrieved: 12.04.2020)
20. www.documents.worldbank.org/curated/en/579941540407455831/pdf/Floating-Solar-Market-Report-Executive-Summary.pdf (last retrieved: 11.02.2020)

E3S Web of Conferences **227**, 06001 (2021) https://doi.org/10.1051/e3sconf/202122706001
GI 2021

GI school projects and their value and transferability to Uzbekistan

Sabine Hennig[1,*], and *Robert* Vogler[2]

[1]University of Salzburg, Department of Geoinformatics, 5020 Salzburg, Austria
[2]University of Salzburg, Department of Geography and Geology, 5020 Salzburg, Austria

Abstract. Geospatial technologies and spatial data have increasingly become a key issue in our world with regard to digitalization. The ability and capability to handle spatial data products is a key competence in the 21st century. Accordingly, school education is an important domain to address these skills, but still lacks behind its actual potential - even if numerous concepts and approaches already exist. A promising means of addressing this issue is to actively involve schools in real-world GI projects. This paper sums up relevant educational approaches and concepts that are important for such initiatives and briefly describes four selected projects that were carried out in Austria. Based on the knowledge gained in these projects, challenges and benefits are presented, and the transferability and the potential of carrying out such projects in Uzbekistan is discussed.

1 Introduction and research questions

With the digitalization in all areas of living, including work and private life, the use of Information and Communication Technologies (ICT) has become a key issue for everyone everywhere on earth. This was compounded by the Covid-19 pandemic [1]. Digital communication and interaction processes in real time based on web 2.0 and mobile technologies are an essential part of our everyday life [2]. This also includes the use of geospatial technologies and data. Online maps, for example, which are omnipresent on the internet today, are an excellent example of the role geospatial technologies and data play for everybody [3]. They help people to orient themselves in physical space, navigate from place to place, plan routes, acquire spatial knowledge, build up cognitive or mental maps, share spatial information, organize information, and participate in different types of public participation initiatives that increasingly integrate geospatial technologies [4-6]. Now, the competent and capable use of spatial products including online maps requires spatial literacy skills on the part of the users. In the literature, some specific skills are emphasized that are important for this. These skills include digital skills as well as the capabilities to handle spatial data and spatial data products [7]. Table 1 gives an overview of these skills.

* Corresponding author: sabine.hennig@sbg.ac.at

Table 1. Overview of spatial literacy skills including general digital skills and spatial representation handling skills [based on 7]

Digital skills	Spatial representations handling skills		
	Map reading/ one way communication	**Map making/ two way communication**	**Basic GI and cartography related concepts and terms**
• Using a digital medium in general • Handling the formal structures of the medium (e.g., browsing, navigation) • Searching, selecting, evaluating information in digital media • Mailing, contacting, creating online identities, draw attention, giving opinions • Making contributions with a plan, design; taking data privacy, intellectual property rights IIPR) into account • Using the digital medium as a means to achieve particular professional and personal goals • Etc.	• Using online maps: e.g., find and open tools, orient and navigate on the map (e.g. zoom, pan), find places switch basemaps, turn layers on/ off, measure) • Accessing/ retrieving information (e.g., available in feature pop-ups) • Output (e.g., print, save, share) • Etc.	• Creating maps, adding features (i.e., markers, lines, areas) • Adding and combining information (e.g., by pop-ups) • Handling spatial data (i.e., import, export, convert, transfer) • Re-using spatial data (find, assess, integrate) • Etc.	• Knowing/ understanding relevant technical terms (e.g., pan, layer, basemap) • Being aware of cartographic design guidelines (e.g., decide upon adequate symbols, basemaps, use/ integrate multimedia) • Critical reflection on the power of maps; use of maps as a powerful mediator of particular interests • Etc.

In society, however, people not only have different levels of skills to use ICT [8, 9], but also to handle geospatial technologies [10, 11]. This is where formal school education can play a key role in building these skills in society. Especially secondary education has become a fundamental domain for geomedia education approaches and concepts [12]. Here, the inclusion of spatial data (products) has already taken place to a certain extent [e.g., 13, 14]. Nevertheless, we are (still) faced with the challenge that many teachers still lack spatial literacy skills [15], and that this topic cannot be/ is not assigned to a single school subject. A promising means of addressing these issues is to include schools in projects that focus on the use of geospatial technologies and data (GI projects) and that require public involvement. In addition to public participation, citizen science, crowd mapping, and non-profit sharing economy initiatives, this refers to research projects aiming at developing user-centered products, learning about ICT and GI laypeople and raising people's awareness of topics related to the use of ICT and GI. The added value of this approach for schools is underpinned by the numerous benefits that refer, among other things, to project-based learning such as increased motivation and sustainability of learning [16].

In developed countries (e.g. the EU) such initiatives have gained increasing interest in recent years (see, e.g., conferences such as Learning by Geoinformation, GI_Forum). With regard to developing countries, there are fewer examples such as such run by the Humanitarian OpenStreetMap Team [see, e.g., 17]. But, what additional benefits does cooperating in GI projects bring for schooling and school students – apart from benefits related to project-based learning? If we distinguish between developed and developing countries, what are the benefits for developing countries and does it make sense to apply this approach in those countries?

Based on a selection of GI projects that took place in Austria in which the cooperation with schools and school students was a key aspect; these questions are discussed in this paper. Section 2 presents approaches and educational concepts that are of relevance for such cooperations. Based on the experience from four projects (section 3), selected challenges and benefits of such cooperation are discussed in section 4 and 5. The value of such projects for developing countries is highlighted using the example of Uzbekistan (section 6). A conclusion and outlook is given in section 7.

2 Approaches and concepts

In recent years, a wide range of approaches using geoinformation technologies in education emerged. Of particular importance for the cooperation of schools in GI projects are (i) science education, (ii) the spatial citizenship concept, and (iii) the general support of learning processes by spatial contextualization with geomedia (learning) environments (e.g., the SELTAS approach). These concepts are briefly described below.

2.1 Science education

Many approaches regarding the use of geoinformation in education stem from the science education domain. These concepts aim to foster spatial thinking on one hand and (technology-driven) analytical thinking on the other. One of the most prominent initiatives in this field is the "Learning to think spatially" concept: "The key to spatial thinking is a constructive amalgam of three elements: concepts of space, tools of representation, and processes of reasoning. It is the concept of space that makes spatial thinking a distinctive form of thinking" [18: 12]. In this context, spatial thinking, analytical thinking and technology-supported digital (geo)visualization abilities are considered as key competences in the 21st century, which can be fostered by an active integration of geoinformation technologies in the classroom [19].

2.2 Spatial Citizenship

On the contrary, the spatial citizenship approach [14] aims at the ability to interact and participate in societal negotiation processes with the support of geomedia. Based on a fundament of critical cartography [20], critical GIScience [21, 22], and a constructivist understanding of space and spatial meaning [23], it links geomedia usage with citizenship education aims. Theoretically, it is linked to an activist citizenship education approach [24] in combination with the emergence of online communities, networks and collaborations [25].

Based on the premise, that geomedia serve as powerful platform for the manifestation of spatial meaning, spatial citizenship focuses on two dimensions: (i) reflection and (ii) communication. Reflection refers to a mature appropriation of space and spatial meaning; and communication is linked to actively communicate own alternative meanings supported by geomedia. For this, spatial citizenship defines three competence fields, which are required to actively act as a spatial citizen: (i) technical and methodologic competences to handle spatial representations and geomedia, (ii) the competence to reflect, appraise and evaluate the use of spatial representations and geomedia, and (iii) the competence to actively communicate and participate with spatial representations and geomedia [14]. When these three competences are consistently fostered in (school) education, school students develop the ability and capability to participate in processes of spatial meaning production.

2.3 SELTAS

Another approach focusing on geomedia and education is SELTAS (Spatially Enabled Learning and Teaching Across Subjects). In contrast to science education approaches and spatial citizenship, SELTAS is a concept that provides benefit to learning through spatial contextualization. Its main aim is the individual and collective generation and documentation of knowledge (including spatial analysis) in the framework of learning environments by adding a (geo)mediated spatial context to the leaning content [26, 27].

SELTAS uses the omnipresence of digital spatial information to support learning while taking into account the communicative and collaborative potential of the geoinformation society. Inspired by the variety of content, to which georeferences have been added in everyday life – from political issues to sustainable consumption to spatialized documentation of leisure activities – the focus lies on learning that uses spatial representations as added value for several subjects beyond purely geographical topics. Examples within this approach are learning about places mentioned in novels, about statistics concerning countries and regions, and about cartographic narrations on product chains. The approach is based on two main references:

i. relational spac, i.e., the idea of the attachment of meaning to mainly physical matter by communication through spatial representations [23] and a supportive spatial contextualization of content
ii. dual coding [28], i.e., the combination of pictorial (maps as picture) and symbolic elements (maps as text) that target different modes of information processing in the learners' reception, as described in mapping theory [29]

Based on the advantages of geomedia, SELTAS "[…] makes use of web-based spatial representations to support interaction, communication and document learning outcomes in educational contexts by connecting complex learning content to a spatial dimension. This can be achieved by contextualization, by cartographic support of arguments in communication processes or simply as a spatialized portfolio that may be either individual or collective or both" [26: 206].

3 Exemplary projects

Over the last couple of years, we carried out or took part in several projects in which we collaborated with schools such as (i) YouthMap5020, (ii) Fairtrade Oberndorf, (iii) the4bees, and (iv) citizenMorph. Even though their aims as well as methods and tools used differ widely, the common thread of all four projects was to address school students' geomedia skills in the context of specific settings: High school students contributed to each project in the context of several hours / days lasting workshops. Common to all projects is also that each one aimed at developing a product to which high school students have made a significant contribution. The four projects are briefly described in Table 2 and below.

Table 2: Overview of the projects

Project	Duration/ Funding	Key topic	Domains involved	Tools used
Youtmap-5020	2013-2014 National/ Austria: FFG	Youth-centred city map	Cartography, geography, geoinformatics, software development	ESRI's ArcGIS Online
Fairtrade Oberndorf	2015 National/ Austria: FFG	Sustainability, responsible consumption	Cartography, geography, geoinformatics, economics	ESRI's ArcGIS Online

the4bees	2015-2019 International/ EU: Interreg Alpine Space	Sustainable energy consumption	Cartography, geoinformatics, sensing technology	ESRI's Storymaps for ArcGIS Online
citzen-Morph	2018-2020 National/ Austria: FWF	Geomorphological features and dynamics, climate change	Geomorphology/ geograph, remote sensing, geoinformatics	ESRIS's Survey 123 for ArcGIS Online

3.1 YouthMap

The YouthMap 5020 project aimed at creating an online map for the city of Salzburg that is tailored to the needs of young people. The map was technically implemented as an ArcGIS online map application. Applying the approach of participatory design, about 120 high school students from several local schools were involved in all kinds of tasks related to the phases of generating the youth-centric Salzburg online city map. The decision to use the participatory design approach arose from the fact that literature highlights it as a strategy that is particularly useful when it comes to working with young people and when there is a need to directly and actively involve them in the development process. Following the participatory design approach, the YouthMap 5020 project team comprised adults as experts in geoinformatics and application development as well as young people as experts for their own requirements including, e.g., map content and design [11].

3.2 FairTrade Oberndorf

This project was a technical spin-off from the YouthMap5020 project. Since it developed its own dynamic, it will be mentioned here as an independent project. Despite the above-described, increasing (geo)digitalization of our world, this project showed that we are still faced with many challenges on the path towards an emancipatory geomedia usage by the general public. These challenges arise mainly from technical hurdles such as different data formats, data standards and systems, and their (at least partly) incompatibilities. To minimize these problems, within the project a workflow was developed that shows how to use a minimalistic and therefore universal technology structure for cross-platform work with spatial data, especially useful in school environments. The workflow development took place in the context of a school initiative, in which high school students created a dynamic and interactive online map application of their school town showing all retail stores where fair-trade products are available [30].

3.3 the4bees

The the4bees project aimed at an interactive teaching approach to give school students a better understanding of energy consumption by the implementation of real time sensors in their school building. It tackled the research question of how school students can get a clear and sustainable understanding about the topic of energy saving. To achieve this, students from three schools in the federal state of Salzburg were directly involved in research into energy saving. The school students participated in the entire research process from data collection to analysis and visualization of the results in digital story maps. Based on the related insights, teaching material and learning modules were designed for supra-regional/ international use [31].

3.4 CitizenMorph

The main objective of the citizenMorph project was to create a technological solution that allows and supports the general public to contribute data on landforms and geomorphological phenomena, the dynamic of which is actually steadily increasing due to the effects of climate change. The technological solution comprises a mobile app for mapping such landforms on-site as well as an extended internet presence consisting of a content management web site (providing all relevant information related to the topic and social networking capabilities) and different social media accounts. To provide a technological solution that is as tailored as possible to the citizens and their needs, representatives of the public (including high school students) were involved in its development in different ways. This included the specification of the citizen requirements as well as the design and implementation of prototypes which finally fed into the final product [32].

4 Selected challenges

Cooperating with young people is a challenging task per se. One of the reasons for this is that the youth is a fairly heterogeneous and complex user group. Regarding their ICT use, they can be described by the following aspects [33]: (i) overconfident of their computer and web abilities, (ii) reduced attention span, long reaction time, (iii) little patience, quick judgment, (iv) reluctances to read on screen and use manuals, (v) low willingness to explore implemented functionalities, and (vi) partly problems to get and keep an overview of complex tasks and workflows. In addition to these aspects, different challenges and problems with regard to working with students were identified in the context of the four projects. These are: lack of work routine and experience, lack of experience in participating in long-term tasks, age-related level of soft skills and technical skills.

This requires additional work and efforts that must be considered when cooperating with young people. Selected aspects identified in the course of the four projects are listed in Table 3. Moreover, conducting projects in cooperation with schools requires certain organization effort. Thus, for instance, to enable school students to participate in project-related workshops lasting several hours or days, teachers usually have to coordinate these activities in advance by clarifying with their school administration and other teachers, if their teaching time will be affected.

Table 3: Additional efforts (adapted from [32, 33])

Categories	Description
Suitable equipment	• Hardware, devices, software applications and methods • Mix of web-based and traditional methods • Devices and tools preferred by young people (e.g., mobile devices/ phones)
Suitable support	• Providing an easy-to-understand big picture: overall project aim, workflow etc. • Clearly structured and defined tasks, sticking to the agreed workflow • Team composition: Support by peers to give appropriate explanations etc. • Availability of comprehensive support: online and offline options • Appropriate supportive material: technical terms, concepts, explanations etc. • In-person meetings and roundups
Enough scope for development/ time to work	• Facing and handling long discussions that offer opportunities to present, exchange and discuss ideas and gain recognition from others • Expecting and facing unreliability, being flexible: failure to meet deadlines, being available for (online) discussions/ support out of company time etc. • Invite to experiment and to think beyond common structures

	• Expecting and dealing with school students' different knowledge and background • Trusting in school students' skills • Helping to carry responsibility • Supporting decision making without bringing in your own ideas and opinions • Helping young people to express themselves (reasoning, presentations) • Involving school students in PR activities: school, family, other peers etc.

5 Selected benefits

On the contrary to these challenges, there are also numerous benefits. In addition to building spatial literacy skills, we have identified many of them in the context of the four projects. Some are highlighted and discussed below.

5.1 Spatial literacy skills

The projects presented have been proven to give school students the opportunity to build spatial literacy skills. This is related to the fact that in all projects school students contributed to the respective product development, which automatically contributed to the development of these skills. Thus, by actively and directly taking part in the various tasks, receiving explanations from the GI experts and/or working through tutorials and supportive material independently, they build these skills.

Here, our experiences are confirmed by the finding of other scholars. This relates, for instance, to De Luca et al. [34] or in another context Kerski [35] who emphasize that the use of spatial data and spatial data products in schools can foster spatial awareness and spatial literacy skills. An overview of the spatial literacy skills addressed by the four projects is given in Table 4.

Table 4: Spatial literacy skills addressed by the four projects

Project	Spatial literacy skills (required/ build)
Youthmap5020	Map reading, mapping, map production
Fairtrade Oberndorf	Map reading, mapping
the4bees	Map reading, mapping, map production
citizenMorph	Mapping

In addition, the school students' interest on spatial data products was aroused across all projects. This includes the relevance for everyday tasks as well as the availability of spatial data and tools (e.g., purposes, potential of crowdsourcing approaches). This corresponds to Sui et al. [36] who argue that such initiatives make it more likely that young people will use spatial data products for their private and school activities and later on in their working life. It also has the potential to spark students' interest in studying geoinformatics.

5.2 Expanding horizons

Cooperating with experts from different fields and working on real world topics in the context extracurricular activities showed the school students how important theory is for practical work and how different school subjects are intertwined and help tosolve problems and find solutions. It has helped to broaden their horizons on aspects that are generally not covered in school subjects as such.

The projects YouthMap 5020 and the4bees – with a focus on software development and geoinformatics as well as more sustainable and reflective energy consumption behaviour – sensitized the school students in particular to STEM subjects (Science, Technology, Engineering, and Mathematics). The projects FairTrade Oberndorf and citizenMorph, which dealt with topics such as sustainability and social responsibility, addressed the improvement of school students' understanding of the relevance of public participation. Both are closely related to aspects such as citizenship responsibilities, which are considered as key educational issues – especially for the social challenges of the 21st century [37]. Regarding all projects, the school students were impressed by the experience and knowledge they were able to gain, as well as the fact that they found interest in topics they previously considered uninteresting and boring. In terms of the YouthMap 5020 project, high school students explained they were surprised that software development is much more than just coding and how many different people from different backgrounds contribute to the development of software applications (e.g., media designer, sociologists, communication and PR experts).

Moreover, the fact of cooperating with experts (who serve as role models for the school students) and, in this context, the discussions with the experts about their work, the study and career opportunities behind it were seen as very useful and interesting by the young people. With regard to the citizenMorph project, for example, the interest of the school students was aroused in particular in topics such as geology and geomorphology. Further, by observing the experts, school students can modify their behavior/ attitude accordingly, as the behavioural change of students (in the context of energy consumption) throughout the 4bees project showed [31]. This is in line with the approach of social learning.

5.3 Soft skills

Soft skills play an important role in school, studies and work. Here, activities like taking part in a project that focuses on collaboration and interaction, as well as being responsible for specific tasks, can help developing various soft skills. This refers to competencies related to teamwork, conflict management, communication and presentation tasks, and responsibility. Moreover, it required high school students to learn and work independently and self-paced and meet deadlines. Responsibility and independent work of the students were addressed by the fact that their personal interest was met by the project topics – which relate to their life and current situation and cover issues that are important to them (Table 2). Prüller et al. [38] also underline this: They highlight that taking local and/ or regional topics and data into account can motivate students to work self-driven and self-paced and to intensify their learning experience. This is also in line with the findings of situated learning approaches in general [39] and for science education in particular [40]. This is driven by the fact that personal relationships, consternation and interest in a topic are seen as key motivational factors for people to take part and get involved in whatever activity [41].

5.4 Spatially enabled society and geo digital society

Technological advances have triggered the development towards a society in which digital information and communication are key concepts, and in which actions are often supported – in some cases even replaced – by digital tools [42, 43]. Due to the advances and omnipresence of geospatial technologies, a spatially enabled society has emerged [44]. This means that almost everyone not only has spatial data products at his/her fingertips (e.g., geo portals, open government data), but should also have a certain level of spatial literacy skills to be able to use spatial data products in a competent and capable manner and, thus, benefit from spatial data. This relates to aspects such as becoming spatially informed, producing, publishing and sharing spatial data (e.g., user generated content, volunteered geographic information), and contributing to space-related questions such as citizen science or public participation projects. In this context, the spatially enabled society can also be understood as a geo-digital society [11]. Here, our projects contribute to building spatially enabled societies and geo digital societies in different ways. This refers, firstly, to building spatial literacy skills across the high school students. Secondly, it relates to getting teachers' interest in becoming (more) familiar with geospatial technologies and spatial data, and being prepared to include such aspects in their teaching activities [45]. Thirdly, school students, who were involved in cooperation with GI projects, can and will spread their knowledge and experience with geospatial technologies, products and data. As in the case of the YouthMap5020 and citizenMorph project this refers to their families but also friends. This is an important aspect, since for young people promotion through word of mouth and referrals is an important way to get informed [46].

6 Transferability to Uzbekistan

Several of the benefits outlined above are of interest for future educational GI projects in Uzbekistan. This Central Asian country is one of several republics from the former Soviet Union. A rough overview using some figures that are of particular interest in connection with this work is given in Table 5.

Table 5: Key figures on Uzbekistan [47]

Area [km^2]	445 711
Population (2018)	32.96 Mio.
School enrollment, secondary (%) (2018)	93
Mobile cellular subscriptions (per 100 people) (2018)	71,5
Individuals using the Internet (% of population) (2018)	55,2
Net migration (thousands) (2018)	-44

In Uzbekistan, the need for a sustainable economic development as well as to respond to climate change is affecting the country (e.g., changes in water distribution, natural hazards, loos of biodiversity) and creates a particular demand for GI experts [38, 48]. As emphasized by GPS [49:12), there is a substantial need to increase "… the number of youth and adults who have relevant skills, including technical and vocational skills, for employment, decent jobs and entrepreneurship …" in the next few years thus "…. that all learners acquire the knowledge and skills needed to promote sustainable development". Considering this, schooling in Uzbekistan can greatly benefit from cooperating in GI projects. Hence, the aspects discussed in section 4 also apply to Uzbekistan, as this can increase the proportion of young people with ICT skills, including spatial literacy skills. Depending on the type and topic of the project, it also increases the extent to which global citizenship education and education for sustainable development are carried out – as requested by GPS [49].

Apart from this, there are additional benefits. For instance, with respect to soft skills, English language proficiency – which are often a problem in Central Asian countries [see e.g. 48], but a key issue in many areas – can be improved and interest in it can be aroused. Cooperating in GI projects, including using different (expert) tools, requires to become familiar with them. Since much of the supporting materials, including e-learning content, are usually available in English (e.g., video tutorials), school students will become aware of the importance of proficiency in English and will be motivated to learn English and improve their English skills. Thus, such projects can help to improve students' language skills in general [48].

Nevertheless, there are also some challenges particular for Uzbekistan regarding such teaching initiatives: For such projects, a good internet connectivity is substantially required. This is often a problem in Uzbekistan [50]. This can be countered by the fact that a large proportion of people around the world now own and use smartphones (incl. developing countries; Uzbek school students). This, along with high internet user penetration rates (i.e., mobile Internet) can free project managers from worrying too much about internet connectivity. In addition, allowing school students to use their mobile devices meets the preferences of this user group (i.e., digital natives) and, thus, can increase the acceptance of the project and the tasks they have to do.

The use of geospatial technologies is generally related to the use of substantial computer resources including software and spatial data [38; 50]. Here, free-of-charge software is suitable for such projects, as there are no or only minimal costs for licensing and great community-based support [38]. Moreover, for many chargeable software products special licenses for the education field, including schools, exist (e.g. ESRI campus program).

7 Conclusion and outlook

The cooperation of schools in GI projects is a useful means to build school students' spatial literacy skills. In addition, it brings some other benefits for young people (expanding their horizon, developing soft skills) as well as for society (development towards a spatially enabled society and geo digital society; citizenship responsibilities). Further, the link with school education also brings benefits for the GI projects: more publicity, awareness rising in society, projects will run smoother and faster, new insights etc. Nevertheless, cooperating with schools and school students is always challenging and requires additional work and effort. While, for instance, there are many examples of such initiative in EU states, the situation is still different in developing countries such as the Central Asian states. Through GI school projects, they can benefit from the opportunities described above.

To carry out such projects, opportunities such as those offered, e.g., by EU funding should be taken into consideration (for instance capacity building initiatives with regard to formal school education). This will promote interesting partnerships between institutions and organization in EU countries and Central Asian states, and bring many opportunities of exchange for school students.

References

1. European Commision, "Digital technologies - actions in response to coronavirus pandemic" [Online]. Available: https://ec.europa.eu/digital-single-market/en/content/digital-technologies-actions-response-coronavirus-pandemic [Accessed June 28 2020].
2. A. Darwish, K.I. Lakhtaria, *The Impact of the New Web 2.0 Technologies in Communication, Development, and Revolutions of Societies*, Journal of advances in information technology **2**, 4, 204-216 (2011).
3. T. Thielmann, L. Van der Velden, F. Fischer, R. Vogler, *Dwelling in the Web: Towards a Googlization of Space*, HIIG Discussion Paper Series **2012**, 03 (2011).
4. A.S. Helal, E. Moore, B. Ramachandran, *Drishti: An Integrated Navigation System for Visually Impaired and Disabled*, 2nd IEEE Annual Conference on Pervasive Computing and Communications, Orlando, March 14–17 (2004).
5. D.R. Montello, S.M. Freundschuh, *Cognition of Geographic Information* (A Research Agenda for Geographic Information Science, edited by R. B. McMaster and E. L. Usery, 61–91. Boca Raton: CRC Press, 2005).

6. J. Panek, *From mental maps to geoparticipation*, The Cartographic Journal **53**, 4, 300-307 (2016).
7. S. Hennig, S., R. Vogler, I. Gryl, *Spatial Education for Different User Groups as a Prerequisite for Creating a Spatially Enabled Society and Leveraging SDI*, International Journal of Spatial Data Infrastructre Research **8**, 2013, 98-127 (2013) .
8. J. Nielsen, "Digital divide: the 3 stages" [Online]. Available: https://www.nngroup.com/articles/digital-divide-the-three-stages [Accessed Dec. 5 2019].
9. J. Van Dijk, *The evolution of the digital divide - the digital divide turns to inequality of skills and usage*, Digital Enlightenment Yearbook 2012, 57-75 (2012).
10. M. Bartling B. Resch, A. Eitzinger, L. Zurita-Arthos, *A multi-national human–computer interaction evaluation of the public participatory GIS GeoCitizen*, GI_Forum **2019**, 1, 18-39 (2019).
11. S. Hennig, R. Vogler, *YouthMap 5020! What should youth-centred web maps look like?*, GI_Forum **2014**, 1, 285-294, (2014).
12. International perspectives on teaching and learning with GIS in secondary schools (Edited by A. Milson, A. Demirci, J. Kerski. New York: Springer, 2012).
13. Using geoinformation in European geography education (Edited by K. Donert. Rome: International Geographic Union, 2010).
14. I. Gryl, T. Jekel, *Re-centring Geoinformation in Secondary Education: Toward a Spatial Citizenship Approach*, Cartographica: The International Journal for Geographic Information and Geovisualization **47**, 1, 18-28 (2012).
15. F.J. Fernandez-Cruz, M.J. Fernandez-Diaz, *Teachers Generation Z and their Digital Skills*. Comunicar, **24**, 46, 97-105 (2016).
16. P. Blumenfeld, E. Soloway, R.W. Marx, J.S. Krajcik, M. Guzdial, A. Palincsar, *Motivating project-based learning: Sustaining the doing, supporting the learning*, Educational Psychologist **26**, 369–398 (1991).
17. F. Delattre "Mapping the Thien Sgan mountains" [Online]. Available: https://www.hotosm.org/updates/2016-05-10_mapping_the_tien_shan_mountains [Accessed June 28 2020].
18. Learning to Think Spatially (Edited by National Research Council. Washington, DC: The National Academies Press, 2006).
19. D. Stuart Sinton, S.W. Bednarz, P. Gersmehl, R.A. Kolvoord, D.H. Uttal, *The People's Guide to Spatial Thinking*, National Council for Geographic Education (2013).
20. J.B. Harley, *Deconstructing the Map,* Cartographica **26**, 2, 1-20 (1989).
21. J.W. Crampton, *Maps as Social Constructions: Power, Communication and Visualization*, Progress in Human Geography **25**, 2, 235-352 (2001)
22. J. Pickles, *Ground Truth 1995–2005*, Transactions in GIS **10**, 5, 763-772 (2006)
23. B. Werlen, *Society action and space: an alternative human geography* (London & New York: Routledge, 2006).
24. S. Elwood, K. Mitchell, *Another Politics Is Possible: Neogeographies, Visual Spatial Tactics, and Political Formation*, Cartographica: The International Journal for Geographic Information and Geovisualization **48**, 4, 275-292 (2013).
25. W.L. Bennett, C. Wells, A. Rank, *Young citizens and civic learning: two paradigms of citizenship in the digital age*, Citizenship Studies **13**, 2, 105-120 (2009).
26. R. Vogler, S. Hennig, T. Jekel, K. Donert (2012). Towards a concept of spatially enabled learning. In T. Jekel, A. Car, J. Strobl, & G. Griesebner (Eds.), GI_Forum 2012: Geovizualisation, Society and Learning, 204-211. Berlin: Wichmann.
27. R. Vogler, T. Jekel, E. Killingseder, *Using the Geographies of Learning. An Exploratory Categorization for Spatially Enabled Learning*, GI_Forum **2018**, 2, 181-192 (2018).

28. A. Paivio, *Mental Representations: A dual coding approach* (New York: Oxford University Press, 1990).
29. D. Wood, *The Power of Maps* (London: Routledge, 1993).
30. R. Vogler, S. Hennig, M. Lindner-Fally, *Shrinking technical Challenges – Zur strukturierten Erstellung digitaler Karten im Unterricht*, GW-Unterricht **137**, 1, 49-60 (2015).
31. B. Vockner, R. Vogler, C. Atzl, M. Mittlböck, *THE4BEES – Energieverständnis durch smarte Gebäude*, AGIT – Journal für Angewandte Geoinformatik, **2018**, 4, 348-353 (2018).
32. S. Hennig, L. Abad, D. Hölbling, *Geo citizen science and the role of participatory design in the field of geomorphology,* InterCarto. InterGIS. GI support of sustainable development of territories: Proceedings of the International conference **26**, 1, 116–126 (Moscow: Moscow University Press, 2020).
33. S. Hennig, R. Vogler, *User-Centred Map Applications Through Participatory Design. Experiences Gained During the 'YouthMap 5020' Project*, The Cartographic Journal **53**, 3, 213-229 (2016).
34. N. De Luca, N. Ferber, H. Atteneder, T. Jekel, Feminist and Queer Approaches to Education for Spatial Citizenship, GI_Forum **2015**, 1, 272-282 (2015).
35. J. Kerski, *Geo-awareness, Geo-enablement, Geotechnologies, Citizen Science, and Storytelling: Geography on the World Stage*, Geography Compass **9**, 14-26 (2015).
36. D.Z. Sui, S. Elwood, M.F. Goodchild, *Crowdsourcing Geographic Knowledge* (Berlin: Springer, 2013).
37. L. Keller, A. Oberrauch, *Can innovative learning settings support Education for Sustainable Development? – Teenagers from Northern and Southern Tyrol perform research on quality of life*, Standards and Research in Geography Education. Current Trends and International Issues. (Edited by D. Schmeinck, & J. Lidstone, J., Berlin: Mensch und Buch Verlag, 81-92, 2013).
38. R. Prüller, J. Scholz, C. Strauß, T. Achleitner, *Teaching GIS in Central Asia*, Vermessung & Geoinformation **1**, 174-178 (2009).
39. J. Lave, E. Wenger, *Situated learning: Legitimate peripheral participation* (Cambridge: Cambrige University Press, 1991).
40. T.D. Sadler, *Situated learning in science education: socio-scientific issues as contexts for practice*, Studies in Science Education **45**, 1, 1-42 (2009).
41. E.G. Clary, M. Snyder, R.D. Ridgr, J. Copeland, A.A. Stukas, J. Haugen P. Miene, *Understanding and assessing the motivations of volunteers: a functional approach*, Journal of Personality and Social Psychology **74**, 6, 1516-1530 (1998).
42. T.F. Dapp, *The digital society. New ways to more transparency, participation and innovation* (Frankfurt: Deutsche Bank Research, 2011).
43. A. Martin, *Digital literacy and the "Digital Society"*, Digital Literacies: Concepts, Policies and Practices (Edited by C. Lankshear, M. Knobel, M. Peters, New York, Oxford: Peter Lang, 151-176, 2008).
44. S. Enemark, A. Rajabifard, *Spatially Enabled Society*, Geoforum Perspektiv – Tidsskrift for Geografisk Information **20**, 3-16 (2011).
45. A. Koller, T. Jekel, R. Vogler, *RECC Geographie und Wirtschaftskunde – Geomedien*, GW Unterricht **151**, 2, 66-68 (2018).
46. NCVO Knowhow "How to engage with young people in your community" [Online]. Available: https://knowhow.ncvo.org.uk/how-to/how-to-engage-with-young-people-in-your-community# [Accessed June 28 2020].
47. Worldbank "Uzbekistan Country Profil" [Online]. Available: https://databank.worldbank.org/views/reports/reportwidget.aspx?Report_Name=Countr

yProfile&Id=b450fd57&tbar=y&dd=y&inf=n&zm=n&country=UZB [Accessed June 28 2020].

48. S. Hennig, A. Nazakulova, *Benefits and challenges of eLearning in Central Asia*, International Journal of Geoinformatics **15**, 4, 43-51 (2019).
49. GPS Global Partership "Education Sector Plan (ESP) of Uzebekistan 2019-2023" [Online]. Available: https://www.globalpartnership.org/sites/default/files/2019-04-gpe-esp-uzbekistan.pdf [Accessed June 28 2020].
50. O. Akbarov, B. Markus, A. Podor, *Development of Advanced Education in Geoinformatics for Enabling Sustainable Development in Uzbekistan* (Proceedings FIG Congress, 1-15, 2014).

Wheat harvesting by combine with GPS receiver and grain sensor

Komil Astanakulov[1*], *Kodir* Shovazov[1], *Atkham* Borotov[1], *Akram* Turdibekov[1] and *Saidmukhammad* Ibrokhimov[2]

[1]TIIAME, Agricultural Machines Department, 100000, 39 Kari Niyaziy str., Tashkent, Uzbekistan
[2]SRDI "UZDAVYERLOYIHA", 100097, 5 Chapanata str., Tashkent, Uzbekistan

Abstract. This research describes the result of study and analyze the working process of the harvester combine with GPS receiver and grain level sensor in wheat harvesting. GPS receiver and grain level sensor showed that the combine harvester was working with what level of velocity in the real time regime and how much grain was being uploaded into grain tank. According to results when the yield increased from 2.62 t/ha to 6.22 t/ha, because of decreasing the work-velocity of the combine harvester, the work efficiency of the combine harvester decreased from 1.96 ha/hour till 1.30 ha/hour, and grain loss rose from 2.03 per cent till 3.76 per cent. In the field where the weeds had covered 16.8 per cent, the work efficiency of the combine was 1.74 ha/hour, in another field where weeds level was 56.7 the work efficiency of the combine decreased almost 1.7 times, made up 1.07 ha/hour, the grain loss was observed that it increased from 2.1 per cent till 7.42 per cent. When the moisture of the cereal decreased it was defined that the combine work efficiency increased, however the rate of grain loss in the header rose too.
Keywords: GIS, harvester combine, GPS receiver, grain level sensor, wheat harvesting, work efficiency.

1 Introduction

In farms the combine harvesters are being used in different conditions, namely they are being used with the same working regime in low or high yield, grassy or lain down fields as well in condition that the moisture of the cereal is lower than measures. As a result, in some areas the grain loss level is reaching till 15-20 per cent, fuel expense is consisting of 35-40 kg for one hectare, and even in some situations 45-50 kg, therefore the effectiveness of the combine harvester is being low.

One of the more perspective ways of increasing effect managing agricultural production is using informative system on the base of geo-information technology. The similar systems allow solving planned agro technical operation, monitoring agro technical operations and condition of crops, forecasting the yield of the crop and evaluation losses, planning, monitoring and analyzing the used technics. Such data must be completed by an accurate

*Corresponding author: komil_uzmei@mail.ru

position on the Earth, specifically using all modern methods: GIS based digital maps, GPS, GNSS satellite systems, etc. [1].

GIS technologies are used for defining the using efficiency of the combine harvesters. Steering system for combine harvester based on machine vision was developed for increasing the efficiency of harvesting works by using GIS technologies [2, 22].

It is important to intrude GPS receivers and yield sensors for using GIS technologies in agricultural machines. Some works have been performed for intruding GPS receivers and sensors as well using correctly them.

In India, an automated yield monitoring system consisting of a yield sensor, global positioning system (GPS), field computer with custom software was mounted on a self-propelled indigenous grain combine harvester for real-time crop yield mapping along with moisture data. By means of optical sensors (light emitter and detector), the height of the grain on the elevator paddles was measured and converted in the grain mass flow. Three rice fields were harvested to evaluate the performance of the yield monitor for grain yield and moisture mapping of harvested grains. The actual yield maps were generated by using ArcGIS software from the data collected for three different rice fields. The total area harvested by using indigenous combine harvester fitted with yield monitor and moisture sensor was 0.84 ha comprising three fields having areas 0.3, 0.22 and 0.32 ha respectively. On an average, yields for all the three fields were 4325.91, 5093.14 and 4287.66 kg ha^{-1} with moisture content of 21.42%, 22.78% and 20.42% (wet basis) recorded respectively [3, 4].

In harvesting of the wheat in autonomous navigation of tracked combine harvester, the state and posture can be obtained from the positioning and inertial sensor measurements. But these sensor measurements have measurement uncertainties, which can be compensated by using the tracked combine harvester kinematic and dynamic model.
On the other hand, the computation of soil parameter and track coefficients are also important for the tracked combine harvester dynamic model. An autonomous tracked combine harvester is developed in the vehicle robotics laboratory, Hokkaido University based on Real Time Kinematic Global Positioning System (RTK-GPS) and Inertial Measurement Unit (IMU) for the harvesting of wheat and paddy [5].

Some researchers focused on degree of lodging at harvest time an alternative of yield. It depends on completion of extension of spikes up to heading stage and too much N and lack of sunshine make lodging large. A relationship between lodging and these are few after the heading stage. Also, excessive N application in paddies leads to lodging and decreased grain quality. They estimated that rough estimates of yield were possible by looking at the degree of lodging [6,7].

Researchers observed to estimate the absolute heading of robot combine harvesters by compensating this gyro measurement bias during non-linear turning maneuvers. Therefore, this estimation method can be used to estimate the absolute heading when the tracked combine harvester will makes turn at high speed and with high order steering commands in order to cut wheat and paddy rice near headlands to calculate the exact crop periphery for the development of a harvesting map [8].

In general, a single sensor is not able to provide enough information, whereas multi-sensor integration can provide more useful information, which is more helpful and informative than what can be observed using a single sensor. This information needs to be fused in a way that reduces sensor uncertainties and the additional task of interpretation must be performed [9]. However, it can be cause combine to become complicated and expensive. Therefore, it is necessary to take important results by using less sensors than usually.

In general, main aim of using new techologies during the harvesting process is to make longer the work time of the combine harvester. Niehaus used spatial data to evaluate the corn harvesting operation on an Iowa grain farm and reported an overall TE of 62.4%; with 16.1% of total time spent in machine idling, 9.1% in in-field or road travel, 9.3% in turning within field headlands, and 2.9% unloading grain while not harvesting [10].

Other scientists compared the practical collection methods of cereal harvesting costs in different agricultural holdings in order to effectively manage combine harvester fleets, make economically reasoned decisions on the exploitation of combine harvesters, reduce harvesting costs and consequently the cost price of cereals. Analyzed this information and provided assessments on the effectiveness of their combine harvester fleet [11].

2 Materials and methods

2.1 Study area

This study was conducted on 18-hectare irrigated, almost float cropland area of Yangiyul district of Tashkent region (Uzbekistan). The district lies in the middle of the Tashkent oasis and covers 431.5 km2 area of region The corresponding time zone is UTC+5 (UZT) (Fig. 1). On the main part of the district, the vegetation period of plants exceeds 180-200 days. Favorable climatic terms allow tilling a cotton plant, winter wheat and other heat-loving cultures, get 2−3 harvests of vegetables in a year. Generally, taking into account weather condition, winter wheat is sowed between middle of autumn and beginning of winter.

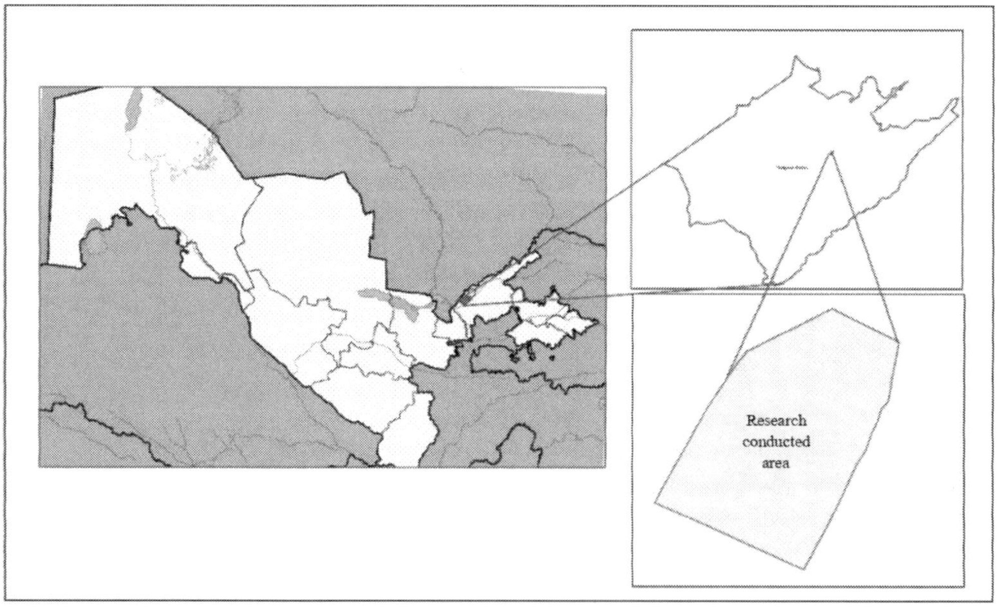

Fig. 1. Study area

2.2 Methodology

For carrying experiments, the experimental harvester combine was prepared at a base of combine Dominator-130 (Fig. 2) according to previous researchers' results in this direction [12-15].

Fig. 2. Experimental cereal combine

Above mentioned Han and others developed method to evaluate auto guidance systems of linear parallel-tracking applications of agricultural machines that equipped with GNSS devices [11]. Keskin used an automatic agricultural vehicle based on low cost GPS receiver for increasing the rate of them [12]. Coen developed a robot combine harvester as the aim more realizing GIS technologies during the harvesting works [13].

A cereal harvester combine was tested that equipped with GPS receiver and grain level indicator for harvesting the wheat as well the experiments for analyzing its work were performed by wheat's Krasnodar-99 sort. Size-mass and yield indicators of the cereal in the field were defined before experiments.

For taking data about work efficiencies of cereal combine Dominator-130 at real time regime the Teltonika FMB920 GNSS/GSM/Bluetooth terminal and Escort DB-2 sensor were installed in the grain tank (Fig. 3).

Teltonika FMB920 GNSS/GSM/Bluetooth terminal collects data about position and working of moving object and it sends the data to server through GSM net-work.

Teltonika FMB920 terminal gives observing opportunity the position of combine during the work in real time regime. The technical characteristics of this device as following: Bluetooth B 3.0 micro CD card (till 32 GB) internal GPS antenna, the internal GPS antenna1, number enter 1, analog enter 1 exit were installed. Battery 170 was provided with mA/hour IP54.

Escort DB-2 grain sensor is installed on the wall of grain tank, it defines the fluctuation of every cereal and other sorts of productions that their moisture does not increase from 25 per cent and sends the data to observing system through artificial satellite. The high level of the defense (IP 67) wide temperature gap (from - 60 till +85) guarantees reliable forking of censor indicator in every climate condition and in different relief. The sensor is suitable for different types of GLONASS/GPS terminals.

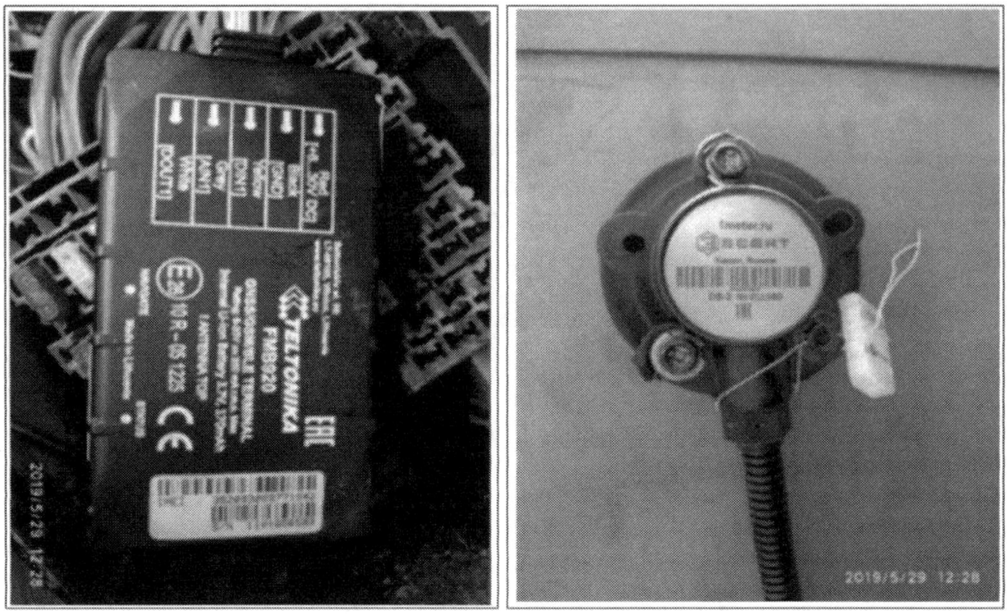

a b

Fig. 3. Teltonika FMB920 GNSS/GSM/Bluetooth terminal (a) and Escort DB-2 sensor (b)

During the experimental-test the work regimes and technological parameters of the combine Dominator-130 are as following: installing height of header - 20 cm; rotation frequency of drum - 1200 rpm; clearance between concave and drum: entering part - 13 mm; exiting part - 3 mm; opening angle of sieve jalousies - 30 degree; rotation frequency of fan - 800 rpm. When the influence of installing height of header to work efficiency indicators was studied, the cutting height was changed from 10 cm till 50 cm with 10 cm step. Also, working speed of the combine harvester was in diapason 4-6 km/hour according to fed cereal mass into thresher and working condition.

In defining work and idle movement of combine in field Oksanen's results were used about coverage path planning algorithms for agricultural field in moving zone of agricultural machines [16].

3 Results and discussion

According to obtained data through GPS receiver, work efficiency of combine decreases while grain yield increases in work process of combine. Grain loss consisted of 1.96 ha/hour when yield became 2.62 t/ha, the rate of grain loss decreased to 1.78 ha/hour when yield became 3.58 t/ha, then with yield increased the grain loss rose too (Table 1).

The grain loss in combine header almost did not change, when yield rose from 4.46 t/ha to 6.24 t/ha, losses behind the combine increased from 2.72 per cent till 3.76 per cent, and work efficiency declined from 1.63 ha/hour till 1.30 ha/hour (Table 1).

According to data that was received from GPS receiver it should be mentioned that, when crop yield increased to 4.46 t/ha and more, after that the travel speed of the combine decreased from 6 km/hour to 4 km/hour respectively.

As well in Grisso's researches: crop yield affects the field efficiency of a combine when standard or typical field speeds are used to calculate theoretical field capacities, with greater yields usually resulting in reduced travel speed due to the heavier weight of grain

[17]. Agricultural machines field efficiencies have a significant effect on the effective field capacities of machinery, which in turn impact the overall cost of production [18].

Table 1. Influence of grain yield to work efficiency of combine Dominator-130

Work quality indicators	**Grain yield (t/ha)**				
	2.62	3.58	4.46	5.63	6.24
Work efficiency in main period (ha/hour)	1.96	1.78	1.63	1.50	1.30
Total grain loss (%):	2.19	2.03	2.72	3.28	3.76
- in combine header	0.60	0.51	0.52	0.56	0.63
- behind the combine	1.59	1.57	2.20	2.72	3.13

When combine Dominator-130 was tested in fields where weeds level is different (Table 2), its work efficiency was 1.74 ha/hour in field where weeds level was 16.8 per cent, the work efficiency decreased almost 1.7 times and constituted 1.07 ha/hour in the field where weeds level was 56.7 per cent. In that conditions the grain loss of the combine was observed to increase to 3.5 times, namely in the first variant it was just 2.10 per cent, in the last condition the index consisted of 7.42 per cent.

In this case, it is important to emphasize that, when grassy level increases, in combine header the grain loss changes slightly, namely it reached from 0.52 per cent till 0.62 per cent. However, in the thresher changing of the grain loss was insignificantly, it reached from 1.58 per cent to 6.8 per cent, namely it rose to 4.3 times.

This situation is described that, when combine harvests the weeds cereal field the process of working units of the combine in threshing part becomes complicated and they cannot thresh, separate the grain.

Table 2. Influence of grassy level fields to work quality indicators of combine Dominator-130

Work quality indicators	**Weeds level fields (%)**				
	16.8	24.2	36.7	43.9	56.7
Work efficiency in main period (ha/hour)	1.74	1.65	1.61	1.36	1.07
Total grain loss (%):	2.10	3.31	4.63	6.18	7.42
- in combine header	0.52	0.58	0.6	0.62	0.62
- behind the combine	1.58	2.73	4.03	5.56	6.80

According to data that was received from GPS receiver, the travel speed of the combine was observed to decrease during crop harvesting in the field where there are many weeds.

According to results of performed experiments about studying the influence of the cereal moisture to combine work quality (Table 3).

Table 3. Influence of cereal moisture to work quality indicators of combine Dominator-130

Work quality indicators	Cereal moisture (%)				
	20.3	16.0	12.4	10.8	7.5
Work efficiency in main period (ha/hour)	1.78	1.73	1.80	1.81	1.96
Total grain loss (%):	2.10	2.21	2.51	2.78	3.26
- in combine header	0.43	0.57	0.89	1.27	1.44
- behind the combine	1.67	1.64	1.62	1.51	1.82

At high working regimes when the cereal moisture was 20.3 per cent the work efficiency in main period made up 1.78 ha/hour, grain loss was equal to 2.10 per cent, when the cereal moisture decreased from 16.0 per cent till 7.5 per cent the work efficiency rose from 1.78 ha/hour till 1.96 ha/hour. In the same time of increasing of the work efficiency the grain loss increased from 2.10 per cent to 3.26 per cent.

This situation is described that, when the cereal moisture decreased, work efficiency of combine increases as well cereal mass following rises, increasing of grain loss is described with spilling as a result its mechanical influence. It can be seen also the grain loss is being taken part mainly in combine header.

Above analyzed experiments showed that, combine harvester should be used at the suitable working regimes depending on the sort of crop, yield and grain moisture. As if like these recommendations were suggested by Pristavka and Aldoshin [19-21].

4 Conclusion

When Dominator-130 for cereal harvesting combine that equipped with GPS receiver and grain level indicator sensor was tested by using GIS technologies, becoming its header installing height - 20 cm; rotation frequency of drum - 1200 rpm; clearance between concave and drum: entering and exiting part - 13 mm and 3 mm respectively; opening angle of sieve jalousies - 30 degree; rotation frequency of fan - 800 rpm, as cereal yield – 3.5-3.6 t/ha, moisture - between 16-20 per cent the cereal can be harvested with little grain loss. When cereal yield, moisture and weeding level change for decreasing of grain loss combine should be used by changing its work regimes.

The study was funded by the Uzbekistan Applied Research Foundation, grant No KHA-3-009.

References

1. V. Voštová, T. Vondráčková, Proceedings of the 4th Int. Conference TECHSTA (2007)
2. E. R. Benson, J. F. Reid, Q. Zhang, ASAE Annual Meeting (2001)
3. M. Singh, R. Kumar, A. Sharma, B. Singh, P. Mishra, K. Sharma, Agricultural Engineering International: CIGR Journal, 15 (3) (2013)
4. M. Singh, J. Singh, A. Sharma, Journal of Agricultural Engineering, 48 (4) (2011)
5. Z. Zhang, N. Noguchi, K. Ishii, L. Yang, C. Zhang, IFAC Proceedings, 46 (2013)
6. E. Makino, T. Sugiyama, T. Ichikawa, K. Hamada, M. Kawanaka, Journal of JSAM. 69 (4), (2007)
7. E. Makino, T. Sugiyama, T. Ichikawa, K. Hamada, M. Kawanaka, Journal of JSAM. 69 (5), (2007)
8. M. Rahman, K. Ishii, Sensors, 18 (2018)
9. R. Adla, B. Yousef, N. Al-Holou, In Proceedings of the International Conference on Parallel and Distributed Processing Techniques and Applications (Las Vegas, 2013)
10. C. R. Niehaus, M.S Thesis, University of Illinois at Urbana-Champaign (2014)
11. J. Olt, K. Kuut, R. Ilves, A. Kuut. Research in Agricultural Engineering, 65 (1) (2019)
12. S. Han, Q. Zhang, H. Noh, B. Sin, Transactions of the ASAE, 47 (1) (2004)
13. M. Keskin, S. M. Say, Computer and Electronics in Agriculture, 54 (2006)
14. T. Coen, A. Vanrenterghem, W. Saeys, J. de Baerdemaeker, Computers and Electronics in Agriculture, 63 (2008)
15. M. Pexa, Research in Agricultural Engineering, 57 (2011)
16. T. Oksanen, A. Visala, Journal of field robotics, 26 (8) (2009)

17. R. D. Grisso, P. J. Jasa, D. E. Rolofson, Applied Engineering in Agriculture, 18 (2) (2002)
18. S. Pitla, N. Lin, S. Shearer, J. Luck, Applied Engineering in Agriculture, 30 (6) (2015)
19. M. Pristavka, M. Bujna, M. Korenko, Journal of Central European Agriculture, 14 (2013)
20. M. Pristavka, K. Krištof, P. Findura, Agronomy Research, 15 (2017)
21. N. Aldoshin, O. Didmanidze, Research in Agricultural Engineering, 64 (2018)
22. G. Shodmonova, U. Islomov, O.Abdisamatov, Sanjar Khikmatullaev, U.Kholiyorov and Sh. Khamraeva, IOP Conf. Series: Materials Science and Engineering, 896 (2020)

Indicators of the combine equipped with GPS receiver in sunflower harvesting

Otabek Ochildiev[1,*], *Golibjon* Fozilov[1], *Shodi* Achildiev[2], *Maxmud* Karimov[3], *Nurali* Ashurov[4]

[1]SRIMA, 110800 Tashkent region, Yangiyul district, Samarkand 41, Uzbekistan
[2]TerSU, Transport Systems and Facilities Department, 190111 Termez, Barkamol avlod 43 Uzbekistan
[3]Termez branch TSTU, Electrotechnical Department, 190111 Termez, Barkamol avlod 26 Uzbekistan
[4]TIIAME, Agricultural Machines Department, 100000 Tashkent, Kari Niyaziy 39 Uzbekistan

Abstract. In sunflower harvesting for decreasing the grain loss and increasing using efficiency of the combines GIS technologies, namely combines equipped with GPS devices and control sensors provides good results. This is explained that by the fact that with a length of 400-500 mm, the cut baskets did not partially fall on the divider-guide, and with a length of 700-800 mm, seed losses increased due to uncut baskets, since with an increase in length the divider-guide bent some stems forward and did not cut them. Analyzing the studies of working process of the adapted grain harvester combine Dominator-130 that equipped with GPS receiver and grain level indicator sensor in sunflower harvesting defined optimal parameters its header, as the following values: the length of the divider-guide (l_T) - 0.6 m, the gap between the divider-guides (S_T) - 0.1 m, the installation angle of the divider-guide relative to the bottom of the header (β_T) - 5 degrees. With these parameters, the combine that has adapted header for harvesting sunflower meets the agro technical requirements for harvesting sunflower.
Keywords: sunflower harvesting, combine, GIS, GPS receiver, grain sensor, header, adaption, guide-divider.

1 Introduction

In modern agriculture, for harvesting sunflower, mainly combine harvesters are being used that equipped with a special header or adapted grain header for harvesting sunflower. In Uzbekistan, the sunflower is grown as a second crop after cereals in relatively small areas of farmers. Therefore, it is economically viable to use a combine harvester by adapting the combine header for harvesting sunflower under these conditions.

One of the important and crucial indicators when harvesting sunflower is harvesting in a short time and with minimal losses. For this, when harvesting sunflower, it is important to

* Corresponding author: otabek_uzmei@mail.ru

implement GIS technologies. For application on GIS technologies, it is necessary to use widely combines equipped with GPS devices and control sensors.

With using GIS technologies the working efficiency of the harvester combines is defined. Y.Cao researched on application of the Internet of things technology in financial leasing of intelligent manufacturing enterprises. In his research, he analyzed of financial leasing risk control with agricultural combine harvester has front end-speed sensor A is arranged at the drive shaft of the combine harvester, fuel consumption sensor B is arranged in the fuel tank, body positioner C, weight sensor D, distance sensor E are arranged, and power-off control element F is arranged at the ignition device. The transfered node, the company's personalized data encryption algorithm is added to ensure remote and secure transmission of information data [1].

Developed dynamic performance evaluation method of agricultural machines that equipped with DGPS devices [2]

Feasibility of low-cost GPS receivers for ground speed measurement was developed for increasing the rate of using an automatic agricultural vehicle based on low cost GPS receiver [3].

As the aim more realizing GIS technologies during the harvesting works an autopilot was developed for a combine harvester. [4].

Tractor-based Real-time Kinematic-Global Positioning System (RTK-GPS) was created for guidance system for geospatial mapping and intra-row weed knife control system [5, 6].

K.Hayashi developed a geographic information system (GIS) for collecting farming information. They decided the time from harvesting to tank discharge and proposed a method to solve the time lag of combine harvester by setting tailing return. Solving the issue of tailing return is required in order to do highly accurate yield mapping using head-feeding combine harvester [7, 8].

Besides of above mentioned the scientific fundamentals of sunflower harvesting are occupied by many researchers, in particular, D.Shafarostov and S.S. Makarov developed and justified the parameters of a four-row sunflower harvesting system [9].

E.I.Sitchenko developed a self-propelled machine for harvesting selective sunflower [10]. S.Sudajan, V.M.Salokhe and K.Triratanasirichai have developed a thresher for sunflower [11]. T.Csanadi and B. Hamphoff have advanced a special header for harvesting sunflower for combine harvesters CLAAS [12].

I.S.Dalmis et al developed a sunflower harvester that has a chopper under the header for chopping stems during harvesting [13,14]. A.H.Mirzabe and G.R. Chehini created a thresher for sunflower that separates grains from baskets with an air stream [15]. X.Li et al developed the ADRC-DMC controller, which for stable it controls the rotation speed of the threshing drum when harvesting sunflower [16].

In above works researches have not been carried out about adaptation and defining indicators of the grain harvester combine equipped with gps receiver and grain level sensor in sunflower harvesting. Besides this full-scale researches in this direction have not been conducted in Uzbekistan.

2 Materials and methods

The justification of technical means for harvesting sunflower based on local conditions is relevant. In this regard, studies were conducted and an adapter was developed to adapt the grain header for harvesting sunflower [17, 18] (Fig.1).

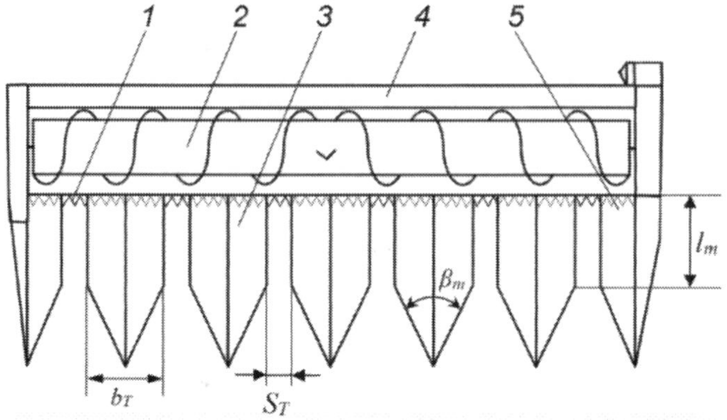

1-segment-finger cutting apparatus; 2-auger; 3-directional dividers; 4-back side of the header; 5-side guide divider.

Fig. 1. Adapted grain header for harvesting sunflower.

The adapted header consists of a cutting unit, auger of header, reel and divider-guides that are mounted on top of a segmented-finger cutting unit.

In the process of harvesting sunflower for taking data about work efficiencies of combine Dominator-130 at real time regime combine was used with GPS receiver Teltonika FMB920 and grain level sensor Escort DB-2 (Fig. 2).

Fig. 2. Outlook of the adapted combine harvester equipped with GPS receiver, grain sensor.

 https://doi.org/10.1051/e3sconf/202122707002

Therefore, in the process of harvesting sunflower seeds, the cut baskets, which are guiding apparatus, should fall on the divider-guide (Fig.1). For this, it is important to determine correctly the length and width of the divider guide. If a divider-guide is long, this can affect to the maneuverability of the combine. If the divider-guide is short, the cut baskets fall forward in the cutting device can fall to the ground (especially with curved stems in the direction of the combine) in front of the divider. In addition, if the gap between the dividers is too wide, then baskets with a diameter less than the gap may fail, and if the width is less than the stems that deviate strongly from the sowing line may not fall and the gap will bend under the header after W_{ii} in increased grain loss. Therefore, it is very important to determine the optimal parameters of the divider-guide.

3 Results and discussion

At first we conducted experimental studies to find the optimal value of the length of the guide-divider. At which the loss of seeds was determined depending on the length of the guide-divider (Fig.3).

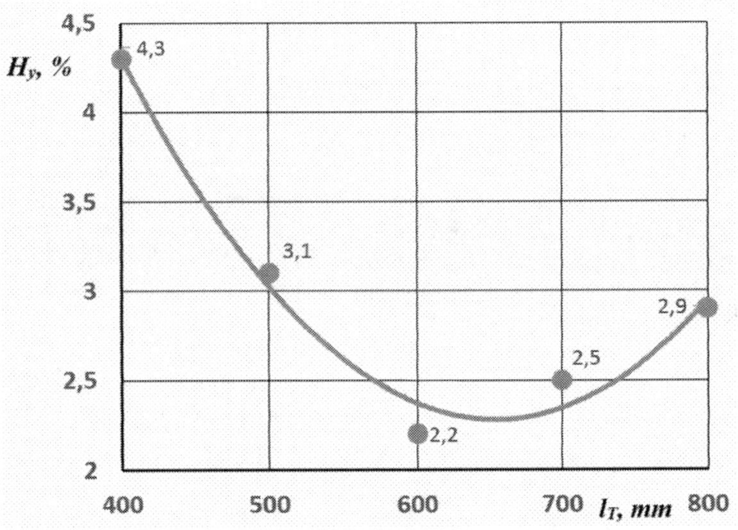

Fig. 3. The influence of the length of the dividers-guides (l_T) on the loss of sunflower seeds (H_y) behind the header.

It can be seen from the graph (Fig. 3) that with the length of the divider guides from 400 mm to 600 mm, the seed loss changed in the direction of decreasing from 4.3 percent to 2.2 percent. Further, with an increase in the length of the guide divider (700 mm), the seed loss increased from 2.5 percent to 2.9 percent (800 mm). This is explained that by the fact that with a length of 400-500 mm, the cut baskets did not partially fall on the divider-guide, and with a length of 700-800 mm, seed losses increased due to uncut baskets, since with an increase in length the divider-guide bent some stems forward and did not cut them.

The width of the divider guide should be less than the minimum distance between the extreme points of the sunflower stalks in the row spacing and greater than or equal to the distance between the fingers of the cutting apparatus. Given that the distance between the fingers is 7.6 cm, with a row spacing of 70 cm, the width of the divider should be greater or equal to 62.4 cm and less than the deviation of the plants from the centerline of the row. Therefore, to determine the optimal width of the divider guide, experimental studies were

 https://doi.org/10.1051/e3sconf/202122707002

performed, namely the influence of the gap between the dividers (divider width) on the loss of sunflower seeds behind the header was determined (Fig.4).

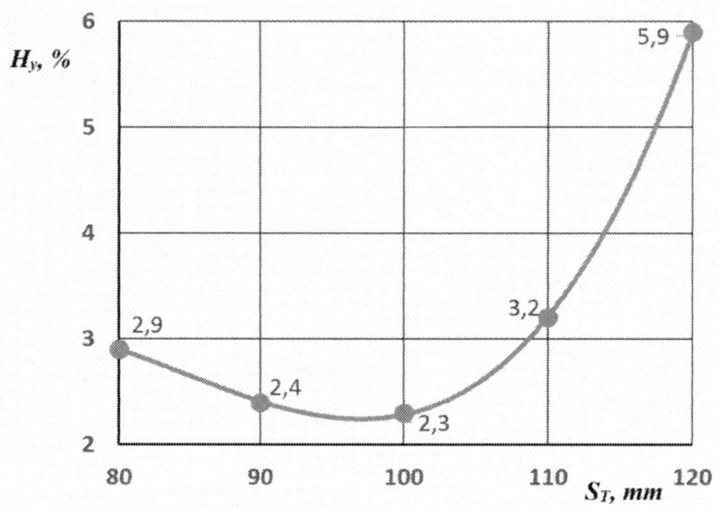

Fig. 4. The influence of the gap between the dividers (S_T) on the loss of sunflower seeds (H_y) behind the header.

As can be seen from the graph (Fig. 4), with a gap of 80 mm, the loss of seeds was 2.9 percent, with a gap of 100 mm, the loss was 2.3 percent, then with an increase in the gap (120 mm) the loss of seeds increased from 3.2 percent to 5, 9 percent (140 mm).

This is because with a gap of 80 mm, some stems that deviate strongly from the sowing line bend under the header, and with a gap of 120 -140 mm, cut baskets with a small diameter fall into the gap.

Next, we conducted studies to determine work efficiencies of combine Dominator-130 with the guide-divider at real time regime using GPS receiver Teltonika FMB920 and grain level sensor Escort DB-2. The results this experiment shown in the table 1.

Table 1. Influence to combine work quality indicators of work velocity changing of the combine Dominator-130

Work quality indicators		**Velocity of the combine harvester, m/s**			
		0.6	**0.9**	**1.2**	**1.5**
Work capacity, ha/hour		0.83	1.13	1.47	1.82
Losses, %	In combine header	0.36	0.40	0.47	0.84
	In combine thresher	0.50	0.68	0.92	1.64

According to data of GPS receiver, when work velocity of the combine was 0.6 m/s, its work capacity in main time became 0.83 ha/hour. As the velocity of the combine increased to 0.9, 1.2 and 1.5 m/s, work capacity of the combine harvester constituted 1.13, 1.47 and 1.82 ha/hour respectively. Seed losses increased too while velocity of the combine increased. In this situation, when velocity of the combine rose, seed losses was significantly in the header of the combine and made up 0.36, 0.40, 0,47 and 0.84 per cent. In thresher of the combine harvester seed losses that are free form were equal to 0.50, 0.86, 0,92 and 1.64 per cent accordingly to velocities. It has been known that, work velocity of the combine harvester should not be no more than 1.2 m/s, this has been founded as optimal.

4 Conclusion

Analyzing the studies of working process of the grain harvester combine Dominator-130 that equipped with GPS receiver and grain level indicator sensor in sunflower harvesting we have taken optimal parameters, as the following values: the length of the divider-guide (l_T) - 0.6 m, the gap between the divider-guides (S_T) - 0.1 m. With these parameters, the combine that has GPS receiver, grain level indicator sensor and adapted header for harvesting sunflower meets the agro technical requirements for harvesting sunflower.

The study was funded by the Uzbekistan Applied Research Foundation, grant No KHA-3-007-2015.

References

1. Y. Cao, The International Journal of Advanced Manufacturing Technology (2019)
2. S. Han, Q. Zhang, H. Noh, B. Sin, Transactions of the ASAE, **47(1)**, (2004)
3. M. Keskin, S. M. Say, Computer and Electronics in Agriculture, **54**, (2017)
4. T. Coen, A.Vanrenterghem, W. Saeys, J. De Baerdemaeker, Computers and Electronics in Agriculture, **63**, (2008)
5. M. Perez-Ruiz, D. C. Slaughter, C. Gliever, S. K. Upadhyaya, Biosystems Engineering, **111**, (2012)
6. M. Perez-Ruiz, D. C. Slaughter, C. Gliever, S. K. Upadhyaya, Computers and Electronics in Agriculture, **80**, (2012)
7. K. Hayashi, Journal of Japanese Society of Agricultural Machinery, **69(5)**, (2007)
8. K. Hayashi, Journal of Japanese Society of Agriculture and Biological Engineering, **6**, (2014)
9. D. V. Shafarostov, S. S. Makarov, ACTA Technological Agriculture, **22(2)**, (2019)
10. E. I. Sitchenko, Research Institute of Oil Crops, 124, (2001) (in Russian)
11. S. Sudajan, V. M. Salokhe, K. Triratanasirichai, **83 (4)**, (2015)
12. T. Csanadi, B. Hamphoff, VDI Berichte, (2007)
13. I. S. Dalmis, B. Kayisoglu, Y. Bayhan, P. Ulger, F. Toruk, F. Durgut, Bulgarian Journal of Agricultural Science, **19(5)**, (2013)
14. I. S. Dalmiş, B. Kayişoğlu, Y. Bayhan, F. Toruk, F. T. Durgut, Tarim Bilimleri Dergisi, **19 (1)**, (2013)
15. A. H. Mirzabe, G. R. Chegini, Agricultural Engineering International: CIGR Journal, **17(3)**, (2015)
16. X. Li, X. Zhang, X. Zhu, Journal of Agricultural Mechanization Research, **7**, (2019)
17. K. Astanakulov, O. Ochildiyev, European science review, **1**, (2018)
18. K. D. Astanakulov, O. Sh. Ochildiyev, In Proceedings of the III Tashkent international innovation forum, (2017)

Monitoring of the Combine with Smart Devices in Soybean Harvesting

Komil Astanakulov[1], *Tulegen* Abdillaev[1], *Abdigappor* Umirov[2], *Golibjon* Fozilov[1*], and *Bobamurod* Hatamov[3]

[1]TIIAME, Agricultural Machines Department, 100000 Tashkent Kari Niyaziy 39 Uzbekistan
[2]TerSU, General Technical Subjects and Technologies Department, 190111 Termez Barkamol avlod 43 Uzbekistan
[3]NamECI, Transport Logistics Department, 160103 Namanagan Islam Karimov street 12 Uzbekistan

Abstract. GIS technology based on real-time smart devices, including GPS devices and sensors will be widely used in agricultural technology. Therefore, this technology is being expected in crop harvesting process. One of the most important processes in soybean cultivation is the efficient use of combines and harvesting without loss of harvest. The grain harvester combine Dominator-130 equipped with GPS receiver and grain level sensor was researched in harvesting the soybean. According to results of experiments, during the harvesting of soybean the grain cleanness in a tank of the combine harvester Dominator-130 made up 95.9 per cent, grain loss in a thresher of combine made up 0.7 per cent, these indexes answer to the demands, however in a reaper of combine harvester the grain loss and damaging of grain were higher than demand. Therefore, it is important to define the optimal technological parameters and working regimes that give opportunity to decrease the loss of grain in the reaper of combine and damaging of grain in the thresher of combine for harvesting the soybean.
Keywords: GIS, soybean harvesting, GPS, combine, grain sensor, real time, working efficiency.

1 Introduction

Significant work is being done in Uzbekistan on large-scale cultivation of soybeans as a primary and secondary crop. However, there are still a number of difficulties in soybean harvesting due to the peculiarities of this crop and the methods and means of harvesting it at the required level, taking into account the conditions of different regions. One of the most important processes in soybean cultivation is the efficient use of combines and harvesting without loss of harvest. At present, in Uzbekistan, the use of combiners is Dominator-130, New Holland TS-5060 and many others, but most are Dominator-130 combines. Recently, agricultural technology has been widely used in GIS technology based on real-time Smart devices, including GPS devices and sensors.

In Japan, several agricultural robots had been developed, the tractor robot working autonomously using RTK-GPS and FOG [1], the tractor robot that followed the path by D-

* Corresponding author: golibjon270784@mail.ru

GPS and Geomagnetism Sensor [2-3], the paddy trans-plant robot using RTK-GPS and IMU [4-6], the weeding robot based on the paddy trans-plant robot using RTK-GPS and IMU on paddy field [7], the head feeding type combine robot that able to go straight and turn using GPS, and the rice combine harvester robot using CAN Bus network with RTK-GPS and GPS compass [8]. Each operation like cultivation, planting and harvesting are developed to work autonomously on the field. An automation of agricultural machinery has been studied in the world in the same way.

However, the sensors used in these agricultural robots are very expensive and it is necessary to find cost-effective ways to develop these robots. According to this M.Saito studied that a robot combine harvester for beans was developed using a conventional bean combine harvester adopting a CAN Bus network, handheld sensing devices. The harvesting efficiency was also developed by harvesting and unloading crops simultaneously [9].

As a result his work a robot combine harvester for beans was developed using CAN Bus network. Two sensors are connected to the CAN Bus; a RTK-GPS to locate the position, and a GPS Compass to detect the heading direction. This robot is programmed to perform autonomous operation such as moving forward, backward, and turning in right angle. When the robot was tested in harvesting soybeans on a 100 m × 30m rectangular field, the deviations between the target path and the robot's traveling path was approximately 0.07m RMS, and the yaw angle error was approximately 1.82° RMS along the long side of the field. Therefore this robot can work autonomously along the target path without overlapping and shortening. In addition, this robot can unload harvested grain to adjacent transport truck during harvesting operation, which improved harvesting capacity to approximately 10% [10].

Much research has been done on soybean harvesting and new methods and techniques have been developed. Nowadays, in world practice, combines are used as grain harvesters equipped with headers that are not adapted for soybean harvesting.

Although there are developments and recommendations on the quality of soybean harvesting worldwide, there are no studies on the quality of soybean harvesting with the GIS technology-based smart combine. That is why it is important to study this combine in soybean collection and to identify performance indicators.

2 Materials and methods

The experiments were carried out in the fields of Agrobioholding LLC, located in Yangiyul district of Tashkent region.

The threshing and cleaning components of the grain harvester remain unchanged and are used only with their technological parameters adjusted and operating modes adjusted. Where soybeans are cultivated in small areas, small techniques have also been developed for its collection [11-13]. Although soybeans are ripe, their stems are high in moisture, making it difficult to harvest from the combine. Therefore, a method for decontamination is recommended [14-15].

The field size where experiment performed was larger than other regions, the moisture of the crop was not higher. Therefore, there was not need to use small technics and desiccation plants.

During the experiments, the width of the Dominator-130 combine was 4.2 m, the drum rotation frequency was 1200 rpm, the hole inclination between the drum and its deck was 13 mm, the exit angle was 3 mm, and the open angle 30 cm. The number of ventilation cycles was 800 rpm. Harvest height was 5-10 cm depending on the mass of the stem and the working speed was 4-5 km/h.

The key performance indicators were the combine's real-time performance, grain losses, grain cleaning and damage in the bunker. The combine's real-time performance was determined by subtracting salts and stops on the GPS receiver and grain level sensor data. Grain losses were obtained separately by combine harvester and crusher. Grain purity and damage were extracted from the grains in the bunker, separating the whole grain and the damaged grain and foreign impurities and found by their ratio.

In experiments at first the condition of crops before harvesting, namely description of the agro background was defined.

According to experiments the height of the soybean stalk that harvested is average 85 cm, average quadratic declination is 13.6 cm, diameter of the stalk in bottom cutting part is average (7.1 ± 0.37) mm. There are average 29.8 pieces of pod on a plant, the lowest pod is located on average 6.2 cm height from the land, there are 2-3 pieces of grain that is 6.3 gram in one pod and the ratio of grain than stalk consists of 1:1.8.

After defining the harvesting condition in the field where combine works, it was started to harvest the soybean (fig.1) and according to research method its work quality indicators were defined.

Fig. 1. Process of harvesting soybean with combine Dominator-130.

Fig. 2. GPS receiver Teltonika FMB920 installed in combine Dominator-130.

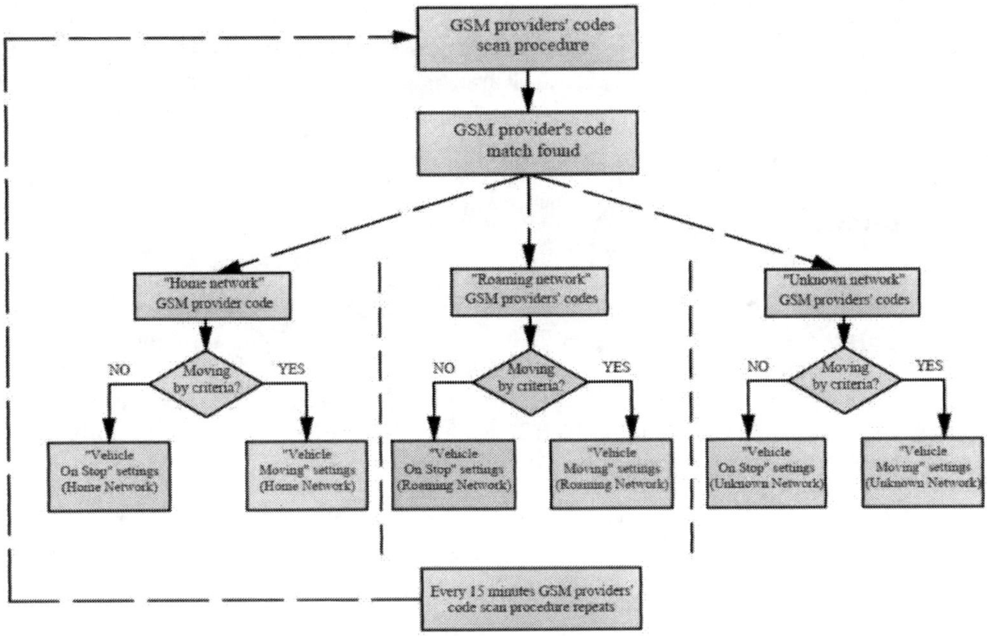

Fig. 3. Regime of GPS recording date.

3 Results and discussion

According to analyzing the results of experiments, work indicators of the combine in soybean harvesting were more different than indicators for harvesting other sorts of crops (look at the table). It was known that from experimental results, the work efficiency of combine Dominator-130 for harvesting the soybean in mainly time consists of 1.1 ha/hour.

The grain loss was observed only in cutting unit of the combine and made up 5.34 per cent. Grain losses on the Dominator-130 combine occurred mainly in the form of legumes that were not cut below the harvest height, resulting in a larger grain loss index.

Table 1. Quality indicators of work on soybean harvesting of combine Dominator-130

Work quality indicators	The amount of work quality indicators
Capacity in main time (ha/h)	1.1
Grain loss (%)	
- in combine header	5.34
- in combine thresher unit	0.7
Grain cleanliness in grain tank (%)	95.9
Grain losses (%)	7.8

It is also important to pay attention to the period of harvesting soybeans, and studies have also shown that early or late harvest can have a significant impact on grain and its quality [16-19].

Worldwide, soybeans are harvested mainly in combines, with the most common loss being grain spills [20, 21]. Grain loss depends on many factors, and one of the main ways to prevent it is to develop appropriate harvesting techniques for each crop. Therefore, every soybean in the US has appropriate recommendations for its harvesting [15].

In addition, the mechanical damage caused by combine harvester blades was revealed as a result of grain spills and death. Grain clearance in the bunker was relatively good at 95.9%, despite the presence of weeds in the field. Grain damage was also significantly higher than the target and was 7.8%.

4 Conclusion

According to the results of experimental studies, grain harvester combine Dominator-130 equipped with GPS receiver and grain level sensor was tested in soybean harvesting showed that grain cleanliness in tank is 95.9 %, and the loss of grain in combine thresher unit is 0.7%. However, the loss of grain in the header combine harvester is 5.34 % and the grain damage is 7.8 %, which is higher than the set requirements. Therefore, further research will need to identify parameters and modes of operation that reduce grain loss and damage to combine harvester and crusher.

The study was funded by the Uzbekistan Applied Foundation of the Ministry Innovative Developed, grant No MV-KH-A-KH-2018-189.

References

1. M. Kise, N. Noguchi, K. Ishii, H. Terao, Journal of JSAM, **63** (5) (2001)
2. O. Yukumoto, Y. Matsuo, N. Noguchi, M. Suzuki, Journal of JSAM, **60** (4) (1998)
3. Y.Matsuo, O.Yukumoto, Y.Irie, N.Ichisugi, Journal of JSAM, **63** (3) (2001)
4. Y.Nagasaka, K.Taniwaki, R.Otani, K.Shigeta, Y.Sasaki, Journal of JSAM, **61** (6) (1999)
5. Y.Nagasaka, N.Umeda, Y.Kanetani, K.Taniwaki, Y.Sasaki, Computers and Electronics in Agriculture, **43** (3) (2004)
6. Y.Nagasaka, H.Saito, H.Kitagawa, K.Kobayashi, M.Seki, K.Tamaki, K.Taniwaki, M.Miyazaki, K.Tateishi, Journal of JSAM, **70** (1) (2008)
7. J.Reid, Q.Zhang, N.Noguchi, Computers and Electronics in Agriculture, **25** (1) (2000)
8. M.Saito, K.Tamaki, K.Nishiwaki, Y.Nagasaka, Journal of JSAM, **74** (4) (2012
9. M.Saito, Y.Nagasaka, K.Tamaki, K.Kobayashi, Journal of JSAM, **72** (3) (2010)
10. M.Saito, K.Tamaki, K.Nishiwaki, Y.Nagasaka, K.Motobayashi, (2013) DOI.10.3182/20130828-2-SF-3019.00058
11. A.Vejasit, V.M.Salokhe, Journal of ASAE, **31** (4) (2004)
12. S.Musoni, R.M.Nazare, L.Mukosera, Journal of AVS, **4** (1) (2013)
13. A.A.Timothy, A.B.Osakpamwan, I.E.Osaivbie Journal of ASAE, **18** (2) (2016)
14. J.L.Griffin, C.A.Jones, L.M.Etheredge, J.J.Boudreaux, D.Y.Lanclos, Louisiana State University Agriculture Research and Extension Center, (2002)
15. A.Sharda, L. Haag, Soybean Production Handbook. (2016).
16. A.C.Santana, M.C.Carrão-Panizzi, J.G.Mandarino, R.S.Leite, J.B.Silva, E.I.Ida, Ciência e Tecnologia de Alimentos, Campinas, **32** (2) (2012)
17. A.Abbasi Surki, F.Sharifzadeh, R.Tavakkol Afshari, African Journal of Agricultural Research, **7** (36) (2012)

18. A. P. Gaikwad, R. W. Bharud, International Journal of Current Microbiology and Applied Sciences, **6** (4) (2017)
19. B. D. Philbrook, E. S. Oplinger, Agronomy Journal, (81) (2014)
20. S. Butzen FIELD FACTS. **9** (18) (2015)
21. C. S. S. Paixão, R. P. Da Silva, M. A. Voltarelli, M. T. Cassia, T. O. Tavares, Australian Journal of Crop Science, **10** (2016)

AUTHOR INDEX